Finite Geometries

Finite Geometries

György Kiss
Tamás Szőnyi

CRC Press
Taylor & Francis Group
Boca Raton London New York

CRC Press is an imprint of the
Taylor & Francis Group, an **informa** business
A CHAPMAN & HALL BOOK

CRC Press
Taylor & Francis Group
6000 Broken Sound Parkway NW, Suite 300
Boca Raton, FL 33487-2742

First issued in paperback 2022

© 2020 by Taylor & Francis Group, LLC
CRC Press is an imprint of Taylor & Francis Group, an Informa business

No claim to original U.S. Government works

ISBN 13: 978-1-03-247538-7 (pbk)
ISBN 13: 978-1-4987-2165-3 (hbk)

DOI: 10.1201/9781315120072

Library of Congress Cataloging-in-Publication Data

Names: Kiss, György (Mathematics professor), author. | Szőnyi, T., author.
Title: Finite geometries / György Kiss and Tamas Szőnyi.
Description: Boca Raton : CRC Press, Taylor & Francis Group, 2020. |
Includes bibliographical references and index.
Identifiers: LCCN 2019013231 | ISBN 9781498721653
Subjects: LCSH: Finite geometries. | Combinatorial geometry.
Classification: LCC QA167.2 .K57 2020 | DDC 516/.11--dc23
LC record available at https://lccn.loc.gov/2019013231

Visit the Taylor & Francis Web site at
http://www.taylorandfrancis.com

and the CRC Press Web site at
http://www.crcpress.com

Contents

Preface

This is an extended and updated version of our earlier textbook *Véges Geometriák* published in 2001 in Hungarian. The above book was used for various courses on finite geometry at Eötvös Loránd University, Budapest, and the University of Szeged. We had different types of courses: for mathematics students we had an overview type course touching the most important results in all the chapters of the book without going into detail, for students interested in finite geometries we had special courses going into more detail for example on arcs, blocking sets, (k, n)-arcs and also higher dimensional analogues. For students specialized in mathematics education we focused on configuration, coordinatization and collineations, because that was related to their courses on classical geometry. Finally, we have a finite geometry seminar (following the tradition of Professor Ferenc Kárteszi), where even more special and recent results are discussed. Since 2010 the book has been used at the University of Primorska, Koper, for the course *Selected Topics in Finite Geometry*.

Although this book is introductory, it also contains some more recent material. In particular, we try to illustrate some methods in more detail, such as the use of polynomials. We also try to indicate recent developments in the topics covered by the book without proofs. Therefore, the book can be used as a textbook for higher level undergraduate and lower level graduate courses by choosing some of the topics appropriately, and also as a reference material in the sense that it contains pointers to recently published works. We tried to limit the background assumed in our book. However, linear algebra (over finite fields), some knowledge of (finite) fields, polynomials, elementary group theory and basic combinatorics is used in many places. Hungarian students typically know some (classical) geometry, so we did not use much and tried to briefly refer to the most important facts in the first chapter.

There are important books on the topic. The authors grew up on Kárteszi's book *Introduction to Finite Geometries* [105] and several chapters follow it. The chapter on coordinatization and collineations follows the book by Hughes and Piper [95] and the appendix by Lombardo-Radice to Segre's book *Lectures on Modern Geometry* [151]. In many cases the books *Projective Geometries over Finite Fields*, [88] *Projective Spaces of Three Dimensions* [87] by Hirschfeld and the book *General Galois Geometries* [93] by Hirschfeld and Thas, are very useful and can be used to check the material only mentioned in our textbook. In the nineties we had a Tempus cooperation with many universities and a lot of teaching material on topics related to the topics treated in the present book were produced in those years. For example we had lectures on cryptography and coding theory, which is why we used the book *Projective Geometries* [21] by Beutelspacher and Rosenbaum in the last chapter.

There are very important recent results by Ball on arcs in higher dimensions and the MDS conjecture; they can be found in his book [8]. That book also contains recent results about the connection of graph theory and finite geometry. The books *Generalized Quadrangles* [140] by Payne and Thas, and *Generalized Polygons* [177] by Van Maldeghem are the monographs related to the topic in Chapter 10 of our book.

We would like to thank our colleagues and (former) students who helped us in reading the different chapters. Here we have to mention Péter Sziklai and András Gács in the case of the Hungarian edition and Zoltán Blázsik, Dániel Lenger and Tamás Héger in the case of the English translation. We are indebted to Tamás Héger for drawing the figures. Throughout the years we worked closely with Aart Blokhuis (Eindhoven), Gábor Korchmáros (Potenza), Leo Storme (Ghent). We have used the discussions with them at many places in this book.

We gratefully acknowledge the support of the NKFIH Grant NN114614 and the editorial work of Miklós Bóna. Finally, we thank our publisher for the technical help in publishing the book.

1

Definition of projective planes, examples

In this book some familiarity with classical geometry will be assumed. The classical results will not be used explicitly, but will just provide some background motivation for some of the results. Probably everyone has learnt about Euclidean planes. The classical projective plane comes from the classical Euclidean plane by introducing ideal (or infinite) elements. Associated to a parallel class of lines we have an ideal (or infinite) point, and the ideal line (or line at infinity) consists of all the infinite points. The advantage of introducing the classical projective plane is that there is no difference between ordinary and ideal points; two lines always intersect. In classical geometry typical theorems state that under some conditions certain lines pass through a point (for example, if we take a triangle, then the angle bisectors pass through a point) or certain points are on a line. In some cases, the classical theorems use metric properties of the plane (distances and angles), in other cases the order of the points on a line, but there are interesting results that just use incidences of points and lines. A notable example for this is the celebrated Theorem of Desargues.

Theorem 1.1. *Let $A_1A_2A_3$ and $B_1B_2B_3$ be two triangles in such a position that the lines A_iB_i pass through a point O. Consider the points $A_iA_j \cap B_iB_j = C_k$, where $\{i,j,k\} = \{1,2,3\}$. Then the points C_1, C_2, C_3 are on a line t.*

Less formally, when the two triangles are in perspective from the point O then they are also in perspective from the line t. More details on Desargues' theorem can be found in Coxeter's book [48], where similar theorems, for example the Theorem of Pappus, are also discussed. These theorems will also occur in our book, mainly in the context of finite planes and spaces. In Chapters 2 and 3 we shall see how particular cases of Desargues' theorem are related to properties of the coordinate structure of the projective plane. We shall also call the configuration of the ten points (A's, B's and C's and O) and the ten lines (the lines A_iB_i, the sides of the two triangles and the line t) a *closed Desargues configuration*.

Let us now start the more systematic study of projective and affine planes.

Definition 1.2. *A triple $(\mathcal{P}, \mathcal{E}, I)$, where $\mathcal{P}$ and $\mathcal{E}$ are two disjoint non-empty sets and $I \subset \mathcal{P} \times \mathcal{E}$ is an incidence relation, is called an* incidence geometry.

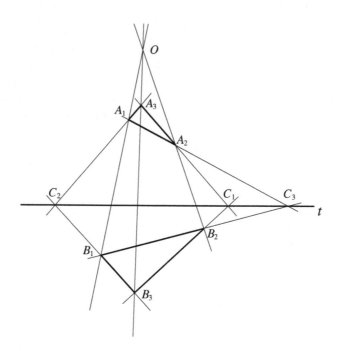

FIGURE 1.1
Desargues configuration

In the classical projective plane (the Euclidean plane extended by infinite points and line) the following facts are known for the incidence of points and lines:

- For any two distinct points there is exactly one line (the line joining them) which is incident with both of them.

- For any two distinct lines there is exactly one point (the common point of the two lines) which is incident with both of them.

The abstract projective plane is defined by requiring these two properties.

Definition 1.3. *An incidence geometry* $\Pi = (\mathcal{P}, \mathcal{E}, \mathrm{I})$ *is called a* projective plane *if it satisfies the following four axioms:*

- **P1.** *For any two distinct elements of* $\mathcal{P}$ *there is precisely one element of* $\mathcal{E}$, *which is in relation* I *with both of them.*

- **P2.** *For any two distinct elements of* $\mathcal{E}$ *there is precisely one element of* $\mathcal{P}$, *which is in relation* I *with both of them.*

- **P3.** *Each element of* $\mathcal{E}$ *is in relation* I *with at least three distinct elements of* $\mathcal{P}$.

- **P4.** *Each element of* $\mathcal{P}$ *is in relation* I *with at least three distinct elements of* $\mathcal{E}$.

A triple Π *is called a* degenerate projective plane *if it satisfies P1 and P2, but either it does not satisfy at least one of P3 and P4, or at least one of the sets* $\mathcal{P}$ *and* $\mathcal{E}$ *is empty.*

Points will be denoted by upper case Latin letters, lines by lower case Latin letters. The usual geometric terminology will also be used in case of abstract projective planes. As in everyday language, if the point P is incident with the line ℓ, then we will say that P is on ℓ; we shall talk about the line joining two points, and the common point (or intersection point) of two lines. AB denotes the line joining the points A and B, $e \cap f$ will be the common point of lines e and f. The axioms P1 and P2 guarantee that AB and $e \cap f$ are well-defined.

The *duality principle* known from classical projective geometry is valid for abstract projective planes, too. Since P1–P2, and P3–P4 are duals of each other, whenever a theorem can be deduced from axioms P1–P4, then its dual can also be deduced. This observation will be used frequently later. Associated to an abstract projective plane one can define its *dual plane* whose points (lines) are the lines (points) of the original plane and incidence is defined as the inverse of the incidence relation I. The dual of a projective plane Π is denoted by Π^*.

Axioms P3 and P4 in Definition 1.3 rule out degenerate projective planes, so they can be replaced by other axioms which guarantee that the plane is not degenerate.

Lemma 1.4. *A triple* $\Pi = (\mathcal{P}, \mathcal{E}, \mathrm{I})$ *is a(n abstract) projective plane if and only if it satisfies axioms P1 and P2 and one of the following two axioms:*

- **P3'.** *There are four points in general position, that is four points no three of which are collinear.*

- **P3".** *Two lines cannot cover the points of the plane; in other words, for every two lines of the plane there is a point which is not incident with any of the two lines.*

Proof. Assume that Π satisfies axioms P1, P2, P3 and P4. Through an arbitrary point P there are three distinct lines by P4. Let us denote these lines by e, f and g. Take four distinct points $E_1 \,\mathrm{I}\, e, E_2 \,\mathrm{I}\, e, F \,\mathrm{I}\, f$ and $G \,\mathrm{I}\, g$. They exist because of P3. At least one of the points E_1, E_2 cannot be on line FG. Without loss of generality we may assume that E_1 is not on FG. Then P, F, G and E_1 are four points in general position, hence Π satisfies axioms P1, P2 and P3'.

Now assume that Π satisfies axioms P1, P2 and P3' but not P3". If two lines e and f contain all the points of the plane, then, by P3', there are points $E_1 \,\mathrm{I}\, e$, $E_2 \,\mathrm{I}\, e$, $F_1 \,\mathrm{I}\, f$ and $F_2 \,\mathrm{I}\, f$, all different from the point $e \cap f$. On the other hand, by P1 and P2, there exists a point $E_1 F_1 \cap E_2 F_2$, which cannot be on lines e or f. This contradiction shows that Π satisfies axioms P1, P2 and P3".

If Π satisfies axioms a P1, P2 and P3", then, by P3", there are at least three lines through any point of the plane, hence P4 is also satisfied. If e is any line, then there is a point P not on e, by P3". There are at least three distinct lines through P. By P2 they meet e, so there are at least three distinct points on e. Therefore Π satisfies axiom P3, too. $\qquad\square$

It is clear that the points and lines of the classical projective plane, together with the incidence relation satisfy axioms P1–P4, so the classical projective plane is also an abstract projective plane. We are going to see some non-classical examples.

Example 1.5. Consider a regular triangle in the Euclidean plane. Let $\mathcal{P}$ be the set of three vertices, three midpoints and the center of the inscribed circle of this triangle. Let $\mathcal{E}$ be the set consisting of the sides, orthogonal bisectors and inscribed circle of the triangle. Let the incidence relation be the containment relation (see Figure 1.2).

It is easy to check that the structure defined in Example 1.5 satisfies axioms P1–P4 of abstract projective planes. This example is called the *Fano plane*. It bears the name of *Fano*, an Italian mathematician, who axiomatized projective spaces in 1892. (Among his axioms there was one that guaranteed that in a projective space there is no substructure isomorphic to the Fano plane.)

The next example also contains seven points and seven lines.

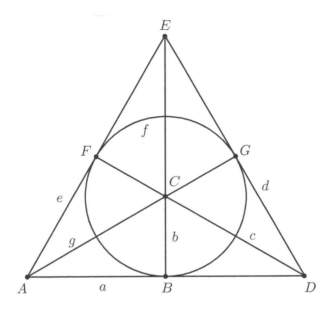

FIGURE 1.2
Fano plane

TABLE 1.1
Incidence table of the Fano plane

	A	B	C	D	E	F	G
a	1	1		1			
b		1	1		1		
c			1	1		1	
d				1	1		1
e	1				1	1	
f		1				1	1
g	1		1				1

Example 1.6. Let $\mathcal{P} = \{A, B, C, D, E, F, G\}$, $\mathcal{E} = \{a, b, c, d, e, f, g\}$, and incidence be defined by Table 1.1

A point is incident with a line if and only if the column of the table corresponding to the point contains a '1' in the row of the table corresponding to the line.

Axiom P1 is satisfied if and only if for every two columns of the table there is precisely one row so that there is a '1' in the row in both columns. Similarly, P2 is satisfied if and only if for every two rows of the table there is precisely one column so that there is a '1' in the column in both rows. This can be checked easily. As there are precisely three '1'-s in each row and column, P3 and P4 are also satisfied.

Comparing this example with the Fano plane shown on Figure 1.2, we see that the structures are essentially the same, since their points and lines can be mapped bijectively in such a way that incidence is also preserved. The bijective correspondence is shown on the figure by labeling the points and lines in the same way. Such planes will be called *isomorphic*. The isomorphism of two planes can be defined more formally. In general, the image of an element x under a mapping φ will be denoted by x^φ throughout the book.

Definition 1.7. *Let* $\Pi = (\mathcal{P}, \mathcal{E}, \mathrm{I})$ *and* $\Pi' = (\mathcal{P}', \mathcal{E}', \mathrm{I}')$ *be projective planes. An* isomorphism (*or* collineation) *between* Π *and* Π' *is a pair* $\varphi = (\varphi_1, \varphi_2)$ *where* $\varphi_1 : \mathcal{P} \to \mathcal{P}'$ *and* $\varphi_2 : \mathcal{E} \to \mathcal{E}'$ *are bijections which preserve the incidence in the sense that for all* $P \in \mathcal{P}$ *and* $e \in \mathcal{E}$

$$P \, \mathrm{I} \, e \iff P^{\varphi_1} \, \mathrm{I}' \, e^{\varphi_2}.$$

Two planes are called isomorphic *if there is an isomorphism between them.*

If we can fill in a table with '1'-s in such a way that for every two columns of the table there is precisely one row so that there is a '1' in the row in both columns and for every two rows of the table there is precisely one column so that there is a '1' in the column in both rows, moreover there are at least three '1'-s in each row and column, then we constructed the so-called *incidence table* (or *incidence matrix*, if the remaining positions are filled in with '0'-s) of the projective plane. It is then natural to ask what the possible sizes of incidence tables are. The next theorem is related to this question.

Theorem 1.8. *If in a projective plane* Π *there is a line which is incident with* $n + 1$ *points, then*

- *every line of* Π *is incident with* $n + 1$ *points,*

- *every point of* Π *is incident with* $n + 1$ *lines,*

- Π *contains* $n^2 + n + 1$ *points and the same number of lines.*

Proof. Let us denote the line containing $n + 1$ points by e. List the points incident with e: they are $P_1, P_2, \ldots, P_{n+1}$. If Q is a point not incident with e, then, by axiom P1, we can take the lines QP_i $(i = 1, 2, \ldots, n+1)$. These lines are pairwise distinct, since Q is not on e. On the other hand, each line through Q intersects e by axiom P2. This common point can only be one of the P_i's, hence there are precisely $n + 1$ lines through Q. Dualizing our argument, we see that if there is a point E which is incident with $n + 1$ lines, then every line not through E is incident with exactly $n + 1$ points.

If f is an arbitrary line different from e, then, by P4, there is at least one line, different from e and f, through the point $e \cap f$. By axiom P3, this line contains a point $R \neq e \cap f$. Since R is not on e, there pass exactly $n + 1$ lines through R. As R is not on f, the line f is incident with $n + 1$ points. This proves the first part of our assertion.

If P is an arbitrary point of the projective plane, then, by axioms P3 and P4, there is a line not through P. By what we have just proved, this line is incident with $n + 1$ points, therefore the number of lines through P is $n + 1$.

The total number of points of the projective plane can be obtained by counting the number of points lying on lines passing through a given point P. The number of lines through P is $n + 1$, and each of these lines contains n points different from P. Hence the total number of points is $1 + (n + 1)n = n^2 + n + 1$. Dualizing this argument we get that the total number of lines is also $n^2 + n + 1$. $\square$

The previous theorem implies the following: if there is a line of a projective plane which is incident with a finite number of points, then every line is incident with the same number of points and this common number is also the number of lines passing through a point. This shows that the following definition is meaningful.

Definition 1.9. *The order of the projective plane Π is n, if Π has a line which is incident with $n + 1$ points.*

Let us remark that the order of a plane is at least 2, and the Fano plane has order 2.

Example 1.10. Let $\mathbb{K}$ be an arbitrary (commutative) field and V be a three-dimensional vector space over $\mathbb{K}$. Moreover, let $\mathcal{P}$ be the set of one-dimensional, $\mathcal{E}$ the set of two-dimensional subspaces of V, and define the incidence relation I as the set theoretical containment relation.

Two distinct one-dimensional subspaces generate a two-dimensional subspace, hence the above defined structure satisfies axiom P1. Two distinct two-dimensional subspaces of a three-dimensional vector space intersect in a one-dimensional subspace, hence axiom P2 is also satisfied. Let $\mathbf{e}_1$ and $\mathbf{e}_2$ form a basis of a two-dimensional subspace V_2. Then V_2 contains the distinct one-dimensional subspaces generated by $\mathbf{e}_1$, $\mathbf{e}_2$ and $\mathbf{e}_1 + \mathbf{e}_2$. This shows that P3 is satisfied. If V_1 is a one-dimensional subspace generated by $\mathbf{e}_1$, then there exist vectors $\mathbf{e}_2$ and $\mathbf{e}_3$ that together with $\mathbf{e}_1$ form a basis of the three-dimensional vector space V. Then $\langle \mathbf{e}_1, \mathbf{e}_2 \rangle$, $\langle \mathbf{e}_1, \mathbf{e}_3 \rangle$ and $\langle \mathbf{e}_1, \mathbf{e}_2 + \mathbf{e}_3 \rangle$ are two-dimensional subspaces of V that are pairwise distinct and contain V_1, hence P4 is also satisfied.

In a three-dimensional vector space one-dimensional subspaces can be described by one of their generating vectors, two-dimensional subspaces can be described by a generating vector of their orthogonal complement. This way one can obtain the following algebraic description of the projective plane coming from the three-dimensional vector space V over a commutative field $\mathbb{K}$:

Points are the triples $\mathbf{x} = (x_0, x_1, x_2)$, where $(x_0, x_1, x_2) \neq (0,0,0)$, and (x_0, x_1, x_2) and $(\lambda x_0, \lambda x_1, \lambda x_2)$ represent the same point if $0 \neq \lambda \in \mathbb{K}$; *lines* are the triples $\mathbf{u} = [u_0, u_1, u_2]$, where $[u_0, u_1, u_2] \neq [0,0,0]$, and $[u_0, u_1, u_2]$ and $[\lambda u_0, \lambda u_1, \lambda u_2]$ represent the same line if $0 \neq \lambda \in \mathbb{K}$. Finally, the point (x_0, x_1, x_2) and the line $[u_0, u_1, u_2]$ are *incident* if and only if $\mathbf{x}\mathbf{u}^T = \sum_{i=0}^{2} x_i u_i = 0$. Another way of saying this is that the *equation of the line* $[u_0, u_1, u_2]$ is $u_0 X_0 + u_1 X_1 + u_2 X_2 = 0$.

These planes will be denoted by $\mathrm{PG}(2, \mathbb{K})$ in what follows. If $\mathbb{K}$ is the finite field with q elements, denoted by $\mathrm{GF}(q)$ or $\mathbb{F}_q$, then the corresponding plane is denoted by $\mathrm{PG}(2, q)$ and it is often called a *Galois plane*. Let us remark that the order of $\mathrm{PG}(2, q)$ in the sense of Definition 1.9 is q (see Exercise 1.9).

The theorems known for the classical projective plane which can be proven by computation with homogeneous coordinates using only the properties of fields (and not the fact that the real numbers are ordered) remain valid for the more general planes $\mathrm{PG}(2, \mathbb{K})$. Examples of such theorems are the Theorem of Pascal and Desargues' theorem. We will often use the collinearity condition stating that three points $\mathbf{x} = (x_0, x_1, x_2)$, $\mathbf{y} = (y_0, y_1, y_2)$ and $\mathbf{z} = (z_0, z_1, z_2)$ of $\mathrm{PG}(2, \mathbb{K})$ are collinear if and only if

$$\begin{vmatrix} x_0 & x_1 & x_2 \\ y_0 & y_1 & y_2 \\ z_0 & z_1 & z_2 \end{vmatrix} = 0.$$

We have to be careful when divisions (by 2, 3 etc.) are used in the proof, so in many cases special attention is needed when the characteristic of $\mathbb{K}$ is 2 (or sometimes if it is 3).

For the sake of completeness and also to show how homogeneous coordinates and the vector space model can be used, we prove Desargues' theorem.

Proof of Theorem 1.1. We use the notation in the statement of Theorem 1.1. If a point is denoted by an upper case Latin letter, then the corresponding boldface lower case Latin letter denotes one of its representing vectors. Let us remark that three points are collinear if and only if their representing vectors are linearly dependent.

If $A_i = B_i$ for some $i \in \{1, 2, 3\}$, then the statement is obvious. From now on we assume that $A_i \neq B_i$. This implies that the vectors $\mathbf{a_i}$ and $\mathbf{b_i}$ form a basis of the 2-dimensional subspace corresponding to the line $A_i B_i$. As the point O is incident with each of the lines $A_i B_i$, there exist $\lambda_i, \mu_i \in \mathbb{K}$ for which

$$\mathbf{o} = \lambda_1 \mathbf{a_1} + \mu_1 \mathbf{b_1}, \quad \mathbf{o} = \lambda_2 \mathbf{a_2} + \mu_2 \mathbf{b_2}, \quad \mathbf{o} = \lambda_3 \mathbf{a_3} + \mu_3 \mathbf{b_3} \qquad (1.1)$$

hold. Subtracting the second equation from the first one we get

$$\lambda_1 \mathbf{a_1} - \lambda_2 \mathbf{a_2} = \mu_2 \mathbf{b_2} - \mu_1 \mathbf{b_1}.$$

On the one hand, this vector is a linear combination of $\mathbf{a_1}$ and $\mathbf{a_2}$; on the other hand, it is a linear combination of $\mathbf{b_2}$ and $\mathbf{b_1}$, and thus it represents a point which is incident with both of the lines $A_1 A_2$ and $B_1 B_2$. These two lines have a unique point of intersection, hence the vector $\lambda_1 \mathbf{a_1} - \lambda_2 \mathbf{a_2}$ is a representing vector of the point $C_3 = A_1 A_2 \cap B_1 B_2$.

In the same way we get that the vectors $\lambda_2 \mathbf{a_2} - \lambda_3 \mathbf{a_3}$ and $\lambda_3 \mathbf{a_3} - \lambda_1 \mathbf{a_1}$ are representing vectors of the points $C_1 = A_2 A_3 \cap B_2 B_3$ and $C_2 = A_3 A_1 \cap B_3 B_1$, respectively. The equality

$$(\lambda_1 \mathbf{a_1} - \lambda_2 \mathbf{a_2}) + (\lambda_2 \mathbf{a_2} - \lambda_3 \mathbf{a_3}) + (\lambda_3 \mathbf{a_3} - \lambda_1 \mathbf{a_1}) = 0$$

means that these three vectors are linearly dependent. This implies that the corresponding points, C_3, C_1 and C_2 are collinear, hence the theorem is proved.
□

If $\mathbb{K}'$ is a subfield of $\mathbb{K}$, V' is a three-dimensional vector space over $\mathbb{K}'$, $\mathcal{P}'$ is the set of one-dimensional, $\mathcal{E}'$ is the set of two-dimensional subspaces of V', incidence is the set theoretical containment relation, then the projective plane coming from V contains naturally the projective plane coming from V'. The latter projective plane is called the subplane of the former one. Now we define the notion of a subplane in full generality.

Definition 1.11. *The projective plane* $\Pi' = (\mathcal{P}', \mathcal{E}', \mathrm{I}')$ *is a subplane of the projective plane* $\Pi = (\mathcal{P}, \mathcal{E}, \mathrm{I})$ *if* $\mathcal{P}' \subset \mathcal{P}$, $\mathcal{E}' \subset \mathcal{E}$, *and* I' *is the same as* I *on the set* $\mathcal{P}' \times \mathcal{E}'$ *(in other words,* I' *is the restriction of* I, *or a point of a subplane is incident with a line of the subplane if and only if they are incident in the original plane* Π).

If the set of points of a subplane of order s is given then the lines of the subplane are those lines of the big plane which meet the given set of points in at least two points. The subplane has $s^2 + s + 1$ lines; if a line of the big plane contains at least two points of the subplane, then it contains exactly $s + 1$ points of the subplane. If each of the two lines of the big plane contains $s + 1$ points of the subplane, then their point of intersection belongs to the subplane.

The next theorem shows that the order of a subplane cannot be large when comparing with the order of the plane.

Theorem 1.12 (Bruck). *If a projective plane* Π_n *of order* n *contains a subplane* Π_s *of order* $s < n$, *then either* $n = s^2$ *or* $n \geq s^2 + s$.

Proof. Let t be a line of Π_n meeting Π_s in a single point P (a tangent to Π_s). Such lines exist because there are $n+1$ lines through each point of Π_n and at most $s+1$ of them belong to Π_s. There are s^2 lines in Π_s that do not contain P. No two of them meet t in the same point, hence $n \geq s^2$.

If $n = s^2$, then the total number of tangents to Π_s is $(s^2 + s + 1)(s^2 - s)$, because there are $n + 1 = s^2 + 1$ lines through each point of Π_n, among them $s + 1$ are lines of Π_s, thus the remaining $s^2 - s$ are tangents to Π_s. There are $s^2 + s + 1$ lines in Π_s, thus the number of tangents to Π_s and secants of Π_s is altogether exactly $s^4 + s^2 + 1 = n^2 + n + 1$, that is the total number of lines in Π_n.

If $n \neq s^2$, then there is a line ℓ in Π_n which does not meet Π_s. This implies that each of the $s^2 + s + 1$ lines of Π_s meets ℓ in distinct points, so $n + 1 \geq s^2 + s + 1$.   □

Subplanes of order $\sqrt{n}$ appear several times in the theory of finite geometries. This motivates that they have a distinguished name.

Definition 1.13. *Let n be a perfect square and Π_n be a projective plane of order n. A subplane of order $\sqrt{n}$ of Π_n is called a* Baer subplane.

As mentioned earlier the incidence table of a projective plane can be regarded as a 0–1 matrix (by filling in the positions that are not '1' with '0') which is called the *incidence matrix* of the plane. If the order of the projective plane is n, then the incidence matrix is a $v \times v$ matrix, where $v = n^2 + n + 1$. Now, we prove two properties of the incidence matrices that are needed later.

Lemma 1.14. *If A is the incidence matrix of a projective plane of order n then*

$$AA^{\mathrm{T}} = J + nI,$$

where the matrices in this equation are of size $(n^2 + n + 1) \times (n^2 + n + 1)$, I is the identity, J is the all-1 matrix.

Proof. Let $A = (a_{ij})$ and $AA^{\mathrm{T}} = C = (c_{ij})$. By the definition of C

$$c_{ij} = \sum_{k=1}^{v} a_{ik} a_{jk},$$

where $v = n^2 + n + 1$. The products in the sum on the right-hand side are either 0 or 1 and we get 1 precisely when both a_{ik} and a_{jk} are 1. This is the number of common points of the i-th and the j-th line, which is 1 or $n + 1$ depending on whether the two lines (or the indices i and j) are distinct or equal. This is precisely what we wanted to prove.   □

Lemma 1.15. *If A is an incidence matrix of a projective plane of order n then*

- *the eigenvalues of AA^{T} are $(n+1)^2$ (with multiplicity one) and n (with multiplicity $n^2 + n$);*

- *A is non-singular over the field of rational numbers.*

Proof. By the previous lemma, $AA^{\mathrm{T}} = J + nI$, so the characteristic polynomial of AA^{T}, $f(\lambda)$, can be written as $f(\lambda) = \det(J + nI - \lambda I)$. This determinant can be computed easily. Let us first subtract the first row from all the other rows and then add all the columns to the first one. Then

$$
f(\lambda) = \begin{vmatrix}
n+1-\lambda & 1 & \cdots & 1 \\
1 & n+1-\lambda & \cdots & 1 \\
\vdots & \vdots & \ddots & \vdots \\
1 & 1 & \cdots & n+1-\lambda
\end{vmatrix}
$$

$$
= \begin{vmatrix}
n+1-\lambda & 1 & \cdots & \cdots & 1 \\
-n-\lambda & n-\lambda & 0 & \cdots & 0 \\
-n-\lambda & 0 & n-\lambda & \cdots & 0 \\
\vdots & \vdots & \vdots & \ddots & \vdots \\
-n-\lambda & 0 & \cdots & 0 & n-\lambda
\end{vmatrix}
$$

$$
= \begin{vmatrix}
(n+1)^2-\lambda & 1 & \cdots & \cdots & 1 \\
0 & n-\lambda & 0 & \cdots & 0 \\
\vdots & 0 & n-\lambda & \cdots & \vdots \\
\vdots & \vdots & \vdots & \ddots & \vdots \\
0 & 0 & \cdots & 0 & n-\lambda
\end{vmatrix}
$$

$$
= \left((n+1)^2 - \lambda\right)(n-\lambda)^{n^2+n}.
$$

This proves the first part of the statement at once.

Substituting $\lambda = 0$ we get $\det(J + nI) = (n+1)^2 n^{n^2+n} \neq 0$. Since $\det A = \det A^{\mathrm{T}}$ this implies $|\det A| = (n+1)n^{(n^2+n)/2} \neq 0$, so the incidence matrix is non-singular over the rationals. $\qquad\square$

The next example, due to *Hall* [77], will be described by constructing its incidence table. In this construction both the number of points and that of lines will be infinite. The example constructed in this way is usually called the *free plane of Hall* and the method is the so-called *free extension process*.

Example 1.16. Let us start from a 6×4 incidence table whose rows we call lines and whose columns we imagine as points. Incidence is defined in such a way that any line be incident with precisely two points, so each row contains two '1'-s and for each pair of columns there is a row where we have a '1' in these columns in the table. This is the upper left corner of Figure 1.3. We shall extend this table in successive rounds, the above starting table being the first round. At this point our "plane" satisfies axiom P1 but there are pairs of lines which do not have a common point. In the next round we make sure that any two lines in the previous round have a common point, so we add new columns. These are columns P_5, P_6 and P_7 in Figure 1.3. After this round our

	P_1	P_2	P_3	P_4	P_5	P_6	P_7	P_8	P_9	P_{10}	P_{11}	P_{12}	P_{13}	P_{14}
e_1	1	1			1			1						
e_2	1		1			1			1					
e_3	1			1			1			1				
e_4		1	1				1				1			
e_5		1		1		1						1		
e_6			1	1	1								1	
e_7					1	1				1	1			
e_8					1		1		1			1		
e_9						1	1	1					1	
e_{10}														

FIGURE 1.3
Hall's free plane after the 4th round

table satisfies axiom P2 but not P1, since there are pairs of points which do not have a line joining them. Now we add new lines for each pair of points which are not yet joined, that is add a new row to our table for each unjoined pair of points and put a '1' in the corresponding columns. The extended table again satisfies P1 but not P2. Continue this extension process: in even rounds new points (columns) are added for pairs of previously disjoint lines (rows which do not have a common '1'), in odd rounds lines (rows) are added for unjoined pairs of points (columns). After performing the odd rounds axiom P1, after the even rounds, axiom P2 will be satisfied. In each round finitely many columns or rows are added but the extension process itself is infinite. The table obtained by this infinite process is the incidence table of a projective plane. Figure 1.3 shows the first four rounds of the extension process.

Using induction on r, where r is the number of rounds performed, it is easy to see that any pair of distinct points is joined by at most one line (and any pair of distinct lines intersect in at most one point). Take two points A and B, where A is defined in round k, B is defined in round $\ell \geq k$. Then in round ℓ they are joined by at most one line. If they are not joined in round ℓ, then the unique line joining them will be defined in round $\ell + 1$. Similarly, one can prove that P2 is also satisfied. P3 and P4 are valid, because when we define a new point (or line), it is the common point of two previously defined lines (or joining line of two previously defined points) but they are not incident with all lines (points) defined in previous rounds. Therefore, in the next round they will be used to define at least one new line (or point).

An interesting property of this plane is that it contains no closed Desargues configuration at all (moreover, it contains no finite configuration in which any point is incident with at least three lines and any line is incident with at least three points). Indeed, list the number of rounds in which the points or lines of the configuration are defined and take the maximum r of these numbers. Then the point(s) or line(s) defined in round r are incident with exactly two lines or points at the moment they are defined. Note also that we can start the above free extension process by Hall with any table which does not violate P1 and P2, not just the 6×4 table we started with.

In the next example so-called *cyclic planes* are constructed.

Example 1.17. Let S be a regular $(n^2 + n + 1)$-gon in the Euclidean plane and O be the center of its circumscribed circle. Let us number the consecutive vertices of S by $0, 1, \ldots, n^2 + n$ clockwise. So the point numbered by i is between the points $i - 1$ and $i + 1$. We shall calculate with these numbers modulo $n^2 + n + 1$. In this representation rotation about O by the angle $-360°/(n^2 + n + 1)$ is just the mapping $x \mapsto x + 1$ modulo $(n^2 + n + 1)$.

If A and B are two disjoint points of S numbered by a and b, $(a < b)$, then the distance of A and B (on the circle) is k if the smaller of the two values $b - a$ and $n^2 + n + 1 - b + a$ is k. This is proportional with the smaller of the angles of $AOB\sphericalangle$ and $BOA\sphericalangle$. This also implies that two points have the same Euclidean distance if and only if they have the same distance (on the circle) in the above defined sense. Among the vertices of S there are $(n^2 + n)/2 = \binom{n+1}{2}$ different distances in this sense. Assume that we can select an $(n + 1)$-gon R from S in such a way that the distances of pairs of vertices of R are pairwise different. As in this case the Euclidean distances of the pairs of vertices are also pairwise different, *Kárteszi* [105] called such an $(n + 1)$-gon a totally irregular $(n + 1)$-gon. Figure 1.4. shows this situation for $n = 3$. Here S is a 13-gon and a suitable 4-gon R is also on the picture. Let P be the set of vertices of S, $\mathcal{E}$ be the sets obtained by rotating R about O by the angle $(-k) \cdot 360°/(n^2 + n + 1)$, $(k = 0, 1, \ldots, n^2 + n)$. Incidence is defined as set theoretical inclusion.

In the above numbering this just means that R can be identified with a set $R \subset \{0, 1, \ldots, n^2 + n\}$ and $\mathcal{E}$ is the set of translates of R, that is the sets of the form $x + R = \{x + r : r \in R\}$. In what follows we use the labels of the vertices.

If a and b are two distinct points whose distance is k then there is precisely one pair of vertices of R, numbered by r_1 and r_2, whose distance is also k. We can order a, b and r_1, r_2 so that $b - a = k$ and $r_2 - r_1 = k$. Now $R + (a - r_1)$ will be the unique translate of R which contains a and b.

If $R_1 = R + u$ and $R_2 = R + v$ are two distinct lines, then the distance k of u and v occurs precisely once among the distances of the elements of R_1; denote these two elements by r_1 and r_1'. Again we may index the lines and the

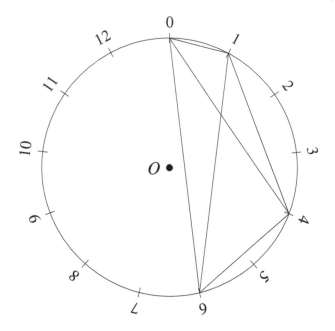

FIGURE 1.4
Cyclic model for $n = 3$

two elements in such a way that $k = v - u$ and $k = r'_1 - r_1$. Then $R_2 = R_1 + k$ and the point r'_1 is the (unique) common point of R_1 and R_2.

For $n \geq 2$ axioms P3 and P4 are clearly satisfied in our model.

For a given $\mathcal{S}$ the totally irregular $(n+1)$-gon $\mathcal{R}$ does not exist necessarily, so we cannot construct a plane for every n with this cyclic model. The key property of R in the additive group $\mathbb{Z}_{n^2+n+1}$ is that every non-zero element g of $\mathbb{Z}_{n^2+n+1}$ can be written uniquely as a difference $g = r - s$, where $r, s \in R$. It holds because the distances of pairs of vertices of $\mathcal{R}$ are pairwise different; hence for any distance $0 < d \leq (n^2 + n)/2$ there is a unique pair $r, s \in R$ such that $d = r - s$; but this means that $-d = s - r$, so $n^2 + n + 1 - d \in \mathbb{Z}_{n^2+n+1}$ can also be written uniquely as a difference of two elements of R. Such sets play an important role in combinatorics. For the sake of completeness we give their general definition.

Definition 1.18. *Let $(G, +)$ be an additive group of order v. A subset $D \subset G$ of k elements is a (v, k, λ)-difference set if every non-zero element g of G can be expressed as a difference $g = d_i - d_j$ of elements of D in exactly λ ways. If $\lambda = 1$, then D is called a* planar difference set.

The reader is referred to the book of *Jungnickel* [98] for the theory of difference sets.

TABLE 1.2

Planar difference sets

n	$n^2 + n + 1$	
2	7	$0, 1, 3$
3	13	$0, 1, 4, 6$
4	21	$0, 1, 4, 14, 16$
5	31	$0, 1, 3, 8, 12, 18$
6	43	does not exist
7	57	$0, 1, 3, 13, 32, 36, 43, 52$
8	73	$0, 1, 3, 7, 15, 31, 36, 54, 63$
9	91	$0, 1, 3, 9, 27, 49, 56, 61, 77, 81$

For small values of n we can just find planar difference sets by exhaustive search. If n is a prime power, then, using the fact that the multiplicative group of finite fields is cyclic, one can explicitly construct difference sets. This method will be discussed in Chapter 2, Theorem 2.15. In Table 1.2 we list planar difference sets for $n < 10$.

If we delete a line of the projective plane $(\mathcal{P}, \mathcal{E}, \mathrm{I})$ together with all its points, then the remaining points and lines form a structure $(\mathcal{P}', \mathcal{E}', \mathrm{I}')$, where I' is the restriction of I onto $\mathcal{P}' \times \mathcal{E}'$. This structure satisfies the following axioms:

- **A1.** For any two distinct points of $\mathcal{P}'$ there is a unique element of $\mathcal{E}'$ which is in relation I' with both points.

- **A2.** If $P \in \mathcal{P}'$ is not in relation I' with the element $e \in \mathcal{E}'$, then there is a unique element of $\mathcal{E}'$ which is in relation I' with P but not in relation I' with any element of $\mathcal{P}'$ being in relation I' with e.

- **A3.** Every element of $\mathcal{E}'$ is in relation I' with at least two elements of $\mathcal{P}'$.

- **A4.** Every element of $\mathcal{P}'$ is in relation I' with at least three elements of $\mathcal{E}'$.

Definition 1.19. *A point-line incidence geometry* $(\mathcal{P}', \mathcal{E}', \mathrm{I})$ *is called an* affine plane *if it satisfies axioms* A1—A4.

Similarly to projective planes, also for affine planes we shall call the elements of $\mathcal{P}'$ points, the elements of $\mathcal{E}'$ lines, and instead of "being in relation I'" we shall use the usual geometric terminology. So a point is on a line, two points are joined by a line etc.

Our procedure also works in the other direction; starting from an affine plane one can extend it to a projective plane by adding one point to each line. This is the generalization of introducing ideal (infinite) elements to the classical Euclidean plane to construct the classical projective plane.

Example 1.20. Let $(\mathcal{P}', \mathcal{E}', \mathrm{I}')$ be an affine plane. Call two elements of $\mathcal{E}'$ (that is, two lines) *parallel* if they do not have a common point or they are the same. Parallelism is an equivalence relation since it is clearly reflexive and symmetric and transitivity follows from axiom A2. Let us call the equivalence classes of lines *ideal points* (or *points at infinity*) and denote them by $\mathcal{P}_\infty$. The *line at infinity* ℓ_∞ will be a new line not in $\mathcal{E}'$. It is also called the *ideal line*. The *projective closure* of the affine plane $(\mathcal{P}', \mathcal{E}', \mathrm{I}')$ will be the projective plane $(\mathcal{P}, \mathcal{E}, \mathrm{I})$, where $\mathcal{P} = \mathcal{P} \cup \mathcal{P}_\infty$ and $\mathcal{E} = \mathcal{E} \cup \{\ell_\infty\}$. Incidence of the elements of the original affine plane is defined by I', incidence of ideal elements is defined as we do it for the classical projective plane: an ideal point is incident with an affine line if and only if the line is in the parallel class corresponding to the ideal pont, an affine point is not incident with the ideal line, and the ideal points are all incident with the ideal line.

The projective closure of any affine plane is a projective plane. The validity of axioms P1 and P2 follow from a case analysis. Two distinct affine points are incident with a unique affine line and the line at infinity is not incident with affine points. Axiom A2 guarantees that for an affine point and an ideal point there is a unique line through the affine point which is in the parallel class corresponding to the ideal point, so in the projective closure there is a unique line joining the affine point and the ideal point. The ideal line does not pass through the affine point. Two distinct ideal points are only incident with the ideal line, because an affine line belongs to precisely one parallel class. This shows that axiom P1 is satisfied. Two distinct affine lines are either parallel or they have a common affine point. In the former case they do not have a common affine point but belong to the same parallel class, hence they have a (unique) common ideal point. In the latter case they have a (unique) common affine point but they are not parallel, so they have no common ideal point. Finally, an affine line and the ideal line have no common affine point and the ideal point of the affine line (the parallel class in which it is contained) will be their unique common ideal point. Thus axiom P2 is satisfied by the projective closure. Axioms P3 and P4 are clearly true.

Every affine plane has a unique projective closure but uniqueness is not true in the other direction: if we delete different lines of a projective plane, then the resulting affine planes are not necessarily isomorphic. (We shall see explicit examples in Chapter 3 when we study translation planes.)

If the projective closure of an affine plane is a projective plane of order n, then Theorem 1.8 implies the next statements immediately.

Theorem 1.21. *If an affine plane $\mathcal{A}$ has a line incident with n points, then*

- *every line of $\mathcal{A}$ is incident with n points,*

- *every point of $\mathcal{A}$ is incident with $n + 1$ lines,*

- *$\mathcal{A}$ has n^2 points and $n^2 + n$ lines in total.*

The number n is called the order *of the affine plane* $\mathcal{A}$.

Example 1.22. If we delete the line with equation $X_0 = 0$ (in other words, the line with coordinates $[1, 0, 0]$) from the projective plane $\mathrm{PG}(2, \mathbb{K})$ together with all of its points, then the resulting affine plane will be denoted by $\mathrm{AG}(2, \mathbb{K})$.

Because of homogeneity of coordinates of the projective plane we may assume that the first coordinate of each point of $\mathrm{AG}(2, \mathbb{K})$ is 1. Therefore, the points of $\mathrm{AG}(2, \mathbb{K})$ can be described by two coordinates, namely (x, y), where $x = x_1/x_0$ and $y = x_2/x_0$.

In this way we get the following algebraic model: points are ordered pairs $(x, y) \in \mathbb{K} \times \mathbb{K}$, lines are those triples $[C, A, B]$, where $(A, B) \neq (0, 0)$ and $[C, A, B]$ is the same line as $[\lambda C, \lambda A, \lambda B]$ if $0 \neq \lambda \in \mathbb{K}$. The point (x, y) and the line $[C, A, B]$ are incident if and only if $Ax + By + C = 0$.

If $\mathbb{K} = \mathrm{GF}(q)$, then $\mathrm{AG}(2, \mathbb{K})$ is denoted by $\mathrm{AG}(2, q)$ and it is often called an *affine Galois plane*.

Examples 1.20 and 1.22 prove the following theorem immediately.

Theorem 1.23. *A projective plane of order n exists if and only if an affine plane of order n exists.*

The points of the classical Euclidean plane can be identified with the field of complex numbers, called the *Gauss-Argand model*. Generalizing this idea we get another algebraic model of $\mathrm{AG}(2, \mathbb{K})$ if $\mathbb{K}$ admits an extension of degree 2 (for example $\mathbb{K} = \mathbb{R}$ and $\mathbb{K} = \mathrm{GF}(q)$).

Example 1.24. Let f be an irreducible polynomial of degree two over the field $\mathbb{K}$; let $\mathbb{F} = \mathbb{K}(i)$ be the quadratic extension of $\mathbb{K}$, where i is a root of f. The elements of $\mathbb{F}$ can be written in the form $a + bi$, where $a, b \in \mathbb{K}$.

Let us map the point (a, b) of $\mathrm{AG}(2, \mathbb{K})$ to the field element $a + bi \in \mathbb{F}$. This map is a bijection between the points of $\mathrm{AG}(2, \mathbb{K})$ and the elements of $\mathbb{F}$. The line $[C, A, B]$ is mapped to the subset $\{x + yi \in \mathbb{F} : x, y \in \mathbb{K}, \ Ax + By + C = 0\}$ of $\mathbb{F}$.

In this model parallelism of lines and collinearity of points can be described easily.

Lemma 1.25. *Consider four points,* $P_j = (a_j, b_j)$ *for* $j = 1, 2, 3, 4$, *of* $\mathrm{AG}(2, \mathbb{K})$ *and the corresponding elements* $a_j + b_j i = z_j \in \mathbb{F}$ *in the model in Example 1.24. Then the lines determined by* $P_1 P_2$ *and* $P_3 P_4$ *are parallel if and only if there is an* $\alpha \in \mathbb{K}$ *so that* $z_4 - z_3 = \alpha(z_2 - z_1)$.

Proof. The equation of the lines $P_1 P_2$, and $P_3 P_4$ in the plane $\mathrm{AG}(2, \mathbb{K})$ is

$$(b_2 - b_1)x + (a_1 - a_2)y + (b_1 a_2 - b_2 a_1) = 0,$$

and

$$(b_4 - b_3)x + (a_3 - a_4)y + (b_3 a_4 - b_4 a_3) = 0,$$

respectively. These two lines are parallel if and only if

$$\begin{vmatrix} b_2 - b_1 & a_1 - a_2 \\ b_4 - b_3 & a_3 - a_4 \end{vmatrix} = 0.$$

The above determinant is zero if and only if the rows of the matrix are linearly dependent. In other words, there is an $\alpha \in \mathbb{K}$, so that

$$a_3 - a_4 = \alpha(a_1 - a_2) \quad \text{and} \quad b_4 - b_3 = \alpha(b_2 - b_1).$$

This happens if and only if

$$z_4 - z_3 = \alpha(z_2 - z_1)$$

is satisfied.   □

Corollary 1.26. *Three distinct points, $P_j = (a_j, b_j)$, $j = 1, 2, 3$, of AG$(2, q)$ are collinear if and only if*

$$(z_2 - z_3)^{q-1} = (z_2 - z_1)^{q-1} \tag{1.2}$$

holds for the elements $a_j + b_j i = z_j \in \mathrm{GF}(q^2)$.

Proof. The three points are collinear if and only if the lines $P_1 P_2$ and $P_2 P_3$ are parallel. Applying the previous lemma in the case $P_4 = P_2$ we get that it happens if and only if there exists $\alpha \in \mathrm{GF}(q)$ so that

$$z_2 - z_3 = \alpha(z_2 - z_1). \tag{1.3}$$

If the points are collinear, then taking the $(q-1)$-st power of this equation we get Equation (1.2), because $\alpha \in \mathrm{GF}(q)$ implies $\alpha^{q-1} = 1$. In the opposite direction, if Equation (1.2) holds, then

$$\left(\frac{(z_2 - z_3)}{(z_2 - z_1)} \right)^{q-1} = 1 \tag{1.4}$$

also holds. It is well-known that in the field $\mathrm{GF}(q^2)$ the roots of the equation $x^{q-1} = 1$ are exactly the non-zero elements of $\mathrm{GF}(q)$. Hence from Equation (1.4) we get that there exists $0 \neq \alpha \in \mathrm{GF}(q)$ so that

$$\frac{z_2 - z_3}{z_2 - z_1} = \alpha,$$

which means that Equation (1.3) also holds.   □

Now two models of infinite projective planes are discussed. Both have natural finite analogs that we shall see in the next chapters.

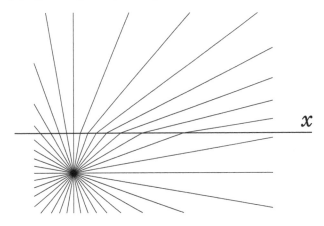

FIGURE 1.5
A pencil of lines in the Moulton plane

Example 1.27. The following example of an (infinite) projective plane is due to *Moulton*. The motivation for constructing this example was to show that Desargues' theorem does not follow from the incidence axioms of projective (or affine) planes. We shall construct the *affine Moulton plane*.

Consider the classical Euclidean affine plane and take a Cartesian system of coordinates. Keep the points and change the definition of "lines" so that the incidence axioms are still valid. The x-axis divides the plane into two halfplanes H_U and H_L. Lines having negative slope, vertical and horizontal lines remain the same, while lines with positive slope are modified above the x-axis. Let ℓ be a line with positive slope and denote by ℓ' the line whose slope is half of the slope of ℓ which intersects ℓ in a point L on the x-axis. We shall associate a distorted line ℓ^*, consisting of two halflines meeting each other in L, to ℓ. The distorted line ℓ^* is incident with the points of ℓ below or on the x-axis, and it is incident with the points of ℓ' above the x-axis. One can easily check that the incidence axioms hold in the Moulton plane. Distinguishing the cases according to their position with respect to the x-axis, one can immediately see that from the line joining them in the Euclidean plane one gets the line joining them in the Moulton plane, except for the case when one of the points (say A) is above, the other one (say B) is below the x-axis and the Euclidean line joining them has positive slope. In this case let A' be the unique point in the upper halfplane whose first coordinate is the same as that of A and whose second coordinate is twice the second coordinate of A. Let ℓ be the Euclidean line joining A' and B. Then the distorted line ℓ^* will be the unique line joining A and B in the Moulton plane. See Figure 1.5.

For the validity of the second axiom, one can show that two lines in the Moulton planes do not intersect if and only if they can be obtained from each other by an Euclidean translation. We leave the proof as an exercise.

Similarly, one can show that Desargues' theorem is not valid in the Moulton plane in general, since one can place an Euclidean Desargues configuration in such a way that most of the figure is below the x-axis; just one of the intersections of two corresponding sides of the perspective triangles are above. One can also achieve that exactly one of the three lines passing through this point above the x-axis has positive slope. Hence this line is modified, so it cannot pass through the point of intersection of the two other lines. Of course, this is just an intuitive sketch of the proof. It can also be seen in Figure 1.6. The details are left as an exercise.

The next example uses the same idea. Points of the Euclidean plane are kept, only the lines are replaced by "new lines".

Example 1.28. Again, let us keep the points of the classical Euclidean plane. We are going to modify the non-vertical lines. Consider a normal parabola $\mathcal{C}$ (the graph of the mapping $x \mapsto x^2$) and call all the translates of $\mathcal{C}$ "new lines". We keep vertical lines as "new lines", too. Then our structure is an (abstract) affine plane. Take two points A and B. If they are on a vertical line, then they cannot be on translated parabolas. If the Euclidean line ℓ joining them is not vertical, then there is a tangent line t to $\mathcal{C}$ which is parallel to ℓ. Let T be the point $\mathcal{C} \cap t$. Move the line t towards the parabola (this means that we continously translate it by a vector perpendicular to t). If we do so, then at each moment the translated line t' will meet $\mathcal{C}$ in two points, say T_1 and T_2. Since the function $x \mapsto x^2$ is continuous, the distance of T_1 and T_2 will be a continuous mapping of the length of the translation vector. The distance can be arbitrary large, therefore, there will be a moment when the distance of T_1 and T_2 is equal to the distance of A and B. This means that the two points A and B can simultaneously be translated to the parabola, and hence there is a translate of $\mathcal{C}$ that passes through A and B. An alternative argument is to translate the parabola $\mathcal{C}$ by the vector $\overrightarrow{AB}$, giving another parabola. The two parabolas have to meet and again we obtain a translate of $\mathcal{C}$ that passes through A and B. Finally, high school analytic geometry can also be used. Let us denote the coordinates of A by (a_1, a_2) and those of B by (b_1, b_2). The translates of $\mathcal{C}$ have equation $y = (x + a)^2 + b$, so we have to solve a system of simple quadratic equations. It is easy to see that the system has a unique solution for $a_1 \neq b_1$. Using either argument one sees that the first axiom of affine planes is valid.

Actually, any of the three arguments above also shows that two "new lines" are parallel if and only if they are vertical translates of each other. This immediately shows that axiom A2 of Definition 1.19 is also valid. Again, the details are left as an exercise.

One should also note that the same construction can be used for any function f whose graph looks like the graph of the parabola (that is f is convex from below and $\frac{f(x)}{x} \to \pm\infty$ if $x \to \pm\infty$). A function that gives an affine plane using the above construction will be called a *planar function*. The high school analytic geometry approach above shows that a function f is planar if and only if $f(x + a) - f(x) = b$ has a unique solution for every $0 \neq a$, and b. So this construction method gives many seemingly different affine planes. It is non-trivial to prove that the resulting function plane is desarguesian if and only if we start from the usual parabola C (or more generally the graph of a quadratic function). Exercises 1.17–1.23 are about these facts.

A *Latin square of order* n is an $(n \times n)$ table whose rows are filled with the numbers $0, 1, \ldots, n - 1$ in such a way that each number occurs precisely once in each row and column. Denote such a Latin square by L, and element in the i-th row and j-th column by $L_{i,j}$. Two Latin squares of order n, L^1 and L^2, are called *orthogonal* if the pairs $(L^1_{i,j}, L^2_{i,j})$, $1 \leq i, j \leq n$ represent exactly once all the n^2 ordered pairs that can be formed from the numbers $0, 1, \ldots, n - 1$. A set of k Latin squares $\{L^1, L^2, \ldots, L^k\}$ is a set of mutually orthogonal Latin squares (MOLS) if any two of them are orthogonal. Such orthogonal Latin squares are used in statistics to design experiments.

It is easy to show that a set of MOLS contains at most $n - 1$ Latin squares. Indeed, we may permute the symbols in the Latin squares independently from the other Latin squares, so we may assume that the first row of each Latin square is $0, 1, \ldots, n - 1$. Consider the elements in the first column of the second row. Since the Latin squares are pairwise orthogonal we cannot get an element (i, i) in this position after putting together two Latin squares. This means that the elements in the first column and second row are pairwise different (and because of the Latin square property they are also different from 0); hence the number of such Latin squares is at most $n - 1$. A set of $n - 1$ mutually orthogonal Latin squares is called a complete set of mutually orthogonal Latin squares (*complete set of MOLS*).

To determine the maximum number of Latin squares in a set of MOLS is an open question. Let us denote this number by $N(n)$. Partial results are known, $N(2) = N(6) = 1$, for $n > 2$ and $n \neq 6$ we have $N(n) \geq 2$, and in general *Erdős, Chowla* and *Straus* [46] showed $N(n) \to \infty$ if $n \to \infty$. For $n = 10$ only pairs of MOLS are known. It is also proved that the known pairs of MOLS of order 10 cannot be extended to a set of three MOLS, but the existence of three MOLS of order 10 is an open problem in general. For $n = p^h$, the next theorem shows that $N(n) = n - 1$. The result also shows the close connection between complete sets of MOLS and projective planes.

Theorem 1.29. *There exists a projective plane of order n if and only if there exists a complete set of MOLS of order n.*

Proof. First suppose that a complete set of MOLS of order n exists. Starting from such a set, $\{L^1, L^2, \ldots, L^{n-1}\}$, we construct an affine plane of order n. Let us call points (elements of $\mathcal{P}$) the ordered pairs (x, y), the symbols (u) and the symbol (∞), lines (elements of $\mathcal{E}$) the ordered pairs $[m, k]$, the symbols $[k]$ and the symbol $[\infty]$, where $x, y, u, m, k \in \{0, 1, \ldots, n - 1\}$. Incidence (relation $I \subset \mathcal{P} \times \mathcal{E}$) is defined in the following way: the point (x, y) is incident with the line $[k]$ if and only if $x = k$; it is incident with the line $[0, k]$ if and only if

$y = k$, and for $m \in \{1, 2, \ldots, n-1\}$ it is incident with the line $[m, k]$ precisely when $k = L^m_{x,y}$; (x, y) is never incident with the line $[\infty]$. Point (u) is incident with line $[m, k]$ if and only if $u = m$; it is never incident with the lines $[k]$, and it is always incident with the line $[\infty]$. Finally, the point (∞) is incident with all lines of type $[k]$ and $[\infty]$ but not with the lines $[m, k]$.

Axioms P1 and P2 are clearly satisfied in those cases when one of the points or lines is not of type (x, y) or $[m, k]$ with $m \neq 0$, respectively. First we show that the lines $[m_1, k_1]$ and $[m_2, k_2]$ have exactly one point in common. If $m_1 = m_2$, then $k_1 \neq k_2$, hence no point of type (x, y) is incident with both lines, and among the points of type (u) only the point (m_1) is incident with the lines. For $0 \neq m_1 \neq m_2 \neq 0$, the Latin squares L^{m_1} and L^{m_2} are orthogonal, therefore the ordered pair (k_1, k_2) appears in exactly one position. Thus there exists a unique row x and a column y so that $k_i = L^{m_i}_{x,y}$ holds for $i = 1, 2$. This means that (x, y) is the unique common point of the two lines.

If (x_1, y_1) and (x_2, y_2) are two distinct points, then consider the n different lines which join the point (x_1, y_1) and points of type (m), and the line $[x_1]$. There are n points on each of these lines different from (x_1, y_1). In total this is $n(n + 1)$ distinct points which is just the number of points different from (x_1, y_1). This implies that the point (x_2, y_2) is on precisely one of these lines.

Since $n \geq 2$, axioms P3 and P4 are clearly satisfied.

In the opposite direction, suppose that $(\mathcal{P}, \mathcal{E}, I)$ is a projective plane of order n. Take a line e (we may think about it as being the line at infinity). Consider two distinct points X, Y on e and list the lines through X, different from e, as $h_1, \ldots, h_n$. Similarly, the lines different from e through Y are $v_1, \ldots, v_n$. So the points of the affine plane obtained by deleting e are the points $v_j \cap h_i$. These points will correspond to the positions of the Latin squares we are about to define. For each point M of $e \setminus \{X, Y\}$ we define a Latin square in the following way: list the lines through M different from e as $m_1, \ldots, m_n$. Put a k in position (i, j) if and only if $h_i \cap v_j \in m_k$. Since m_k meets each h_i and v_j in exactly one point, this defines a Latin square. Do this for the $n - 1$ points of $e \setminus \{X, Y\}$. Then we get a complete set of MOLS. (A more detailed proof of this direction can be found in Chapter 5 of [95].)   □

In the last part of this chapter we summarize results about the existence of projective planes of given order. In our previous examples all the finite planes were of prime power order. Indeed, the number of elements of a finite field is a prime power; hence the order of the planes constructed are prime powers. Conversely, for every prime power q there is a finite field with q elements, so for any prime power q we can construct a projective plane of order q.

For small orders n, the planes are unique up to isomorphism. It is really easy to check that any projective plane of order 2 is isomorphic to the Fano-plane. It is also easy to check that there is a unique projective plane of order 3, and it is also known that planes of order n are unique, when $n = 4, 5, 7, 8$, see [122] and [80]. If $n = 9$, then there are precisely four different projective planes of order 9, we shall say more about them in Chapter 3. We note that

TABLE 1.3
Existence of planes of small order

n	$n^2 + n + 1$	number of planes
2	7	1
3	13	1
4	21	1
5	31	1
6	43	no plane
7	57	1
8	73	1
9	91	4
10	111	no plane

no computer free proof is known for the fact that there are no more planes of order 9. For the reader's convenience we summarize these results in Table 1.3.

We shall return to the cases $n = 6$ and 10 after the proof of Bruck-Ryser's theorem.

For larger orders both the existence and uniqueness is an open question. All the known projective planes have prime power order and the so-called *prime power conjecture* (PPC) states that this is always the case. However, the proof of PPC seems hopeless at present. There are additional conditions which imply PPC for special classes of projective planes (for example, translation planes, see Chapters 2 and 3).

Our next theorem states the only general known necessary condition for the existence of a projective plane of order n. The non-existence of a projective plane of order 10 (see the above table) shows that the condition is not sufficient.

Before the proof let us see two number-theoretic lemmas and an identity, due to Lagrange. The number theoretic tools used in the proof can be found in [134].

Lemma 1.30. *A positive integer n is the sum of at most two perfect squares if and only if each prime divisor of the form $4k + 3$ has an even exponent in the prime factorization of n.*

A variant of this lemma states that an integer is a sum of two rational squares if and only if it is the sum of two perfect squares.

Lemma 1.31 (Lagrange). *Every positive integer is the sum of four perfect squares.*

The proof also uses the next identity (due also to Lagrange)

$$(b_1^2 + b_2^2 + b_3^2 + b_4^2)(x_1^2 + x_2^2 + x_3^2 + x_4^2) = y_1^2 + y_2^2 + y_3^2 + y_4^2. \qquad (1.5)$$

The matrix form of the variables in Lagrange's identity is $\mathbf{y} = \mathbf{x}B$, where

$$
\begin{aligned}
y_1 &= b_1 x_1 - b_2 x_2 - b_3 x_3 - b_4 x_4, \\
y_2 &= b_2 x_1 + b_1 x_2 - b_4 x_3 + b_3 x_4, \\
y_3 &= b_3 x_1 + b_4 x_2 + b_1 x_3 - b_2 x_4, \\
y_4 &= b_4 x_1 - b_3 x_2 + b_2 x_3 + b_1 x_4.
\end{aligned}
$$

The intuitive meaning of this identity is that the norm of the product of two quaternions is the product of their norms. This also explains that the linear transformation $\mathbf{y} = \mathbf{x}B$ (which is intuitively multiplication with the quaternion $b_1 + b_2 i + b_3 j + b_4 k$) is non-singular. As the elements of the matrix B are integers, its inverse matrix B^{-1} has rational elements. This observation will be used in the proof.

Theorem 1.32 (Bruck, Ryser). *Assume that $n \equiv 1, 2 \pmod 4$ and there is a projective plane of order n. Then n is the sum of two perfect squares.*

Proof. Let $A = (a_{ij})$ be an incidence matrix of the projective plane. Then A is a $0 - 1$ matrix of size $v \times v$, where $v = n^2 + n + 1$. As $n \equiv 1, 2 \pmod 4$, we have $v \equiv 3 \pmod 4$. Let $x_1, x_2, \ldots, x_v$ be indeterminates, let

$$\mathbf{X} = (x_1, x_2, \ldots, x_v) \quad \text{and} \quad \mathbf{L} = \mathbf{X}A,$$

then

$$L_i = \sum_{j=1}^{v} a_{ji} x_j, \quad (i = 1, \ldots, v).$$

The elements L_i satisfy the following identity:

$$L_1^2 + \ldots + L_v^2 = n(x_1^2 + \ldots + x_v^2) + (x_1 + \ldots + x_v)^2, \qquad (1.6)$$

because for

$$\mathbf{L} = (L_1, L_2, \ldots, L_v),$$

by Lemma 1.14, we have

$$\sum_{k=1}^{v} L_k^2 = (\mathbf{X}A)(\mathbf{X}A)^T = \mathbf{X}(AA^T)\mathbf{X}^T = \mathbf{X}(nI + J)\mathbf{X}^T =$$

$$n(x_1^2 + \ldots + x_v^2) + (x_1 + \ldots + x_v)^2.$$

We are going to reduce this identity using Lemma 1.31 and the Lagrange-identity mentioned afterwards.

First introduce a new indeterminate x_{v+1} and add nx_{v+1}^2 to both sides of (1.6). Then

$$L_1^2 + \ldots + L_v^2 + nx_{v+1}^2 = n(x_1^2 + \ldots + x_v^2 + x_{v+1}^2) + (x_1 + \ldots + x_v)^2. \quad (1.7)$$

By Lemma 1.31 and identity (1.5) express $n(x_i^2 + x_{i+1}^2 + x_{i+2}^2 + x_{i+3}^2)$ as a sum of four perfect squares. Let $n = b_1^2 + \ldots + b_4^2$ and for $i = 1, 5, 9, \ldots, v-2$

$$(b_1^2 + b_2^2 + b_3^2 + b_4^2)(x_i^2 + x_{i+1}^2 + x_{i+2}^2 + x_{i+3}^2) = y_i^2 + y_{i+1}^2 + y_{i+2}^2 + y_{i+3}^2.$$

For the sake of simplicity introduce $x_1 + \ldots + x_v = w$. Then

$$L_1^2 + \ldots + L_v^2 + nx_{v+1}^2 = y_1^2 + \ldots + y_{v+1}^2 + w^2, \quad (1.8)$$

where the L_i's and the y_j's are linear combinations of the x_i's with integer coefficients. Since the matrix B appearing in (1.5) is non-singular, the mapping $\mathbf{X} \to \mathbf{Y}$ is bijective. So we can use the indeterminates y_j's instead of the x_i's and express everything in (1.8) as a linear combination of the y_j's. Since the matrix B^{-1} has rational elements, the coefficients in these linear combinations are rational. We are going to substitute such a linear combination of the remaining indeterminates in place of y_1 which "shortens" (1.8). We do not just want to eliminate y_1, but also want to get rid of one-one term on both sides of (1.8). To do so look at L_1 as a linear combination of $y_1, \ldots, y_{v+1}$ and find the coefficient of y_1. If it is not 1, then determine y_1 so that $L_1 = y_1$. Of course, the solution will be a linear combination of the remaining variables with rational coefficients. Substituting it in (1.8) we get

$$L_2^2 + \ldots + L_v^2 + nx_{v+1}^2 = y_2^2 + \ldots + y_{v+1}^2 + w^2, \quad (1.9)$$

where all the L_i's are still linear combinations of $y_2, \ldots, y_{v+1}$ with rational coefficients. If the coefficient of y_1 in L_1 was 1, then instead of $L_1 = y_1$ we use $L_1 = -y_1$ and solve it for y_1. Also in this case substituting the result in (1.8) we obtain Equation (1.9).

For the shortened equation we repeat the same trick until everything becomes a constant multiple of the last indeterminate x_{v+1}. So we solve repeatedly one of the equations $L_2 = \pm y_2$, $\ldots$, $L_v = \pm y_v$ for the indeterminate y_2, then for $y_3, \ldots, y_v$. The solutions are linear combinations of the remaining variables with rational coefficients. Finally, we arrive at the equation

$$nx_{v+1}^2 = y_{v+1}^2 + w^2,$$

where w and x_{v+1} are rational multiples of y_{v+1}. This shows that n is the sum of two rational numbers. The variant of Lemma 1.30 mentioned right after it implies that n is the sum of two perfect squares.

We can also use Lemma 1.30 directly. Dividing by y_{v+1}^2 and multiplying by a common multiple of the denominators we get that the equation $nu^2 = v^2 + z^2$ has a non-trivial integer solution. The right-hand side is the sum of two perfect squares, so every prime of the form $4k + 3$ dividing it appears with an even exponent. As all the exponents in u^2 are even, the same holds for n. By Lemma 1.30 this implies that n is the sum of two perfect squares. $\qquad\square$

Since $n = 6$ is not the sum of two perfect squares, there is no projective plane of order 6. This was already "known" to *Euler* because there is no pair of orthogonal Latin squares, hence Theorem 1.29 implies the non-existence of planes of order 6. So far $n = 10$ is the only value which satisfies the condition in the theorem of Bruck–Ryser yet it is known that there is no projective plane of order 10. This was shown in the 1980's by *Lam, Swiercz* and *Thiel* [114] using computers.

Exercises

1.1. A $2 - (v, k, \lambda)$-design consists of points and blocks. It has a set of v points, and the blocks are certain k-element subsets of the points. The key property is that any two distinct points are contained in exactly λ blocks.

   Determine the number of blocks containing a given point (usually denoted by r) and the total number of blocks (usually denoted by b).

1.2. Show that a $2 - (n^2 + n + 1, n + 1, 1)$-design is a projective plane if $n \geq 2$. (Note that this can be regarded as an alternative definition of finite projective planes.)

1.3. Show that a $2 - (n^2, n, 1)$ design is an affine plane if $n \geq 2$.

1.4. Consider the defining axioms for projective planes and the numerical properties in Theorem 1.8. Try to find other possible sets of axioms for projective planes. Do the same for affine planes.

1.5. Determine AA^T for an incidence matrix of a $2 - (v, k, \lambda)$-design.

1.6. Show that for $k < v$ one has $b \geq v$ for a $2 - (v, k, \lambda)$-design. (Hint: use the matrix AA^T.)

1.7. Prove that any finite projective plane of order at least 3 contains a (closed) Desargues configuration.

1.8. Show that there is a unique projective plane of order 4 and 5.

1.9. Prove that the order of $\mathrm{PG}(2, q)$ is q.

1.10. (Kárteszi) Determine the size of the incidence tables obtained in the rounds of Hall's free extension process.

1.11. Show that the reverse process indicated after Example 1.29 indeed gives a complete set of MOLS.

1.12. Use the process indicated in the previous exercise to construct explicitly a complete set of MOLS, starting from the plane $\mathrm{PG}(2, q)$.

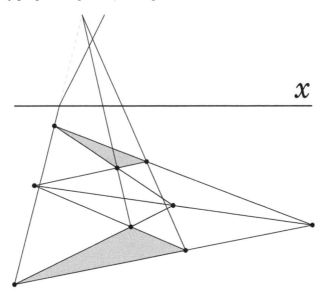

FIGURE 1.6
The Moulton plane is not Desarguesian

1.13. A $2 - (v, k, \lambda)$-design is called square (or symmetric) if it has as many points as blocks (formally: $b = v$). Show that a design is square if and only if $r = k$. Show that in a square design two blocks intersect in exactly λ points. (Hint: Show that $AA^T = A^T A$.)

1.14. Assume that v is even and there is a square $2 - (v, k, \lambda)$-design. Show that $k - \lambda$ is a square. This criterion is due to *Schützenberger*. Hint: compute that $\det AA^T = (\det A)^2 = k^2(k - \lambda)^{v-1}$, for example by determining the eigenvalues of the matrix AA^T.

1.15. Prove the following extension of the theorem of Bruck–Ryser to square designs: assume that v is even, then the Diophantine equation

$$x^2 = (k - \lambda)y^2 + (-1)^{(v-1)/2}\lambda z^2$$

has a non-trivial integer solution (different from $(0, 0, 0)$). (Hint: Try to copy the the proof of the theorem of Bruck-Ryser; for $v \equiv 3 \pmod 4$ introduce the new indeterminate x_{v+1} while for $v \equiv 1 \pmod 4$ separate the last variable and use Lagrange's identity for four consecutive terms as in the original proof.) This result is called the Theorem of Bruck–Chowla–Ryser.

1.16. Show that the Moulton plane (see Example 1.27) is indeed an affine plane.

1.17. Prove that the Moulton plane (see Example 1.27) is not Desarguesian.

1.18. Prove that the parabola plane (see Example 1.28) is indeed an affine plane.

1.19. Prove that the parabola plane is Desarguesian.

1.20. Prove that the function plane coming from the graph of $x \mapsto x^4$ is indeed a plane. In other words, prove that $x \mapsto x^4$ is a planar function.

1.21. Prove that the function plane coming from the graph of $x \mapsto x^4$ is not Desarguesian.

1.22. Generalize the result of the previous two exercises for the function $x \mapsto x^{2k}$, where k is a positive integer.

1.23. Describe planar funcions over the reals.

1.24. Prove that the function plane coming from the graph of $f : x \mapsto f(x)$ is Desarguesian if and only if f is a quadratic function.

2

Basic properties of collineations and the Theorem of Baer

In this chapter the most important definitions and theorems about collineations are collected. We assume that the reader is familiar with the basic results and definitions of group theory, in particular the theory of permutation groups. For an introduction to these notions we refer to the book [174].

Definition 2.1. *Let* $\Pi = (\mathcal{P}, \mathcal{E}, \mathrm{I})$ *and* $\Pi' = (\mathcal{P}', \mathcal{E}', \mathrm{I}')$ *be two projective or affine planes. A* collineation *between* Π *and* Π' *is a pair* $\varphi = (\varphi_1, \varphi_2)$ *where* $\varphi_1 : \mathcal{P} \to \mathcal{P}'$ *and* $\varphi_2 : \mathcal{E} \to \mathcal{E}'$ *are bijections which preserve the incidence in the sense that for all* $P \in \mathcal{P}$ *and* $e \in \mathcal{E}$

$$P \, \mathrm{I} \, e \iff P^{\varphi_1} \, \mathrm{I}' \, e^{\varphi_2}.$$

If Π *is a projective plane and* $\Pi' = \Pi^*$ *is the dual plane of* Π, *then a collineation* $\varphi : \Pi \to \Pi^*$ *is called a* correlation.

For the sake of brevity usually the letter φ (without any subscript) denotes the action of a collineation both on the set of points and set of lines. As it is easily seen, if a bijection $\varphi_1 : \mathcal{P} \to \mathcal{P}'$ has the property that any three points of Π are collinear if and only if their images in Π' are collinear, then φ_1 induces a bijection $\varphi_2 : \mathcal{E} \to \mathcal{E}'$ and the pair $\varphi = (\varphi_1, \varphi_2)$ becomes a collineation. So it is enough to give the action of a collineation only on the set of points.

If φ is a collineation between two affine planes Π and Π', then φ maps parallel lines of Π to parallel lines of Π'. Hence φ can be extended to a collineation $\overline{\varphi}$ between the projective closures of Π and Π'. Thus any collineation between affine planes can be considered as a collineation between projective planes, too. So in the rest of this chapter only collineations between projective planes are considered.

If $\Pi' = \Pi$, then a collineation $\varphi : \Pi \to \Pi$ is called a collineation of Π. The set of all collineations of Π with the composition (subsequent application) as operation form a group, called the *full collineation group of the plane*. This group is denoted by $\mathrm{Aut}(\Pi)$, and the operation will be denoted as a multiplication. Any subgroup of $\mathrm{Aut}(\Pi)$ is called a *collineation group of* Π.

In the rest of this chapter ι denotes the identity element of any collineation group.

First let us look at some examples.

Example 2.2. If Π is the cyclic plane constructed in Example 1.17 and we identify the vertices of the polygon $\mathcal{S}$ with the elements of the additive group $\mathbb{Z}_{n^2+n+1}$, then the mapping $i \mapsto i+1 \pmod{n^2+n+1}$ is obviously a collineation.

Example 2.3. Linear transformations are collineations of the classical projective plane. The planes $\mathrm{PG}(2,\mathbb{K})$ have the same vector space model as the classical projective plane. As expected, due to this property, linear transformations are collineations of these planes, too. Let A be a 3×3 nonsingular matrix over $\mathbb{K}$. Then the mappings

$$\varphi_1 : \mathbf{x} \mapsto \mathbf{x}A \quad \text{and} \quad \varphi_2 : \mathbf{u} \mapsto \mathbf{u}A^{-1T}$$

give a collineation of the projective plane $\mathrm{PG}(2,\mathbb{K})$. These mappings are permutations of the set of points and of the set of lines of $\mathrm{PG}(2,\mathbb{K})$, because $\det A \neq 0$ and if $0 \neq \lambda \in \mathbb{K}$, then for any vector $\mathbf{v}$ the equality $(\lambda \mathbf{v})A = \lambda(\mathbf{v}A)$ holds. These permutations preserve incidence, because

$$\mathbf{x}\mathbf{u}^{\mathrm{T}} = 0 \iff (\mathbf{x}A)(\mathbf{u}(A^{-1})^{\mathrm{T}})^{\mathrm{T}} = \mathbf{x}(AA^{-1})\mathbf{u}^{\mathrm{T}} = 0.$$

Similarly to the case of the classical plane, one can prove (see Theorem 4.16) that the group of linear transformations is transitive on the set of ordered quadruples formed by four points in general position. Dually, the group of linear transformations is transitive on the set of ordered quadruples of lines formed by four lines in general position.

Another important class of collineations comes from automorphisms of $\mathbb{K}$.

Example 2.4. Let σ be an automorphism of $\mathbb{K}$. For any vector $\mathbf{v} = (v_0, v_1, v_2)$ in the 3-dimensional vector space over $\mathbb{K}$ let $\mathbf{v}^\sigma = (v_0^\sigma, v_1^\sigma, v_2^\sigma)$. Then the mappings

$$\varphi_1 : \mathbf{x} \mapsto \mathbf{x}^\sigma \quad \text{and} \quad \varphi_2 : \mathbf{u} \mapsto \mathbf{u}^\sigma$$

give a collineation of the projective plane $\mathrm{PG}(2,\mathbb{K})$. These mappings are evidently permutations of the point-set and of the line-set of $\mathrm{PG}(2,\mathbb{K})$, and they preserve the incidence, because

$$\mathbf{x}\mathbf{u}^T = 0 \iff \sum_{i=0}^{2} x_i u_i = 0 \iff \sum_{i=0}^{2} x_i^\sigma u_i^\sigma = 0 \iff \mathbf{x}^\sigma \mathbf{u}^{\sigma T} = 0.$$

In the next part of this section we consider the elements fixed by a collineation.

Definition 2.5. *Let φ be a collineation of the plane Π. A point F is a fixed point of φ if $F = F^\varphi$, a line e is a fixed line of φ if $e = e^\varphi$. Let $\mathcal{F}_\varphi$ denote the set of fixed points and lines of a collineation φ.*

Definition 2.6. *A set of points and lines of a projective plane is called a* closed configuration *if it contains the line joining any two of its points and the point of intersection of any two of its lines.*

Lemma 2.7. *Let φ be a collineation of a projective plane Π. Then the line joining two points of $\mathcal{F}_\varphi$ is in $\mathcal{F}_\varphi$ and the point of intersection of two lines of $\mathcal{F}_\varphi$ is also in $\mathcal{F}_\varphi$.*

Proof. Let A and B be any two distinct points of $\mathcal{F}_\varphi$. As φ preserves incidence, $(AB)^\varphi = A^\varphi B^\varphi$. Since $A, B \in \mathcal{F}_\varphi$, $A^\varphi = A$ and $B^\varphi = B$, hence $(AB)^\varphi = AB$, so $AB \in \mathcal{F}_\varphi$. Dually, we get that the point of intersection of any two lines of $\mathcal{F}_\varphi$ is in $\mathcal{F}_\varphi$. □

Corollary 2.8. *Let φ be a collineation of a projective plane Π. Then $\mathcal{F}_\varphi$ is a closed configuration. If $\mathcal{F}_\varphi$ contains four points in general position, then $\mathcal{F}_\varphi$ is a subplane of Π.*

If the fixed points and fixed lines of a collineation φ of a finite plane form a subplane, then φ has as many fixed points as fixed lines. It is true for any collineation of a finite plane. To prove this theorem we need some preparatory notions and statements.

Definition 2.9. *A 0–1 matrix is called a* permutation matrix *if each of its rows and columns contains exactly one entry '1'.*

Theorem 2.10. *Let Π_n be a finite projective plane of order n. If A is an incidence matrix of Π_n and φ is a collineation of Π_n, then there exist permutation matrices P and Q for which $PA = AQ$ holds.*

Proof. Let $v = n^2 + n + 1$, and let $P_1, P_2, \ldots, P_v$, and $\ell_1, \ell_2, \ldots, \ell_v$ denote the points and lines of Π_n, respectively. If the incidence matrix is $A = (a_{ij})$, then $a_{ij} = 1$ if $P_i \, I \, \ell_j$, and $a_{ij} = 0$ otherwise. Let us define the matrices $P = (p_{ij})$ and $Q = (q_{ij})$ in the following way:

$$p_{ij} = \begin{cases} 1 & \text{if } P_i^\varphi = P_j, \\ 0 & \text{if } P_i^\varphi \neq P_j, \end{cases} \qquad q_{ij} = \begin{cases} 1 & \text{if } \ell_i^\varphi = \ell_j, \\ 0 & \text{if } \ell_i^\varphi \neq \ell_j. \end{cases}$$

Then P and Q are permutation matrices because φ is a collineation. Consider the matrix $PA = (c_{ij})$. Then $c_{ij} = \sum_{k=1}^{v} p_{ik} a_{kj} = a_{xj}$, where x is uniquely determined by the equation $p_{ix} = 1$, because it holds for exactly one x, namely the one for which $P_i^\varphi = P_x$. In the same way we get that if $AQ = (d_{ij})$, then its entry $d_{ij} = \sum_{k=1}^{v} a_{ik} q_{kj} = a_{iy}$, where y is uniquely determined by the equation $\ell_y^\varphi = \ell_j$. We claim that $c_{ij} = d_{ij}$. As φ is a collineation, we have

$$c_{ij} = a_{xj} = 1 \iff P_x \, I \, \ell_j \iff P_i^\varphi \, I \, \ell_y^\varphi \iff P_i \, I \, \ell_y \iff 1 = a_{iy} = d_{ij},$$

which proves the statement. □

Theorem 2.11. *Let φ be a collineation of a finite projective plane Π. Then the number of fixed points of φ is equal to the number of fixed lines of φ.*

Proof. Let A be an incidence matrix of Π. According to the previous theorem φ can be represented by two permutation matrices P and Q for which $PA = AQ$ holds. If $P = (p_{ij})$, then $p_{ii} = 1$ if and only if $P_i^\varphi = P_i$, thus if and only if P_i is a fixed point of φ. Hence the number of fixed points of φ is equal to the trace of P. In the same way we can prove that the number of fixed lines of φ is equal to the trace of Q. According to Lemma 1.15, A is non-singular, hence from $PA = AQ$ we get $P = AQA^{-1}$, which implies that the trace of P is equal to the trace of Q. $\qquad\qquad\square$

Let us recall some basic notions concerning permutations. Let $\mathcal{H}$ be a set of $n \geq 1$ elements. A *permutation* of $\mathcal{H}$ is a bijection from $\mathcal{H}$ to itself. The composition of any two permutations of $\mathcal{H}$ is again a permutation of $\mathcal{H}$. The *symmetric group* on $\mathcal{H}$ is the set of all permutations of $\mathcal{H}$ with the composition as operation. This group is denoted by $\mathrm{Sym}(\mathcal{H})$, it consists of $n!$ elements. Subgroups of $\mathrm{Sym}(\mathcal{H})$ are called *permutation groups* on $\mathcal{H}$. If Γ is a permutation group on $\mathcal{H}$ and $x \in \mathcal{H}$ is any element, then the *orbit* of x under Γ is the set

$$x^\Gamma = \{x^\gamma : \gamma \in \Gamma\},$$

and the *stabilizer* in Γ of x is the set

$$\Gamma_x = \{\gamma \in \Gamma : x^\gamma = x\}.$$

The set of all orbits of Γ forms a partition of $\mathcal{H}$. If Γ has a single orbit, then Γ is said to be *transitive* on $\mathcal{H}$.

The next result is crucial in the theory of permutation groups. Usually it is called *Burnside's lemma*, but the name *Cauchy-Frobenius lemma* would be more correct because of historical reasons. It states that the number of orbits of a permutation group is equal to the average number of objects fixed by the elements of the group. For a more detailed discussion of the use of this lemma we refer to the books [144] by *Pólya* and *Read* and [158] by *Stanley*.

Lemma 2.12. *Let Γ be a permutation group on a set $\mathcal{H}$. For each $\gamma \in \Gamma$ let $f(\gamma)$ denote the number of elements $\mathcal{H}$ fixed by γ, and let t denote the number of orbits of Γ on $\mathcal{H}$. Then*

$$t|\Gamma| = \sum_{\gamma \in \Gamma} f(\gamma).$$

Proof. We count in two different ways the number of ordered pairs $(x, \gamma) \in \mathcal{H} \times \Gamma$ for which $x^\gamma = x$ holds. On the one hand, if Γ_x denotes the stabilizer of x in Γ then for a given $x \in \mathcal{H}$ there are $|\Gamma_x|$ elements of Γ fixing it. On the other hand, a given $\gamma \in \Gamma$ fixes $f(\gamma)$ elements of $\mathcal{H}$, so

$$\sum_{\gamma \in \Gamma} f(\gamma) = \sum_{x \in \mathcal{H}} |\Gamma_x|.$$

Hence it is enough to show that

$$\sum_{x \in \mathcal{H}} |\Gamma_x| = t|\Gamma|.$$

For each $x \in \mathcal{H}$ the order of Γ_x is equal to the order of Γ divided by the length of the orbit of x, hence if x^Γ denotes the orbit of x, then

$$\sum_{x \in \mathcal{H}} |\Gamma_x| = \sum_{x \in \mathcal{H}} |\Gamma|/|x^\Gamma|.$$

In the sum $\sum_{x \in \mathcal{H}} |\Gamma|/|x^\Gamma|$ the orbit x^Γ appears as many times as the number of its elements, thus this sum is equal to $t|\Gamma|$. This proves the statement. □

Theorem 2.13. *Let Γ be a collineation group of a finite projective plane. Then Γ has equally many point and line orbits.*

Proof. Let t_1 and t_2 denote the number of point and line orbits of Γ, respectively. For each $\gamma \in \Gamma$ let $f(\gamma)$ denote the number of fixed points of γ. Then, by Theorem 2.11, we have that the number of fixed lines of γ is also $f(\gamma)$. As Γ acts both on the points and lines of the plane Lemma 2.12 gives

$$t_1|\Gamma| = \sum_{\gamma \in \Gamma} f(\gamma) = t_2|\Gamma|,$$

hence $t_1 = t_2$. □

Corollary 2.14. *A collineation group is transitive on the points of a projective plane if and only if it is transitive on the lines of the plane.*

The equality of the number of point and line orbits does not imply that the structures of the two orbits are the same. We will present a simple example after the proof of Theorem 2.33.

Now we are ready to prove an important property of the collineation group defined in Example 2.2. This group is generated by the mapping $i \mapsto i+1$ and it acts regularly on both of the points and the lines of the plane. Collineation groups having these properties are called *Singer groups*. The next theorem shows that a finite Galois plane always has a Singer group.

Theorem 2.15. *The plane $PG(2,q)$ possesses a cyclic Singer group. This group is isomorphic to the additive group $\mathbb{Z}_{q^2+q+1}$. The set of points on any line of the plane corresponds to a planar difference set in the group $\mathbb{Z}_{q^2+q+1}$.*

Proof. Let $\mathbb{K} = GF(q)$ and $\mathbb{F} = GF(q^3)$. Then $\mathbb{F}$ can be considered as a 3-dimensional vector space over $\mathbb{K}$. Hence the points of $PG(2,q)$ can be identified with the elements of $\mathbb{F}^*$ modulo $\mathbb{K}^*$, that is $f_1 \in \mathbb{F}^*$ and $f_2 \in \mathbb{F}^*$ represent the same point of $PG(2,q)$ if and only if $f_1/f_2 \in \mathbb{K}^*$.

For each $f \in \mathbb{F}^*$ define the mapping $R_f : \mathbb{F} \to \mathbb{F}$ as $R_f(x) = fx$. Then R_f is a non-singular linear transformation of the vector space $\mathbb{F}$ because for all $x, y \in \mathbb{F}$ and for all $k, m \in \mathbb{K}$ the equality

$$R_f(kx + my) = kR_f(x) + mR_f(y)$$

obviously holds. The set of mappings $\{R_f : f \in \mathbb{F}^*\}$ with the operation of composition forms a group which is isomorphic to the multiplicative group $\mathbb{F}^*$ by definition.

Now consider the action of R_f on the points of $\mathrm{PG}(2, q)$. As R_f is non-singular, this is a collineation. The mapping R_f acts on the points of $\mathrm{PG}(2, q)$ as the identity if and only if for all $x \in \mathbb{F}^*$ the element $R_f(x) = fx$ represents the same 1-dimensional subspace as x does. This happens if and only if there exists $k \in \mathbb{K}^*$ such that $fx = kx$ holds, hence $f \in \mathbb{K}^*$.

The group formed by the mappings of type R_f induces a collineation group Σ of $\mathrm{PG}(2, q)$. By the previous remark, this group is isomorphic to the factor group $\mathbb{F}^*/\mathbb{K}^*$ of the multiplicative group $\mathbb{F}^*$. It is well-known that the multiplicative group $\mathbb{F}^*$ is cyclic, hence Σ is also cyclic. This means that Σ is isomorphic to the cyclic group $\mathbb{Z}_{q^2+q+1}$.

Apparently, the group Σ is transitive on the points of $\mathrm{PG}(2, q)$. Consider the stabilizer of the point corresponding to the element $1 \in \mathbb{F}$. As $R_f(1) = f$, it consists of those elements for which the points corresponding to f and 1 are the same, which means that $f \in \mathbb{K}^*$. Hence in Σ the stabilizer of a point consists of the identity only, so Σ acts regularly on the points of $\mathrm{PG}(2, q)$. One can prove with a direct computation that Σ acts regularly on the lines of $\mathrm{PG}(2, q)$, too. We omit the long calculation, because this statement follows directly from Theorem 2.13.

If we identify the points of $\mathrm{PG}(2, q)$ with the elements of $\mathbb{Z}_{q^2+q+1}$ then any line of the plane corresponds to a subset of $q + 1$ elements of $\mathbb{Z}_{q^2+q+1}$. Let $\ell_0 = \{d_1, d_2, \ldots, d_{q+1}\}$ be one of these subsets. As the mapping $i \mapsto i + 1 \pmod{q^2 + q + 1}$ generates the cyclic group $\mathbb{Z}_{q^2+q+1}$, the lines of the plane correspond to the subsets $\ell_a = \{d_1 + a, d_2 + a, \ldots, d_{q+1} + a\}$ with $a = 0, 1, \ldots, q^2 + q$. Let $0 \neq b \in \mathbb{Z}_{q^2+q+1}$ be an arbitrary element. The lines ℓ_0 and ℓ_b have a unique point of intersection. If the corresponding element is c, then $c = d_i = d_j + b$ with suitable i and j (and the subscripts are uniquely determined). Hence b can be uniquely expressed as $b = d_i - d_j$, which proves the last part of the theorem.

Let us remark that the elements of a difference set correspond to the vertices of the polygon $\mathcal{R}$ in Example 1.17.   □

Among the fixed elements of a collineation may appear some "more fixed" ones. A line could be fixed pointwise and a point could be fixed linewise. These more fixed elements play an important role in the study of projective planes.

Definition 2.16. *Let φ be a collineation of a projective plane Π. A point C is a* center *of φ if each line through C is a fixed line of φ. A line t is an* axis *of φ if each point on t is a fixed point of φ.*

The center and axis are dual notions. Each of the next theorems has its dual pair. We shall often use the dual statements without proof.

Lemma 2.17. *Let φ be a collineation of a projective plane Π. If φ has two distinct axes, then φ is the identity.*

Proof. Suppose that both of the lines t_1 and t_2 are axes of φ. Let $P = t_1 \cap t_2$. Then $P = P^\varphi$ is obvious. If $R \neq P$ is any point of Π, then there are at least two lines through R which are distinct from PR. If e and f are lines distinct from PR, then they meet the axes in distinct points, hence by Lemma 2.7 $e, f \in \mathcal{F}_\varphi$. Then, again by Lemma 2.7, the point $e \cap f$ is also in $\mathcal{F}_\varphi$. Therefore φ fixes all points, hence it also fixes all lines, so it is the identity. □

Lemma 2.18. *Let φ be a collineation of a projective plane Π. Suppose that the line t is an axis of φ and the point F is a fixed point of φ not on t. Then F is a center of φ.*

Proof. Let f be any line through F and let $T = f \cap t$. This point is well-defined and distinct from F because F is not on t. Each point on t is fixed, hence $T \in \mathcal{F}_\varphi$. By Lemma 2.7 the line joining F and T is also fixed, so F is a center of φ. □

Lemma 2.19. *Let φ be a collineation of a projective plane Π.*

- *Suppose that the point C is a center of φ. Then for any point $P \neq C$ the line CP contains the point P^φ.*

- *Suppose that the line t is an axis of φ. If a point $P \notin \mathcal{F}_\varphi$, then the line PP^φ is in $\mathcal{F}_\varphi$.*

Proof. The line CP is fixed because C is a center. Hence $CP = (CP)^\varphi = C^\varphi P^\varphi = CP^\varphi$, which proves the first part of the statement.

For proving the second part let e denote the line PP^φ and consider the point $T = e \cap t$. Then $e^\varphi = (PT)^\varphi = P^\varphi T^\varphi = P^\varphi T = e$, hence $e \in \mathcal{F}_\varphi$. □

Theorem 2.20. *Let φ be a collineation of a projective plane Π. If φ has an axis, then it has a center, too.*

Proof. Let t be the axis of φ. If F is a fixed point of φ not on t, then, by Lemma 2.18, F is a center of φ.

Now suppose that all points of $\mathcal{F}_\varphi$ are on t. Let $P \notin \mathcal{F}_\varphi$ be any point. Take the line $e = PP^\varphi$; let $R \notin (\mathcal{F}_\varphi \cup \{e\})$ be any point, let f denote the line RR^φ, and finally, let C be the point $e \cap f$. We claim that C is a center of φ. Let g be any line through C. If $g \in \{t, e, f\}$, then $g \in \mathcal{F}_\varphi$. If $g \notin \{t, e, f\}$, then let $Q \neq C$ be a point on g. As $Q \notin \mathcal{F}_\varphi$, by the second part of Lemma 2.19, we get $QQ' = g \in \mathcal{F}_\varphi$. Hence C is a center of φ. □

Theorems 2.17 and 2.20 together with their duals give that a non-identity collineation has either one center and one axis, or it has no center and no axis. If φ is the identity collineation, then each point is a center of φ and each line is an axis of φ.

Definition 2.21. *A collineation φ is called a* central-axial collineation *if it has both a center and an axis. If C is a center and t is an axis of φ, then φ is also called a (C, t)-perspectivity.*

A (C, t)-perspectivity is an elation *if $C\,\mathrm{I}\,t$, otherwise it is called a* homology. *The identity can be considered as both an elation and a homology.*

Lemma 2.22. *If α is a (C, t)-perspectivity, then α^{-1} is also a (C, t)-perspectivity.*

If the central-axial collineations α_1 and α_2 have the same center C or the same axis t, then $\alpha_1\alpha_2$ also has center C or axis t, respectively.

Corollary 2.23. *Let C be an arbitrary point and t be an arbitrary line of a projective plane Π. Then each of*

- *the set of all collineations of Π having center C,*

- *the set of all collineations of Π having axis t,*

- *the set of all (C, t)-perspectivities of Π*

forms a subgroup of $\mathrm{Aut}(\Pi)$.

These collineation groups are denoted by Γ_C, Γ_t and $\Gamma_{C,t}$, respectively.

Lemma 2.24. *If α is a (C, t)-perspectivity and β is an arbitrary collineation of a projective plane Π, then $\beta^{-1}\alpha\beta$ is a (C^β, t^β)-perspectivity of Π.*

Proof. We claim that each point of the line t^β belongs to $\mathcal{F}_{\beta^{-1}\alpha\beta}$. Any point on the line t^β can be considered as P^β with a suitable point $P\,\mathrm{I}\,t$. Hence $P^\alpha = P$, therefore

$$(P^\beta)^{\beta^{-1}\alpha\beta} = P^{\alpha\beta} = P^\beta.$$

So t^β is an axis of $\beta^{-1}\alpha\beta$.

Dually, we get that each line through C^β is fixed by $\beta^{-1}\alpha\beta$, hence C^β is a center of $\beta^{-1}\alpha\beta$.   □

Lemma 2.25. *Let $C_1 \neq C_2$ be two points and let t be a line of a projective plane Π. Suppose that for $i = 1, 2$ there exist a (C_i, t)-perspectivity $\alpha_i \neq \iota$. Then the collineation $\beta = \alpha_1\alpha_2$ is a (C, t)-perspectivity with a suitable point C. This point is on the line C_1C_2, but it is distinct from both of C_1 and C_2.*

Proof. By Lemma 2.22, $\beta \neq \iota$ because $C_1 \neq C_2$. The line t is obviously an axis of β, hence Theorem 2.20 gives that β has a center. Let C denote this point.

Suppose to the contrary that C is not on the line C_1C_2. The line C_1C_2 is fixed by both α_1 and α_2, so $C_1C_2 \in \mathcal{F}_\beta$. Because of the dual of Lemma 2.18, this means that C_1C_2 is an axis of β. There are two possibilities. If C_1C_2 and t are distinct lines, then β has two distinct axes, thus $\beta = \iota$ follows from Lemma 2.17, which is a contradiction. On the other hand, if $C_1C_2 = t$, then C is not on the line t. But $C^\beta = C$ gives $C^{\alpha_1} = C^{\alpha_2^{-1}}$. The point C^{α_1} is on the line C_1C, while the point $C^{\alpha_2^{-1}}$ is on the line C_2C. These two lines are distinct, their point of intersection is C, hence $C = C^{\alpha_1}$. Thus, because of Lemma 2.18, $\alpha_1 = \iota$, giving a contradiction again.

So the point C is on the line C_1C_2. If C were C_1, then by Lemma 2.22, the point C_1 would be another center of the collineation $\alpha_1^{-1}\beta = \alpha_2 \neq \iota$, while if C were C_2, then, again by Lemma 2.22, the point C_2 would be another center of the collineation $\beta\alpha_2^{-1} = \alpha_1 \neq \iota$. So $C_1 \neq C \neq C_2$ also hold. $\square$

Corollary 2.26. *Let t be an arbitrary line of a projective plane Π. Then the set of all elations of Π having axis t forms a subgroup of $\mathrm{Aut}(\Pi)$.*

Proof. By Lemma 2.25, the product of two elations with axis t is an elation with axis t. Because of Lemma 2.22 the inverse of an elation with axis t is also an elation with axis t and multiplication is obviously associative. $\square$

The collineation group formed by the elations having a fixed axis t is denoted by $\Gamma_{t,t}$.

If we extend a translation of the classical Euclidean plane to a collineation of the classical projective plane, then we get an elation. Its axis is the line at infinity and its center is the point at infinity corresponding to the direction of the translation. The translations of the Euclidean plane form an abelian group, thus the extended collineations also form an abelian group. Our next theorem generalizes this property.

Theorem 2.27. *Suppose that a projective plane Π contains a line t and two distinct points, C_1 and C_2, on t such that both collineation groups $\Gamma_{C_1,t}$ and $\Gamma_{C_2,t}$ contain at least two elements. Then the group of elations $\Gamma_{t,t}$ is abelian.*

Proof. Because of Corollary 2.26, the only thing we have to prove is the commutativity of the operation.

First we show that any two elations with common axis but distinct centers commute. Let $\alpha \neq \iota$ be a (P,t)-perspectivity, $\beta \neq \iota$ be a (Q,t)-perspectivity and suppose that $P \neq Q$. Consider their commutator

$$\gamma = \beta^{-1}(\alpha^{-1}\beta\alpha) = (\beta^{-1}\alpha^{-1}\beta)\alpha.$$

By Lemma 2.24, $\alpha^{-1}\beta\alpha$ is a (Q^α, t^α)-perspectivity. Here $Q^\alpha = Q$ and $t^\alpha = t$, so $\alpha^{-1}\beta\alpha$ is a (Q,t)-perspectivity. This implies that $\gamma = \beta^{-1}(\alpha^{-1}\beta\alpha)$ is a (Q,t)-perspectivity, too. In the same way, we get that $\beta^{-1}\alpha^{-1}\beta$ has center $P^\beta = P$ and axis t, thus $\gamma = (\beta^{-1}\alpha^{-1}\beta)\alpha$ is a (P,t)-perspectivity. Hence γ has two distinct centers, Q and P, so $\gamma = \iota$. This means that α and β commute.

Now, we have to show that two elations with the same center commute. Suppose that $\gamma \neq \iota$ is a (P,t)-perspectivity, too. Consider the elation $\gamma\beta$. If R denotes its center, then, by Lemma 2.25, $P \neq R \neq Q$. Thus, because of the first part of the proof, $\gamma\beta$ commutes with α and $\gamma\alpha$ commutes with β. Hence

$$(\alpha\gamma)\beta = \alpha(\gamma\beta) = (\gamma\beta)\alpha = (\beta\gamma)\alpha = \beta(\gamma\alpha) = (\gamma\alpha)\beta.$$

Multiplying from the right by β^{-1} we get $\alpha\gamma = \gamma\alpha$, so the theorem is proved.
$\square$

Theorem 2.28. *Suppose that a finite projective plane Π_n of order n contains a line t and two distinct points, C_1 and C_2, on t such that both collineation groups $\Gamma_{C_1,t}$ and $\Gamma_{C_2,t}$ contain at least two elements. Then in the group $\Gamma_{t,t}$ each element has the same order. This order is a prime p, and n is divisible by p. Moreover, if C is any point on t, then the order of $\Gamma_{C,t}$ divides n and the order of $\Gamma_{t,t}$ divides n^2.*

Proof. According to the previous theorem $\Gamma_{t,t}$ is an abelian group and it has finite order because Π_n is finite. This means that $\Gamma_{t,t}$ contains an element of prime order. Let α denote an element whose order is a prime, say p. Denote the center of α by P. Let $\beta \in \Gamma_{t,t}$ be an element whose center is distinct from P. As $\Gamma_{t,t}$ is an abelian group we have

$$(\alpha\beta)^p = \alpha^p\beta^p = \beta^p. \tag{2.1}$$

From Lemma 2.25 we get that $\alpha\beta$ and β have distinct centers. So Equation (2.1) implies that β^p has at least two centers, hence $\beta^p = \iota$. This means that the order of β is also p. If an elation γ has the same center as α, then apply the previous argument for the pair (β,γ) instead of (α,β). This gives that the order of γ is p, so all elements of $\Gamma_{t,t}$ have the same order, p.

We claim that $p|n$. Let $e \neq t$ be a line through P. The elation α generates a cyclic group of order p. Consider the action of this group on the point-set $\mathcal{A} = e \setminus \{P\}$. This set contains n points, because Π_n has order n. If $r < p$, then α^r has no fixed point in $\mathcal{A}$, because if $A \in \mathcal{A}$ and $A^{\alpha^r} = A$, then, by Lemma 2.18, we get $\alpha^r = \iota$ contradicting the fact that the order of α is p. Therefore, each point-orbit of α on the set $\mathcal{A}$ has length p, so p divides the cardinality of $\mathcal{A}$, that is n.

The group $\Gamma_{C,t}$ can be considered as a group of permutations of the points of $e \setminus \{C\}$. If $\iota \neq \alpha \in \Gamma_{C,t}$, then α has no fixed point on $e \setminus \{C\}$. Hence the cardinality of each orbit of points is $|\Gamma_{C,t}|$. The union of the orbits is $e \setminus \{C\}$, its cardinality is n, thus $|\Gamma_{C,t}|$ divides n. In the same way, $\Gamma_{t,t}$ is a group of permutations of the n^2 points of the plane not on t. Again, the cardinality of each orbit of points is $|\Gamma_{t,t}|$, hence this number divides n^2. This completes the proof of the theorem.
$\square$

If a point-line pair (C,t) is given in a plane, then the number of (C,t)-perspectivities cannot be too large.

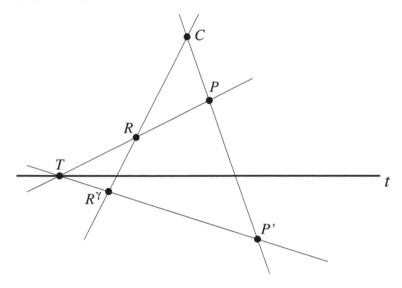

FIGURE 2.1
(C, t)-perspectivity

Proposition 2.29. *Let (C, t) be a point-line pair in a projective plane Π. If a pair of points (P, P') is given such that $P \neq C \neq P'$, none of P and P' is on t and C, P and P' are collinear, then there is at most one element $\gamma \in \Gamma_{C,t}$ for which $P^\gamma = P'$.*

Proof. Suppose that a (C, t)-perspectivity γ maps P to P'. Let R be a point neither on the line CP nor on t. By Lemma 2.19, R^γ must be on the line CR. If $T = PR \cap t$, then the points T, P' and R^γ are collinear, hence R^γ is uniquely determined as the point of intersection of the two lines CR and TP'.

If a collineation is uniquely defined on all points of Π except the points on the line CP, then it is also unique on the points of CP, because the image of any line of Π is uniquely determined. $\square$

The proposition above leads to the following definition.

Definition 2.30. *A projective plane Π is (C, t)-transitive, if for any pair of points (P, P') such that $P \neq C \neq P'$, none of P and P' is on t and C, P and P' are collinear, then there exists a unique element $\gamma \in \Gamma_{C,t}$ for which $P^\gamma = P'$ holds.*

The next statement obviously follows from the definition.

Lemma 2.31. *Let C be a point and t be a line in a projective plane of order n, Π_n. If Π_n is (C, t)-transitive, then*

$$|\Gamma_{C,t}| = \begin{cases} n, & \text{if } C \,I\, t, \\ n-1, & \text{otherwise.} \end{cases}$$

If the full collineation group of a finite projective plane contains a "lot of" elations, then the order of the plane cannot be arbitrary.

Definition 2.32. *A projective plane* Π *is a* translation plane *with respect to the line* t *if for any two points* A *and* B, *not on* t, *there exists an elation with axis* t *mapping* A *to* B.

In this case t *is called a* translation line *of* Π.

Theorem 2.33. *The order of any finite translation plane is a power of a prime number.*

Proof. Let Π_n be a finite projective plane of order n, and suppose that Π_n is a translation plane with respect to the line t. Theorem 2.28 gives that the collineation group $\Gamma_{t,t}$ is abelian, its non-identity elements have the same prime order p, and $p|n$.

Take a point A not on t. It follows from our assumptions that for an arbitrary point B not on t and $A \neq B$, there exists an elation $\gamma \in \Gamma_{t,t}$ such that $A^\gamma = B$. We claim that γ is unique. It is an elation, hence its center must be the point $AB \cap t$. According to Proposition 2.29, the center, the axis and a point and its image uniquely determine a (C, t)-perspectivity. We can choose B from the point-set of the affine plane $\Pi_n \setminus \{t\}$, the only condition is $B \neq A$. Thus we have $n^2 + n + 1 - (n + 1) - 1 = n^2 - 1$ choices for B. There is a one-to-one correspondence between the choices and the non-identity elements of $\Gamma_{t,t}$. Each of these choices defines a different element of $\Gamma_{t,t}$. The group also contains ι (this corresponds to the choice $B = A$), so $\Gamma_{t,t}$ has n^2 elements.

Now apply the classical Theorem of Cauchy: if the order of a group Γ is divisible by a prime number k, then Γ contains an element of order k. In our case the order of $\Gamma_{t,t}$ is n^2 and each of its non-identity elements has order p, hence the only prime divisor of n^2 is p. This means that n^2 is a power of p, so n is also a power of the prime p. $\square$

Corollary 2.34. *A finite projective plane of order* n *is a translation plane with respect to the line* t *if and only if* $|\Gamma_{t,t}| = n^2$.

Let Π_n be a finite projective plane of order n which is a translation plane with respect to the line t. Consider the point and line orbits of $\Gamma_{t,t}$. There are $n + 2$ point orbits. One of them has size n^2, because $\Gamma_{t,t}$ is transitive on the points of the affine plane $\Pi_n \setminus t$, while each of the $n + 1$ points on t is an orbit of size 1. The number of line orbits is also $n + 2$ (according to Theorem 2.13), but only one of them has size 1; this orbit consists of the line t. If T is any point on t, then the lines of the pencil with carrier T, except t itself, form a line orbit of $\Gamma_{t,t}$, so all other line orbits have size n.

The following two theorems show that existence of all elations with a fixed axis t is a consequence of the existence of some special elations having axis t.

Theorem 2.35. *Let* Π_n *be a finite projective plane of order* n. *Suppose that there exists an integer* $h > 1$ *and a line* t *on* Π_n *such that for all points* $C \, I \, t$, *the group $\Gamma_{C,t}$ has order h. Then n is a translation plane with respect to the line t.*

Proof. We know from Theorem 2.28 that the cardinality of $\Gamma_{t,t}$ divides n^2. So there exists an integer m such that $m|\Gamma_{t,t}| = n^2$. By Corollary 2.34, we have to prove that $m = 1$. Each non-identity element of the group $\Gamma_{t,t}$ belongs to exactly one of the groups $\Gamma_{C,t}$. There are $n + 1$ points on t, hence we get

$$\frac{n^2}{m} - 1 = (n+1)(h-1), \quad \text{so} \quad n^2 = m(n(h-1)+h),$$

$$\text{and} \quad n^2 - 1 = (n+1)hm + (m-1).$$

On the one hand, this implies $m < n$, because $h > 1$. On the other hand, $m - 1$ is divisible by $n + 1$, thus $m = 1$. $\qquad\qquad\square$

Theorem 2.36. *Let P and Q be distinct points and let t be a line of a finite projective plane of order n, Π_n. If Π_n is (P,t)-transitive and (Q,t)-transitive, then Π_n is (R,t)-transitive for any point R on the line PQ.*

Proof. If $n \le 4$, then Π_n is a Galois plane, hence it is (R,t)-transitive for any point-line pair. From now on we suppose that $n > 4$.

Let A and B be two distinct points such that none of them is on t and the line AB contains R. We claim that there exists an (R,t)-perspectivity mapping A to B. Let $C = PA \cap QB$. We distinguish three cases.

First, suppose that C is not on t. Then, as Π is (P,t)-transitive, there exists a (P,t)-perspectivity α mapping A to C, and from the (Q,t)-transitivity of Π it follows that there exists a (Q,t)-perspectivity β mapping C to B. Then $\alpha\beta$ maps A to B and it has axis t. The center of $\alpha\beta$ must be $PQ \cap AB = R$, hence $\alpha\beta$ is an (R,t)-perspectivity and $A^{\alpha\beta} = B$.

If C is on t, then take the point $D = PB \cap QA$. If D is not on t, then we can replace C by D and the previous argument shows the existence of an (R,t)-perspectivity again.

If $CD = t$, then let $M = AB \cap CD$, and take a point E on the line AB such that $E \notin \{A, B, R, M\}$. Such a point exists because $n > 4$. Then none of the points $C' = PE \cap QB$ and $C'' = PE \cap QA$ is on t, because $t \cap QB = C \ne C'$ and $t \cap QA = D \ne C''$. So applying the method of the previous paragraph, we can construct an (R,t)-perspectivity γ mapping A to E and another (R,t)-perspectivity δ mapping E to B. Then the (R,t)-perspectivity $\gamma\delta$ maps A to B.

The three cases together prove that Π is (R,t)-transitive. $\qquad\qquad\square$

Corollary 2.37. *Let P and Q be distinct points and let t be a line of a finite projective plane Π. If Π is (P,t)-transitive and (Q,t)-transitive, then Π is a translation plane with respect to the line t.*

Applying the dual of Theorem 2.36 we get the following statements.

Corollary 2.38. *Let t_1 and t_2 be two distinct lines of a projective plane Π. If Π is a translation plane with respect to the lines t_1 and t_2, then Π is a translation plane with respect to any line through the point $t_1 \cap t_2$.*

Corollary 2.39. *If a projective plane Π is a translation plane with respect to three non-concurrent lines, then Π is a translation plane with respect to any of its lines.*

The proof of the following theorem (its algebraic version is called the *Zorn-Levi theorem*) goes beyond the framework of this book. It can be found in [55, 3.4.22], or in [174, Theorem 5.1.31].

Theorem 2.40. *If a finite projective plane Π of order q is a translation plane with respect to all of its lines, then Π is isomorphic to the desarguesian plane $PG(2, q)$.*

In the rest of this chapter we investigate the geometric conditions corresponding to (C, t)-transitivity.

Definition 2.41. *A projective plane Π is (C, t)-desarguesian if for any two triangles which are in perspective from C and two pairs of their corresponding sides meet in points of t, the two triangles are in perspective from t.*

Theorem 2.42 (Baer). *Let (C, t) be any point-line pair in a projective plane Π. Then Π is (C, t)-transitive if and only if Π is (C, t)-desarguesian.*

Proof. First, suppose that Π is (C, t)-transitive. For $i = 1, 2$ let $A_i B_i C_i$ be two triangles which are in perspective from C and the points of intersections $A_1 B_1 \cap A_2 B_2$ and $A_1 C_1 \cap A_2 C_2$ are on the line t. Let α be the (C, t)-perspectivity mapping A_1 to A_2. Then $B_1^\alpha = B_2$ and $C_1^\alpha = C_2$ follows from the construction described in the remark after Theorem 2.28. Thus the line $B_1 C_1$ is mapped to $B_2 C_2$. This means that the point $D = B_1 C_1 \cap t$ is on the line $B_2 C_2$. This point is fixed, because it is on the line t. Hence the point $B_1 C_1 \cap B_2 C_2$ is on t. So the two triangles $A_1 B_1 C_1$ and $A_2 B_2 C_2$ are in perspective from the line t, thus Π is (C, t)-desarguesian, see Figure 2.2.

Now, suppose that Π is (C, t)-desarguesian. Let A_1 and A_2 be two points such that none of them is on the line t, $A_1 \neq C \neq A_2$ and the line $A_1 A_2$ contains C. Let us define a mapping α on the points of Π in the following way:

(1) $C^\alpha = C$,

(2) if $Q \mathrel{I} t$, then $Q^\alpha = Q$,

(3) $A_1^\alpha = A_2$,

(4) if $B \notin t \cup A_1 A_2$, then let $M = A_1 B \cap t$ and $B^\alpha = CB \cap A_2 M$,

(5) if $D \in A_1 A_2$, then let $B \notin t \cup A_1 A_2$ be an arbitrary point and let $D_B^\alpha = CA_1 \cap B^\alpha (DB \cap t)$.

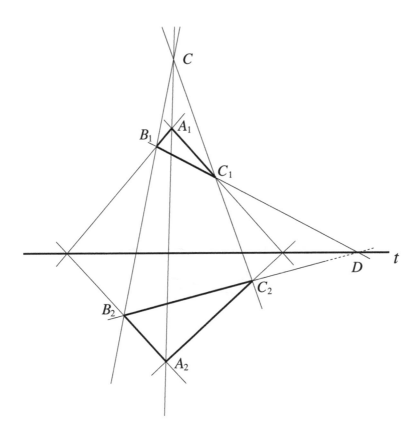

FIGURE 2.2
Theorem of Baer, I

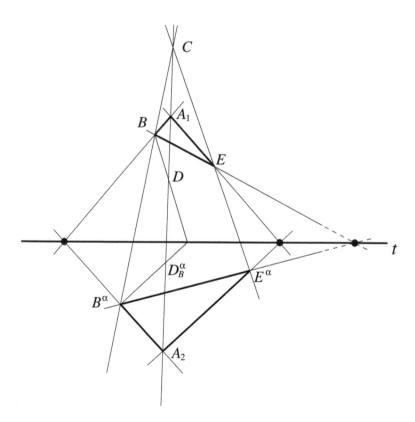

FIGURE 2.3
Theorem of Baer, II

We claim that D_B^α does not depend on the choice of B, and α is a (C,t)-perspectivity.

If $E \notin (t \cup A_1A_2)$ is an arbitrary point, then it follows from the definition of E^α that the triangles A_1BE and $A_2B^\alpha E^\alpha$ are in perspective from C, and the points $A_1B \cap A_2B^\alpha$ and $A_1E \cap A_2E^\alpha$ are on t. Thus the point $BE \cap B^\alpha E^\alpha$ is also on t, because Π is (C,t)-desarguesian, see Figure 2.3.

On the one hand, if $D \in A_1A_2$ and we construct its image according to (5) using three distinct points B_1, B_2 and $B_3 = B_1D \cap B_2D$, then the resulting points $D_{B_i}^\alpha$ coincide for $i = 1$ and 3, and also coincide for $i = 2$ and 3. Thus $D^\alpha = D_{B_i}^\alpha$ does not depend on the choice of B, so α is a well-defined permutation on the points of Π.

On the other hand, if none of the three points K, L and M is on t, then they are collinear if and only if each of the three lines KL, LK and MK meets t in the same point, say T. It happens if and only if each of the three lines $K^\alpha L^\alpha, L^\alpha K^\alpha$ and $M^\alpha K^\alpha$ also meets t in T, hence if and only if the points K^α, L^α and M^α are collinear.

If some of the three points K, L and M are on t, then they are obviously collinear if and only if K^α, L^α and M^α are collinear, hence α is a collineation. It follows from (2) that t is an axis of α, and (4) implies that C is a center of α. Finally, (3) guarantees that α maps A_1 to A_2. Hence Π is (C,t)-transitive. $\square$

Lenz and *Barlotti* have classified projective planes according to the configuration formed by the set of points C and lines t for which the plane is (C,t)-transitive. This list is called *Lenz–Barlotti classification* (see [55], Chapter 3.1). It contains 45 possible classes. A lot of research has been done to find examples with a given configuration or to show that a plane of certain Lenz–Barlotti type cannot exist. In many cases the existence problem is still open.

Exercises

2.1. Let $\alpha_i \neq \iota$ be a (C_i, t_i)-perspectivity for $i = 1, 2$ and suppose that $C_1 \neq C_2$ and $t_1 \neq t_2$. Prove that

- if F is a fixed point of $\alpha_1\alpha_2$, then either $F = t_1 \cap t_2$ or F is on the line C_1C_2;

- if $\alpha_1\alpha_2$ is a (C,t)-perspectivity, then $C_1 \mathrel{I} t_2$, $C_2 \mathrel{I} t_1$, $C = t_1 \cap t_2$ and t is the line C_1C_2.

2.2. Let $\alpha_i \neq \iota$ be a (C_i, t_i)-perspectivity for $i = 1, 2$. Show that $\alpha_1\alpha_2 = \alpha_2\alpha_1$ if and only if $C_1 \mathrel{I} t_2$ and $C_2 \mathrel{I} t_1$.

2.3. A collineation φ of a projective plane Π is called an *involution* if φ has order 2 in $\mathrm{Aut}(\Pi)$.

Show that if a (C,t)-perspectivity interchanges two points, then it is an involution.

2.4. Let φ be an involution. Show that

- if P is a point and $P \neq P^\varphi$, then PP^φ is a fixed line of φ;
- φ has some fixed points;
- φ has at least one fixed point on each line of the plane.

2.5. Let φ be an involution of a projective plane Π. Prove that

- if φ has four fixed points in general position, then $\mathcal{F}_\varphi$ is a Baer subplane of Π;
- if all fixed points of φ are collinear, then φ is an elation and the order of Π is even;
- if $\mathcal{F}_\varphi$ contains three non-collinear points but it does not contain four points in general position, then φ is a homology and the order of Π is odd.

2.6. How many Baer subplanes are there on the projective plane of order 4?

2.7. Determine the number of involutions in the full collineation group of the projective planes of order 2, 3 and 4.

3

Coordinatization of projective planes

In this chapter we explain how one can associate an algebraic structure to any projective plane that coordinatizes the plane. We study how algebraic properties of this coordinate structure are related to collineations of the projective plane. The method explained here is due to *M. Hall Jr.*; a detailed description of the method can be found in Chapter 20 of [77].

Let $\Pi = (\mathcal{P}, \mathcal{E}, I)$ be an arbitrary projective plane, and X, Y, O, E be a set of four points no three of which are collinear. Let $\mathcal{R}$ be a set of symbols containing 0 and 1, but not containing ∞. Assume that the cardinality of $\mathcal{R}$ is one less than the number of points on a line. For coordinatizing our plane elements of $\mathcal{R}$ will be used in the following way: let the coordinates of O be $(0,0)$, the coordinates of E be $(1,1)$, and the coordinates of the points on the line OE different from O, E and $OE \cap XY$ be of the form (a, a) in such a way that a is different from 0 and 1, and the coordinates of different points are different. This can be done because of our assumption on the cardinality of $\mathcal{R}$. When P is a point not on the line XY, then we define the coordinates of P as (a, b), where $YP \cap OE=(a, a)$ and $XP \cap OE=(b, b)$.

If P is a point of the line XY, different from Y, then let the coordinate of P be (m), where $YE \cap OP=(1, m)$. Finally, Y will have coordinate (∞).

If e is a line that does not pass through Y, then the coordinates of e are defined as $[m, k]$, where $e \cap XY=(m)$ and $e \cap OY=(0, k)$. If e goes through Y but it is different from XY, then it will have coordinate $[c]$, where $e \cap OX=(c, 0)$. Finally, let XY have coordinate $[\infty]$.

In particular, if Π is the classical projective plane, $\mathcal{R} = \mathbb{R}$ and we choose XY as the line at infinity, then on the affine plane $\Pi \setminus XY$ the lines OX and OY give a Cartesian coordinate system whose unit point is E.

Definition 3.1. *Let us define* a ternary operation $F(x, m, k)$ *on the set* $\mathcal{R}$ *in the following way:*

$$F(x, m, k) = y \iff (x, y) I [m, k].$$

Lemma 3.2. *For the ternary operation* $F(x, m, k)$ *defined above the following properties hold:*

47

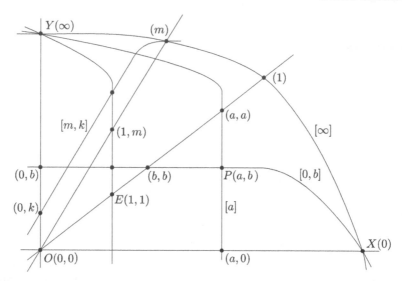

FIGURE 3.1
Definition of coordinates

(1) *For every* $a, b, c \in \mathcal{R}$

$$F(a, 0, c) = F(0, b, c) = c,$$

$$F(1, a, 0) = F(a, 1, 0) = a.$$

(2) *If* $a, b, a', b' \in \mathcal{R}$ *are given and* $a \neq a'$, *then there exists a unique pair* $(m, k) \in \mathcal{R} \times \mathcal{R}$ *for which* $F(a, m, k) = b$ *and* $F(a', m, k) = b'$.

(3) *For every triple* $a, b, m \in \mathcal{R}$ *there exists a unique* $k \in \mathcal{R}$ *for which* $b = F(a, m, k)$.

(4) *If* $m, k, m', k' \in \mathcal{R}$ *are given and* $m \neq m'$, *then there exists a unique* $x \in \mathcal{R}$ *for which* $F(x, m, k) = F(x, m', k')$.

Proof. The assertions of the lemma are just translations of the incidence axioms of projective planes in terms of the function F.

Assertion (1) describes the incidences

$$(a, c) \, \mathrm{I} \, [0, c], \quad (0, c) \, \mathrm{I} \, [b, c], \quad (1, a) \, \mathrm{I} \, [a, 0] \quad \text{and} \quad (a, a) \, \mathrm{I} \, [1, 0]$$

that follow from the definition of coordinates.

Assertion (2) ensures that the points (a, b) and (a', b') have a unique line of type $[m, k]$ joining them. Similarly, assertion (3) means that the points (a, b) and (m) have a unique line joining them.

Assertion (4) simply means that the lines $[m, k]$ and $[m', k']$ have a unique point in common.

$\square$

Definition 3.3. *The algebraic structure* $(\mathcal{R}, F)$ *is called a* ternary ring *if* F *satisfies properties* (1)–(4) *of the previous lemma.*

Reversing the process of introducing coordinates the following lemma can be deduced easily.

Lemma 3.4. *For every ternary ring there exists a projective plane that is coordinatized by the ternary ring.*

Proof. Let $(\mathcal{R}, F)$ be a ternary ring. We define a point-line incidence geometry $\Pi = (\mathcal{P}, \mathcal{E}, \mathrm{I})$. Let

$$\mathcal{P} = \{(a, b) \colon a, b \in \mathcal{R}\} \cup \{(d) \colon d \in \mathcal{R}\} \cup \{(\infty)\},$$

$$\mathcal{E} = \{[m, k] \colon m, k \in \mathcal{R}\} \cup \{[c] \colon c \in \mathcal{R}\} \cup \{[\infty]\}.$$

Incidence is defined as follows:

- $(a, b) \mathrm{I} [m, k] \iff b = F(x, m, k)$, $(a, b) \mathrm{I} [c] \iff a = c$, and (a, b) is never incident with $[\infty]$;

- $(d) \mathrm{I} [m, k] \iff d = m$, (d) is never incident with $[c]$, and $(d) \mathrm{I} [\infty]$ for all d;

- (∞) is never incident with $[m, k]$, $(\infty) \mathrm{I} [c]$ for all c, and $(\infty) \mathrm{I} [\infty]$.

We leave it as an exercise to prove that Π satisfies axioms P1, P2 and P3'. $\square$

It is quite possible that the same projective plane is coordinatized by non-isomorphic ternary rings since the coordinate structures may depend on the four base points of the coordinate system. If the plane has "many" collineations, then the coordinatizing ternary rings are essentially the same. This is the content of the next lemma.

Definition 3.5. *The ternary rings* $(\mathcal{R}_1, F_1)$ *and* $(\mathcal{R}_2, F_2)$ *are* isomorphic, *if there is a mapping* $\alpha \colon \mathcal{R}_1 \to \mathcal{R}_2$ *for which* $\alpha(F_1(a, b, c)) = F_2(\alpha(a), \alpha(b), \alpha(c))$ *holds for every* $a, b, c \in \mathcal{R}_1$.

Lemma 3.6. *Let* $(\mathcal{R}_1, F_1)$ *and* $(\mathcal{R}_2, F_2)$ *be two ternary rings coordinatizing the projective plane* Π, *whose base points are* X_1, Y_1, O_1, E_1 *and* X_2, Y_2, O_2, E_2, *respectively.* $(\mathcal{R}_1, F_1)$ *and* $(\mathcal{R}_2, F_2)$ *are isomorphic if and only if there is a collineation of* Π *which maps the 4-tuple* (X_1, Y_1, O_1, E_1) *to* (X_2, Y_2, O_2, E_2).

Proof. Let us denote the coordinates of points and lines with an index 1 or 2 according to whether the coordinates belong to the ternary ring $(\mathcal{R}_1, F_1)$ or $(\mathcal{R}_2, F_2)$.

Assume that ϕ is a collineation which maps (X_1, Y_1, O_1, E_1) to (X_2, Y_2, O_2, E_2). Define the mapping $\alpha : \mathcal{R}_1 \to \mathcal{R}_2$, using the points of the lines $O_1 E_1$ and $O_2 E_2$, in the following way:

$$\alpha(a_1) = a_2 \iff (a_1, a_1)^\phi = (a_2, a_2).$$

Since ϕ preserves incidence, for every point (a_1, b_1) and line $[m_1, k_1]$ we have that $(a_1, b_1)^\phi = (\alpha(a_1), \alpha(b_1)) = (a_2, b_2)$ and $[m_1, k_1]^\phi = [\alpha(m_1), \alpha(k_1)] = [m_2, k_2]$. Therefore,

$$
\begin{aligned}
y = F_1(x, m, k) &\iff (x, y) \,\mathrm{I}\, [m, k] \iff (x, y)^\phi \,\mathrm{I}\, [m, k]^\phi \\
&\iff (\alpha(x), \alpha(y)) \,\mathrm{I}\, [\alpha(m), \alpha(k)] \\
&\iff \alpha(y) = F_2(\alpha(x), \alpha(m), \alpha(k)).
\end{aligned}
$$

On the other hand, if $\alpha(F_1(a, b, c)) = F_2(\alpha(a), \alpha(b), \alpha(c))$ for every $a, b, c \in \mathcal{R}_1$, then the mapping

$$
\begin{array}{lllll}
(x, y) & \mapsto & (\alpha(x), \alpha(y)), & (m) & \mapsto & (\alpha(m)), & (\infty) & \mapsto & (\infty), \\
[m, k] & \mapsto & [\alpha(m), \alpha(k)], & [k] & \mapsto & [\alpha(k)], & [\infty] & \mapsto & [\infty],
\end{array}
$$

will be a collineation which maps (X_1, Y_1, O_1, E_1) to (X_2, Y_2, O_2, E_2). $\square$

Now some binary operations are defined using the ternary ring.

Definition 3.7. *In the ternary ring $(\mathcal{R}, F)$ the sum and product of two elements are defined by*

$$a + b := F(a, 1, b),$$

and

$$a \cdot b := F(a, b, 0)$$

(see Figure 3.2). The product $a \cdot b$ often will be written as ab.

We recall a definition from algebra.

Definition 3.8. *Let H be a non-empty set, $*$ be a binary operation on H. The algebraic structure $(H, *)$ is called a loop (or quasigroup with unity) if it has a unit element (that is there exists $e \in H$ such that $e * x = x * e = x$ for all $x \in H$) and the equations $a * x = b$ and $x * a = b$ have a unique solution x for every $a, b \in H$.*

The following observation follows from the definitions at once.

Lemma 3.9. *A loop with an associative operation is a group.*

Lemma 3.10. *If $(\mathcal{R}, F)$ is a ternary ring, $\mathcal{R}^* = \mathcal{R} \setminus \{0\}$, then $(\mathcal{R}, +)$ and $(\mathcal{R}^*, \cdot)$ are loops with unit elements 0 and 1, respectively.*

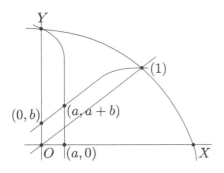 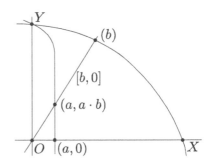

FIGURE 3.2
Sum and product

Proof. The identities $a + 0 = F(a, 1, 0) = a$, $0 + a = F(0, 1, a) = a$, $a \cdot 1 = F(a, 1, 0) = a$ and $1 \cdot a = F(1, a, 0) = a$ follow from assertion (1) of Lemma 3.2 for every $a \in \mathcal{R}$. This shows that 0 and 1 are indeed unit elements.

If $a, b \in \mathcal{R}$ are given, then the unique solvability of equation $a + x = F(a, 1, x) = b$ follows from assertion (3) of Lemma 3.2. As $x + a = F(x, 1, a)$ and $b = F(x, 0, b)$, the unique solvability of equation $x + a = b$ is equivalent to the unique solvability of equation $F(x, 1, a) = F(x, 0, b)$ which follows from assertion (4) of Lemma 3.2. Similarly, the unique solvability of equation $x \cdot a = b$ for $a \neq 0$ follows from assertion (4), because $x \cdot a = F(x, a, 0)$. Finally, as $a \cdot x = F(a, x, 0)$ and $0 = F(0, x, 0)$, the unique solvability of equation $a \cdot x = b$ for $a \neq 0$ is a consequence of assertion (2) of Lemma 3.2 when we apply it for a, b and $a' = b' = 0$. □

Definition 3.11.

- *The ternary ring $(\mathcal{R}, F)$ is called* linear*, if $F(a, b, c) = ab + c$ for every $a, b, c \in \mathcal{R}$.*

- *The ternary ring $(\mathcal{R}, F)$ is called a* (right) quasifield*, if the following properties are satisfied:*

   1. *$(\mathcal{R}, F)$ is linear,*
   2. *$(\mathcal{R}, +)$ is a group,*
   3. *$(a + b)c = ac + bc$ for every $a, b, c \in \mathcal{R}$.*

- *The quasifield $(\mathcal{R}, F)$ is called a* left quasifield *if 1. and 2. above and the other distributivity, that is $a(b + c) = ab + ac$, hold.*

- *The quasifield $(\mathcal{R}, F)$ is said to be* associative *if multiplication is associative.*

- *The quasifield* $(\mathcal{R}, F)$ *is said to be* commutative *if multiplication is commutative.*

- *The quasifield* $(\mathcal{R}, F)$ *is said to be* distributive *if both distributivity laws hold.*

For the sake of completeness, we mention an alternate, purely algebraic definition of quasifields.

Definition 3.12. *A triple* $(Q, \oplus, \odot)$, *where* Q *is a set and* $\oplus$ *and* $\odot$ *are binary operations on* Q, *is called a* right quasifield *if the following hold:*

(1) $(Q, \oplus)$ *is an abelian group with identity* 0;

(2) $(Q \setminus \{0\}, \odot)$ *is a group with identity* 1;

(3) $(a \oplus b) \odot c = a \odot c \oplus b \odot c$ *for every* $a, b, c \in Q$;

(4) *for every* $a \neq b, c \in Q$ *there is a unique* $x \in Q$ *satisfying the equation*

$$x \odot a - x \odot b = c.$$

We leave the proof of the next statements as exercises.

Proposition 3.13. *In the previous definition if* Q *is finite, then Property* (4) *follows from Properties* (1)–(3).

Proposition 3.14. *Let* $(Q, +, \odot)$ *be a right quasifield and for all* $x \in Q$ *let* $-x$ *denote the additive inverse of* a. *Then* $-a = (-1) \odot a$, $-(a+b) = (-a) + (-b)$ *and* $a \odot (-b) = (-a) \odot b = -(a \odot b)$ *hold for every* $a, b \in Q$.

In what follows we shall be interested in geometric properties of the plane corresponding to the above mentioned algebraic properties of the ternary ring coordinatizing the plane. In the remaining part of this chapter Π denotes a projective plane with distinguished base points X, Y, O, E. The corresponding ternary ring will be denoted by $(\mathcal{R}, F)$ throughout.

Theorem 3.15. *The plane* Π *is* (Y, XY)-*transitive if and only if* $(\mathcal{R}, F)$ *is linear and* $(\mathcal{R}, +)$ *is a group.*

Proof. First assume that $(\mathcal{R}, F)$ is linear and $(\mathcal{R}, +)$ is a group. Let A and B be distinct points not on XY so that the line AB passes through Y. We need to define a (Y, XY)-elation which maps A to B. Let $A = (x_1, y_1)$ and $B = (x_1, y_2)$. There is a unique a for which $y_2 = y_1 + a$. Here $a \neq 0$, since A and B are distinct points. Define the mapping α_a in the following way:

$$
\begin{array}{llll}
(x, y) & \mapsto & (x, y + a), \qquad (m) \mapsto (m), \qquad (\infty) \mapsto (\infty), \\
[m, k] & \mapsto & [m, k + a], \qquad [k] \mapsto [k], \qquad [\infty] \mapsto [\infty].
\end{array}
$$

We will show that α_a is the elation mapping A to B. It is clear by the definition of α_a that each point of the line XY is fixed and each line through Y is also fixed. This shows that α_a is indeed a (Y, XY)-elation. We need to show that

$$(x, y)\, \mathrm{I}\, [m, k] \iff (x, y)^{\alpha_a}\, \mathrm{I}\, [m, k]^{\alpha_a}.$$

Using the linearity of F and the associativity of addition we immediately get

$$\begin{aligned}
(x, y)\, \mathrm{I}\, [m, k] &\iff F(x, m, k) = y \iff xm + k = y \\
&\iff (xm + k) + a = y + a \iff xm + (k + a) = y + a \\
&\iff F(x, m, k + a) = y + a \iff (x, y + a)\, \mathrm{I}\, [m, k + a] \\
&\iff (x, y)^{\alpha_a}\, \mathrm{I}\, [m, k]^{\alpha_a},
\end{aligned}$$

which is what we wanted to prove.

To prove the other direction of the theorem, first, the Theorem of Baer (see Chapter 2, Theorem 2.42) is used. Since Π is (Y, XY)-transitive, it is also (Y, XY)-desarguesian. We are going to show that $F(a, b, c) = ab + c$ for every $a, b, c \in \mathcal{R}$. This linearity is straightforward if $a = 0$, $b = 0$, $b = 1$ or $c = 0$. Hence we may assume that none of these cases occurs. Let $C_1 = (0, 0)$, $C_2 = (0, c)$, $A = (a, 0)$, $B = (b)$ and $U = (1)$. With the help of these points we construct two triangles that are in perspective from both Y and XY.

Let $D_1 = C_1 B \cap YA := (a, ab)$, $F_1 = C_1 U \cap D_1 X := (ab, ab)$, $F_2 = C_2 U \cap YF_1 := (ab, ab + c)$ and $D_2 = C_2 B \cap YA := (a, y)$, where $y = F(a, b, c)$, because D_2 is incident with the line $[b, c]$ (see Figure 3.3).

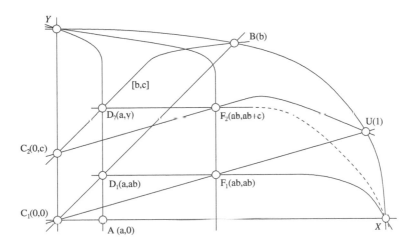

FIGURE 3.3
Linearity of the ternary ring

The triangles $C_1D_1F_1$ and $C_2D_2F_2$ are in perspective from Y, moreover, $C_1D_1 \cap C_2D_2 = B$ and $C_1F_1 \cap C_2F_2 = U$. By the (Y, XY)-Desargues' theorem, the two triangles are also in perspective from XY. This implies that the line D_2F_2 goes through the point $XY \cap D_1F_1 = X$. This means $y = F(a, b, c) = ab + c$, hence the ternary ring is linear.

Now we show that $+$ is associative. By Lemmas 3.9 and 3.10, this implies that $(\mathcal{R}, +)$ is a group.

Let α_a be the (Y, XY)-elation which maps the point $(0, 0)$ to $(0, a)$. Then α_a maps an arbitrary point (x, y) to (x, z), where z is independent from x. Hence $(x, y)^{\alpha_a} = (x, y^{\beta_a})$, where β_a is a permutation of the elements of $\mathcal{R}$ such that $0^{\beta_a} = a$. Let m, k, x and y be arbitrary, then

$$xm + k = y \iff F(x, m, k) = y \; iff \, (x, y) \; \mathrm{I} \; [m, k] \iff (x, y)^{\alpha_a} \; \mathrm{I} \; [m, k]^{\alpha_a}.$$

As the points of type (m) are fixed, $(0, k)$ is mapped to $(0, k^{\beta_a})$, we have that $[m, k]^{\alpha_a} = [m, k^{\beta_a}]$, and for all x, m, k and a

$$(x, y^{\beta_a}) \; \mathrm{I} \; [m, k^{\beta_a}] \iff F(x, m, k^{\beta_a}) = y^{\beta_a}$$
$$\iff xm + k^{\beta_a} = y^{\beta_a}$$
$$\iff xm + k^{\beta_a} = (xm + k)^{\beta_a}.$$

By substituting $k = 0$ and $m = 1$, $x + 0^{\beta_a} = x^{\beta_a}$ follows, hence $x^{\beta_a} = x + a$ for all x and a. Now substitute $m = 1$. We get $x + k^{\beta_a} = (x + k)^{\beta_a}$, that is

$$x + (k + a) = (x + k) + a \qquad\qquad \forall \, x, k, a \in \mathcal{R},$$

hence addition is associative. $\qquad\qquad\qquad\qquad\qquad\qquad\qquad\qquad\qquad\quad\square$

Example 3.16. Let $(\mathbb{K}, +, \cdot)$ be an ordered field, $0 < k \neq 1$ be a fixed element of $\mathbb{K}$. We define a new multiplication on the elements of $\mathbb{K}$ in the following way:

$$a * b = \begin{cases} akb, & \text{if } a \geq 0 \text{ and } b \geq 0, \\ ab, & \text{otherwise.} \end{cases}$$

Let $F(a, b, c) = (a * b) + c$. It is easy to check that $(\mathbb{K}, F)$ is a linear ternary ring. Clearly, $(\mathbb{K}, +)$ is a group, multiplication is commutative and associative, but none of the two distributive laws is satisfied. So this structure is the coordinate structure of a projective plane, but $(\mathbb{K}, +, *)$ is not a quasifield.

For $\mathbb{K} = \mathbb{R}$, and $k = 1/2$, the plane arising from the ternary ring is isomorphic to Moulton's example of a non-Desarguesian plane. Its detailed description is given in Example 1.27.

Theorem 3.17. *The projective plane Π is (X, XY)-transitive if and only if $(\mathcal{R}, +)$ is a group and for every $a, b, c \in \mathcal{R}$ we have $F(a, b, cb) = (a + c)b$.*

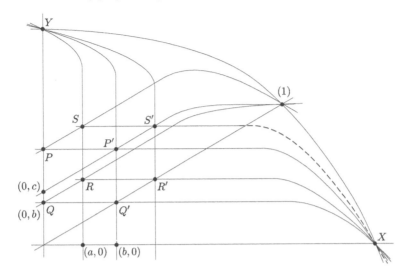

FIGURE 3.4
(X, XY)-transitivity

Proof. Assume that Π is (X, XY)-transitive. Let $Q = (0, b)$, $Q' = (b, b)$, $P = (0, b+c)$, $P' = (b, b+c)$, $R = (a, a+b)$, $R' = (a+b, a+b)$, $S = (a, a+(b+c))$ and $S' = (a + b, (a + b) + c)$ (see Figure 3.4).

Let α be an elation with center X, axis XY for which $Q^\alpha = Q'$. Then $P^\alpha = P'$, $R^\alpha = R'$ and $S^\alpha = S'$, so S, S' and X are collinear. This implies that $+$ is associative and hence $(\mathcal{R}, +)$ is a group.

Now let α_a be the (X, XY)-elation which maps $(0, 0)$ to $(a, 0)$. The elation α_a maps an arbitrary point (x, y) to (z, y), where z does not depend on x. Therefore, $(x, y)^{\alpha_a} = (x^{\beta_a}, y)$, where β_a is a permutation of $\mathcal{R}$, for which $0^{\beta_a} = a$. Suppose that $(x, y)\,\mathrm{I}\,[m, k]$. If $[m, k]^{\alpha_a} = [m, n]$, then, because of $(0, k)\,\mathrm{I}\,[m, k]$, we get $(a, k)\,\mathrm{I}\,[m, n]$ since $0^{\beta_a} = a$. Thus $k = F(a, m, n)$.

For the element n, defined by the above equality,

$$y = F(x, m, k) \iff y = F(x^{\beta_a}, m, n).$$

If $m = 1$, it implies $n = -a + k$ and, as $(\mathcal{R}, +)$ is a group,

$$x^{\beta_a} = y - n = (x + k) - (-a + k) = (x + k) + (-k + a) = x + a.$$

Thus

$$F(x, m, k) = F(x + a, m, n),$$

for every $x, a, m, k \in \mathcal{R}$. Substituting $n = 0$ and $k = am$ gives

$$F(x, m, am) = (x + a)m \qquad\qquad \forall\, x, m, a \in \mathcal{R},$$

and this is what we wanted to prove.

To prove the other direction assume that $(\mathcal{R}, +)$ is a group and for every $a, b, c \in \mathcal{R}$ we have $F(a, b, cb) = (a + c)b$.

Let us define the mapping α_a for $a \in \mathcal{R}$ in the following way:

$$
\begin{array}{llll}
(x, y) & \mapsto & (x + a, y), & (m) & \mapsto & (m), & (\infty) & \mapsto & (\infty), \\
[m, k] & \mapsto & [m, n], & [k] & \mapsto & [k + a], & [\infty] & \mapsto & [\infty],
\end{array}
$$

where n is defined by $k = F(a, m, n)$.

We are going to show that α_a is the desired elation. Obviously, α_a fixes each point of the line XY. It also fixes each line through X, because $k = F(a, 0, k)$ by definition, hence α_a maps $[0, k]$ to $[0, k]$ for all $k \in \mathcal{R}$. We need to show that

$$(x, y) \, \mathrm{I} \, [m, k] \iff (x, y)^{\alpha_a} \, \mathrm{I} \, [m, k]^{\alpha_a}.$$

If $(x, y) \, \mathrm{I} \, [m, k]$, then $(x, y)^{\alpha_a} = (x + a, y) \, \mathrm{I} \, [m, n]$. Let $\bar{n}$ be the unique solution of equation $xm = n$. Then $k = F(a, m, n) = F(a, m, \bar{n}m) = (a + \bar{n})m$ and hence

$$
\begin{aligned}
y = F(x, m, k) &= F(x, m, (a + \bar{n})m) \\
&= (x + a + \bar{n})m = F(x + a, m, \bar{n}m) \\
&= F(x + a, m, n).
\end{aligned}
$$

Therefore, α_a preserves incidence. Since a is an arbitrary element of $\mathcal{R}$, this means that Π is (X, XY)-transitive. $\qquad\square$

The previous two theorems can be summarized using the definition of quasifields.

Theorem 3.18. *The projective plane Π is (X, XY)-transitive and (Y, XY)-transitive, if and only if $(\mathcal{R}, F)$ is a quasifield. If $(\mathcal{R}, +, \cdot)$ is a quasifield, then $F(x, m, k) = x \cdot m + k$. In more detail, it means that the points of the affine plane $\Pi \setminus XY$ are the elements of $\mathcal{R} \times \mathcal{R}$ and the lines not passing through Y are the sets of points of the form $\{(x, x \cdot m + k) : x \in \mathcal{R}\}$, where m and k are elements of $\mathcal{R}$.*

The so-called *Hall quasifields* are constructed in the following example.

Example 3.19. Let $\mathbb{K}$ be an arbitrary field and suppose that the quadratic polynomial $f(x) = x^2 - ax - b$ is irreducible over $\mathbb{K}$. Moreover, let $\mathcal{V}$ be a two-dimensional (left) vector space over $\mathbb{K}$ with basis $1 \, (\in \mathbb{K})$ and $\alpha \, (\notin \mathbb{K})$. Define addition (denoted by $+$) on the elements of $\mathcal{V}$ as the vector space addition and let us introduce a new multiplication (denoted by $*$) in the following way:

$$(x + y\alpha) * (z + w\alpha) = xz - y^{-1}wf(x) + (yz - xw + aw)\alpha,$$

if $y \neq 0$, and

$$x * (z + w\alpha) = xz + (xw)\alpha.$$

It is not difficult (but tedious) to see that the algebraic structure $(\mathcal{V}, +, *)$ is a quasifield, which is not associative and not distributive when $\mathbb{K} \neq \mathrm{GF}(2)$ and $\mathbb{K} \neq \mathrm{GF}(3)$.

If a plane Π is (X, XY) and (Y, XY)-transitive, then, by Theorem 3.18, it is a translation plane with respect to the line XY and its coordinate structure is a quasifield.

Since elations having the same axis form an abelian group by Theorem 2.27, and addition in the quasifield can be identified with (Y, XY)-elations we can state the following theorem.

Theorem 3.20. *The additive group of a quasifield is an abelian group.*

Now some important substructures of quasigroups are introduced. They play a crucial role in describing quasifields.

Definition 3.21. *The* kernel *of the quasifield* $(\mathcal{R}, +, \cdot)$ *is the set* K *of those elements of* $\mathcal{R}$, *for which the following two identities hold for every* $k \in K$ *and* $x, y \in \mathcal{R}$:
(i) $k(x + y) = kx + ky$,
(ii) $k(xy) = (kx)y$.

Let us remark that in the literature the word "nucleus" is also used instead of the word "kernel".

Theorem 3.22. *The kernel of any quasifield is a field and the quasifield is a vector space over its kernel. Hence the order of a finite quasifield is a prime power.*

Proof. It is clear that $0, 1 \in K$. First we show that $(K, +)$ is a subgroup of $(\mathcal{R}, +)$. For this, it is enough to show that K is closed under subtraction. Let l and k be arbitrary elements of K, x and y be arbitrary elements of $\mathcal{R}$. Then

$$(k - l)(x + y) = k(x + y) - l(x + y) = kx + ky - lx - ly$$
$$= kx - lx + ky - ly = (k - l)x + (k - l)y,$$

(here we used Proposition 3.14) and

$$(k - l)(xy) = k(xy) - l(xy) = (kx)y - (lx)y = (kx - lx)y = ((k - l)x)y.$$

We still have to show that $(K^*, \cdot)$ is also a group. As associativity is immediate, it is enough to show that K^* is closed under multiplication and each element of K^* has a multiplicative inverse in K^*. The fact that K is closed under multiplication comes immediately from the identities

$$(kl)(x + y) = k(l(x + y)) = k(lx + ly) = k(lx) + k(ly) = (kl)x + (kl)y,$$

and

$$(kl)(xy) = k(l(xy)) = k((lx)y) = (k(lx))y = ((kl)x)y.$$

As $(\mathcal{R}^*, \cdot)$ is a loop, for each $k \in K$ there is a unique k', for which $kk' = 1$. We wish to show that $k' \in K$. For this, consider the elements $k'(x + y)$ and $k'x + k'y$. Multiplying them by k from the left, we get the elements $k(k'(x+y))$ and $k(k'x+k'y)$. Using the fact that k belongs to K, we get $(kk')(x+y) = x+y$ and $k(k'x)+k(k'y)$, respectively. The latter element is $(kk')x+(kk')y = x+y$, so after multiplying the elements $k'(x+y)$ and $k'x+k'y$ from the left by k, the same element of $\mathcal{R}$ is obtained. Since $(\mathcal{R}^*, \cdot)$ is a loop, the original elements must be the same. This means that k' satisfies property (i) in the definition of the kernel. Checking (ii) is similar.

Considering the addition in $\mathcal{R}$ as addition in a vector space and multiplication with elements of K as multiplication with scalars, the defining properties of K just show that $\mathcal{R}$ is a (left) vector space over K. If this vector space has dimension d, then $|\mathcal{R}| = |K|^d$, hence $|\mathcal{R}|$ is indeed a prime power. $\qquad\square$

Using the previous theorem we can give an alternative proof of Theorem 2.33 of Chapter 2.

Corollary 3.23. *The order of a finite translation plane is a prime power.*

Proof. A translation plane can be coordinatized by a quasifield. The order of a quasifield is a prime power and the order of the plane is equal to the order (number of elements) of the ternary ring coordinatizing it. $\qquad\square$

Using quasifields one can construct projective planes, the so-called *Hughes planes*, which cannot be coordinatized by a linear ternary ring.

Example 3.24. Let q be an odd prime, N be an associative quasifield with kernel $K = \mathrm{GF}(q)$ and assume that N is two-dimensional over K. Let $\mathcal{V}$ be a three-dimesional left vector space over N. Let A be a 3×3 matrix with elements from K, such that $A^{q^2+q+1} = kI$ for some $k \in K$ and assume that $q^2 + q + 1$ is the smallest exponent with this property. This property means that the collineation $x \mapsto Ax$ permutes the points in one cycle, so it represents a collineation of type R_f introduced in the proof of Theorem 2.15.

Let the points of the new plane $\mathcal{H}$ (of order q^2) be the one-dimensional subspaces of $\mathcal{V}$. Points are represented by vectors different from $(0,0,0)$, contained in the subspace.

For any $t \in N$ let $L(t)$ be the set of those vectors $\mathbf{x} = (x_1, x_2, x_3) \in \mathcal{V}$ for which $x_1+x_2t+x_3 = 0$ holds. This happens if and only if $kx_1+(kx_2)t+kx_3 = 0$ for some $0 \neq k \in N$. Therefore, $L(t)$ can be considered as a set of points in the plane $\mathcal{H}$. The lines of the new plane are the sets of points of type

$$L(t)A^m = \{\mathbf{x}A^m : \mathbf{x} \in L(t)\},$$

where $0 \leq m \leq q^2 + q$ and $t \in \{1\} \cup (N \setminus K)$. Incidence is defined in the usual way as set theoretical inclusion.

At the end of this chapter we shall see an example of an associative quasi-field used in the construction of Hughes planes. Indeed, the quasifields in Example 3.29 for $r = 2$ are such examples. This is a special case of *André quasifields*.

The number of points of the plane $\mathcal{H}$ is

$$(|\mathcal{V}| - 1)/(|N| - 1) = (q^6 - 1)/(q^2 - 1) = q^4 + q^2 + 1.$$

The number of lines is the same, since the parameter m in the definition of lines can take $q^2 + q + 1$ different values and t can take $1 + (q^2 - q)$ different values. A simple computation gives that for $m = 0$ the number of points of $L(t)$ is $q^2 + 1$, and as A maps different points to different points, every line contains $q^2 + 1$ points. A tedious computation gives that two different lines have exactly one common point. The computation is left as an exercise. These properties guarantee that $\mathcal{H}$ is a projective plane of order q^2. The detailed proof, together with the fact that the ternary rings coordinatizing the Hughes planes cannot be linear, can be found in [95], Chapter 9.6.

The order of the smallest Hughes plane is 9. This plane can be constructed by the above construction from the André quasifield described in Example 3.29 if we take $p = 3$ and $r = 2$. The same Example 3.29 also gives a translation plane as described in Theorem 3.18. A seemingly different translation plane can be constructed by a Hall quasifield, see Example 3.19. However, it can be shown that all translation planes of order 9 are isomorphic. One can also show that this translation plane is not isomorphic to its dual plane. By computer search it was shown in the late 1980's that there are precisely four non-isomorphic projective planes of order 9. These are PG$(2, 9)$, the translation plane, the dual of the translation plane and the Hughes plane.

Let Π^* be the dual plane of Π. Then Π^* and Π have the same order, hence Π^* can be coordinatized by the elements of the same set $\mathcal{R}$ which is the underlying set of the coordinate structure of Π. Let us denote coordinates in the dual plane by $(x, y)'$, $[m, k]'$, $(m)'$ and $[c]'$. Choose the base points in such a way that

$$\begin{aligned}
(0, 0)' &= [0, 0], & (0)' &= [0], & (\infty)' &= [\infty], \\
[0, 0]' &= (0, 0), & [0]' &= (0), & (1)' &= [1],
\end{aligned}$$

and map the elements of $\mathcal{R}$ onto the points of $[1, 0]'$ so that $(a, a)' = [a, a]$ for every $a \in \mathcal{R}$. (From the practical point of view this means that coordinates in the dual plane are obtained by changing round brackets to square ones, and vice versa.) Define the ternary function F' using incidences in the dual plane. This means that

$$F'(x, m, k) = y \iff (x, y)' \, \mathrm{I}' \, [m, k]' \iff (m, k) \, \mathrm{I} \, [x, y] \iff F(m, x, y) = k.$$

Define the sum $\oplus$ and product $\odot$ of two elements in the ternary ring $(\mathcal{R}, F')$ in the same way as in Definition 3.7:

$$a \oplus b := F'(a, 1, b) \quad \text{and} \quad a \odot b := F'(a, b, 0).$$

Then the following theorem can be formulated.

Theorem 3.25. *If $(\mathcal{R}, F)$ is linear and addition is associative, then $(\mathcal{R}, F')$ is also linear and its addition is associative, too.*

Proof. The conditions on $(\mathcal{R}, F)$ ensure that Π is (Y, XY)-transitive, which implies that Π^* is $(Y', (XY)')$-transitive. Since $Y' = XY$ and $(XY)' = Y$, Theorem 3.15 gives that $(\mathcal{R}, F')$ is linear and its addition is associative. $\square$

If both $(\mathcal{R}, F)$ and $(\mathcal{R}, F')$ are linear, then we have the following relations between the additions and multiplications of the two structures:

$$c = a \oplus b \Longleftrightarrow b = a + c, \quad \text{because} \quad c = F'(a, 1, b) \Longleftrightarrow b = F(1, a, c),$$
$$c = a \odot b \Longleftrightarrow 0 = ba + c, \quad \text{because} \quad c = F'(a, b, 0) \Longleftrightarrow 0 = F(b, a, c).$$

If $(\mathcal{R}, +, \cdot)$ is a quasifield, then Π is a translation plane with respect to the line XY. Hence Π^* is a dual translation plane with respect to the point $(XY)' = Y$. This means that Π^* is (Y, XY)- and (Y, OY)-transitive, since $X' = OY$. The (Y, XY)-transitivity of Π^* implies that $(\mathcal{R}, \oplus, \odot)$ is linear and $(\mathcal{R}, \oplus)$ is an abelian group. From the identity $0 = ba + a \odot b$ we have

$$u(v + w) + (v + w) \odot u = uv + uw + v \odot u + w \odot u \qquad \forall u, v, w \in \mathcal{R}.$$

Hence in the ternary ring $(\mathcal{R}, F)$ distributivity from one side is satisfied if and only if distributivity is satisfied in the ternary ring $(\mathcal{R}, F')$ from the other side. Therefore, combining Theorems 3.17 and 3.25 and the definition of a distributive quasifield, we get the following theorem.

Theorem 3.26. *For the projective plane Π the following are equivalent:*

(1) *The plane is (Y, XY)-transitive, (X, XY)-transitive and (Y, OY)-transitive.*

(2) *The plane is (P, XY)-transitive for every point $P \in XY$ and (Y, ℓ)-transitive for every line ℓ passing through Y.*

(3) *The ternary ring $(\mathcal{R}, F)$ is a distributive quasifield.*

If a ternary ring is linear and addition is a group then we have seen above that the left distributivity law and the (Y, OY)-transitivity of the plane are equivalent. This can be proven without referring to the dual plane (as we did). The direct proof is similar to that of Theorem 3.15 and the details can be found in Chapter 20 of [77].

The associativity of multiplication also has a geometric translation.

Theorem 3.27. *The plane Π is (X, OY)-transitive if and only if $(\mathcal{R}, F)$ is linear and $(\mathcal{R}^*, \cdot)$ is a group.*

Proof. Assume that $(\mathcal{R}, F)$ is linear and multiplication is associative. It is enough to construct an (X, OY)-homology which maps $(1,0)$ to $(a,0)$. Consider the mapping α_a:

$$
\begin{array}{llll}
(x,y) & \mapsto & (xa,y), & \quad (m) \mapsto (a^{-1}m), \quad (\infty) \mapsto (\infty), \\
[m,k] & \mapsto & [a^{-1}m,k], & \quad [k] \mapsto [ka], \quad\quad\ [\infty] \mapsto [\infty].
\end{array}
$$

We are going to show that α_a is the desired (X, OY)-homology. It is clear that the points of the line OY and the lines through Y are fixed. The point $(1,0)$ is mapped to $(a,0)$ indeed, so it only remains to be checked that α_a preserves incidence. This is true for points of type (x,y) and lines of type $[m,k]$, because

$$
\begin{aligned}
(x,y)\,\mathrm{I}\,[m,k] &\iff F(x,m,k) = y \iff xm + k = y \\
&\iff (xa)(a^{-1}m) + k = y \iff F(xa, a^{-1}m, k) = y \\
&\iff (xa,y)\,\mathrm{I}\,[a^{-1}m,k] \iff (x,y)^{\alpha_a}\,\mathrm{I}\,[m,k]^{\alpha_a}.
\end{aligned}
$$

In other cases it is even simpler to check that α_a preserves incidence.

For proving the other direction, assume that Π is (X, OY)-transitive. This ensures the existence of an (X, OY)-homology α_a which maps $(1,0)$ to $(a,0)$. This homology fixes XY and OX, hence it maps points of type (m) to (m^{β_a}) and points of type $(x,0)$ to $(x^{\gamma_a}, 0)$, where β_a and γ_a are permutations of $\mathcal{R}$ that satisfy $0^{\beta_a} = 0$, $1^{\gamma_a} = a$ and $0^{\gamma_a} = 0$. Therefore, the action of α_a can be described in the following way:

$$
\begin{array}{llll}
(x,y) & \mapsto & (x^{\gamma_a}, y), & \quad (m) \mapsto (m^{\beta_a}), \quad (\infty) \mapsto (\infty), \\
[m,k] & \mapsto & [m^{\beta_a}, k], & \quad [k] \mapsto [k^{\gamma_a}], \quad\ \ [\infty] \mapsto [\infty].
\end{array}
$$

For all $m, x, y \in \mathcal{R}$ we have

$$
\begin{aligned}
F(x,m,k) = y &\iff (x,y)\,\mathrm{I}\,[m,k] \\
&\iff (x,y)^{\alpha_a}\,\mathrm{I}\,[m,k]^{\alpha_a} \iff (x^{\gamma_a}, y)\,\mathrm{I}\,[m^{\beta_a}, k] \\
&\iff F(x^{\gamma_a}, m^{\beta_a}, k) = y.
\end{aligned}
$$

Substituting $k = 0$ gives

$$
xm = x^{\gamma_a} m^{\beta_a} \qquad\qquad \forall\, m, x \in \mathcal{R}, \tag{3.1}
$$

and for $x = 1$ it simplifies to $m = am^{\beta_a}$. By substituting now $m = a$ implies $a = a \cdot a^{\beta_a}$. The equation $a = xa$ has a unique solution, namely $x = 1$, hence $a^{\beta_a} = 1$. By substituting this in (3.1), we get $x^{\gamma_a} = xa$ for every x. Putting this back again in (3.1),

$$
xm = (xa)m^{\beta_a} \qquad \forall\, m, x \in \mathcal{R}
$$

is obtained. Now, substituting au in place of m, where u is arbitrary, gives us

$$
x(au) = (xa)(au)^{\beta_a} \qquad \forall\, u, x \in \mathcal{R}.
$$

As $au = a(au)^{\beta_a}$ is also true and the unique solution of equation $au = ax$ is $x = u$, we have $u = (au)^{\beta_a}$, and hence

$$x(au) = (xa)u \qquad\qquad \forall\, a, u, x \in \mathcal{R}.$$

(The above proof makes use of $a \neq 0$, however, for $a = 0$ the assertion is immediate.) This proves the associativity of multiplication. Hence, by Lemmas 3.9 and 3.10, $(\mathcal{R}, \cdot)$ is a group.

It remains to show linearity of the ternary operation. We have seen $m = am^{\beta_a}$, which is the same as $m^{\beta_a} = a^{-1}m$. This implies

$$F(x, m, k) = F(xa, a^{-1}m, k) \qquad\qquad \forall\, a, x, m, k \in \mathcal{R}.$$

For $m = a$ this means

$$F(x, a, k) = F(xa, 1, k) = xa + k, \qquad\qquad \forall\, x, a, k \in \mathcal{R},$$

hence $(\mathcal{R}, F)$ is linear. $\qquad\qquad\qquad\qquad\qquad\qquad\qquad\qquad\qquad\qquad\qquad$ $\square$

One can prove similarly that Π is (Y, OX)-transitive if and only if $(\mathcal{R}^*, \cdot)$ is a group (so multiplication is associative) and $F(a, b, cb) = (a + c)b$ for every $a, b, c \in \mathcal{R}$. This proof can also be found in Chapter 20 of [77], and we leave it to the reader.

Again, we can summarize the proven results.

Theorem 3.28. *A projective plane Π is (X, OY)-transitive and (P, XY)-transitive for every $P \in XY$ if and only if its coordinatizing ternary ring $(\mathcal{R}, F)$ is an associative quasifield.*

Example 3.29. Let $\mathcal{R} = \mathrm{GF}(p^r)$, where p is an odd prime, $r \geq 2$ and let σ be an automorphism of $\mathrm{GF}(p^r)$. Define a new multiplication on the elements of $\mathcal{R}$:

$$a * b = \begin{cases} ab & \text{if b is a square,} \\ a^{\sigma}b & \text{if b is a non-square.} \end{cases}$$

Let $F(a, b, c) = (a * b) + c$. It is easy to check that $(\mathcal{R}, F)$ is a linear ternary ring, addition is an abelian group, multiplication is a loop and also the (right) distributive law is valid. Hence this is a quasifield but not a field, not even a distributive quasifield.

In particular, if $r = 2$, then we get a so-called *Dickson quasifield*. In this case $\sigma\colon x \mapsto x^p$ and this implies that multiplication is associative, too. In the next example we define *André quasifields* in general.

Example 3.30. Let $\mathcal{R} = \mathrm{GF}(q^n)$, Γ be the group of automorphisms of $\mathrm{GF}(q^n)$ which fix each element of $\mathrm{GF}(q)$. Then Γ consists of the mappings $x \mapsto x^{q^i}$, $(i = 1, \ldots, n)$ and Γ is a cyclic group of order n. Let Φ be an arbitrary mapping from GF(q)* to Γ for which $\Phi(1)$ is the neutral element of Γ (so the identity automorphism). Finally, for any $x \in \mathcal{R} \setminus \{0\}$ let $\alpha(x)$ be the Φ-image of the norm of x, that is

$$\alpha(x) = \Phi\left(x^{1+q+\ldots+q^{n-1}}\right).$$

A new multiplication is defined on the elements of $\mathcal{R}$ as follows:

$$a * b = a^{\alpha(b)}b, \quad \text{if} \quad b \neq 0; \quad a * 0 = 0.$$

It can be proven that $(\mathcal{R}, +, *)$ is a quasifield which is not associative and not distributive in general. The mapping Φ can be chosen in n^{q-2} different ways since the only condition is that $\Phi(1)$ be the identity, the remaining $q - 2$ elements of GF(q)* can be mapped to an arbitrary element of Γ. One can also show that $*$ is associative if and only if Φ is a homomorphism from GF(q)* to Γ.

Let us recall the notions of division algebra and field. According to Definition 3.11, a *division algebra* (also called as a *skewfield*) is an associative and distributive quasifield, and a field is a division algebra with commutative multiplication, that is an associative, commutative and distributive quasifield. The next theorem describes finite projective planes coordinatized by a field in terms of (C, t)-transitivities.

Theorem 3.31. *If Π is a finite projective plane, then the next five properties are equivalent:*

(1) Π *is* (Y, XY)*-transitive,* (X, XY)*-transitive,* (Y, OY)*-transitive and* (X, OY)*-transitive.*

(2) *The ternary ring* $(\mathcal{R}, F)$ *is a field.*

(3) *Desargues' theorem is valid in* Π.

(4) Π *is* (P, l)*-transitive for every point-line pair.*

(5) *Every ternary ring coordinatizing* Π *is a field.*

Proof. If Π satisfies the transitivity properties in (1), then, by Theorems 3.15, 3.17 and 3.27 the ternary ring $(\mathcal{R}, F)$ is an associative and distributive quasifield, namely a division algebra. According to a theorem of Wedderburn, the multiplication in any finite division algebra is commutative, so $(\mathcal{R}, F)$ is a finite field.

Property (2) implies (3), because Desargues' theorem can be proven by computing with coordinates and using the field axioms as we have seen in Chapter 1. Now, according to Baer's theorem, (4) is also true. This means that no matter how the base points X, Y, O, E of a coordinate system are chosen, (1) will always be satisfied. This proves (5). From Property (5), by using Theorems 3.15, 3.17 and 3.27 one gets (1). $\square$

In the finite case the commutativity of multiplication in a division algebra is a consequence of the remaining axioms. This is not true for infinite structures (we may just think of the quaternions). According to this, Desargues' theorem does not imply commutativity of multiplication, so there are infinite planes which are (P, l)-transitive for every point-line pair but their coordinate structure is not a field. Such an infinite plane can be constructed for example using quaternions. The commutativity of the multiplication is related to Pappus' theorem, which holds in $PG(2, \mathbb{K})$ for any filed $\mathbb{K}$.

Theorem 3.32 (Pappus). *Let ℓ and ℓ' be two intersecting lines in a projective plane $PG(2, \mathbb{K})$ where $\mathbb{K}$ is a commutative field. If F, G, H are distinct points on ℓ, and F', G', H' are distinct points on ℓ' such that all of them are different from the point $\ell \cap \ell'$, then the points*

$$FG' \cap F'G, \quad FH' \cap F'H \quad \text{and} \quad GH' \cap G'H$$

are collinear.

Theorem 3.33. *If the Theorem of Pappus holds in a projective plane Π, then the multiplication is commutative in any ternary ring coordinatizing Π.*

Proof. Let $(\mathcal{R}, F)$ be a ternary ring which coordinatizes Π. Let O, E, X and Y be the base points of $(\mathcal{R}, F)$. We show that $F(a, b, 0) = F(b, a, 0)$ holds for all $a, b \in \mathcal{R}$. If $a = 0, 1$ or $b = 0, 1$, then the statement follows from Part 1 of Lemma 3.2. Suppose that $0 \neq a \neq 1$, $0 \neq b \neq 1$ and let $A = (a, a)$, $B = (b, b)$, $K = EY \cap AX$ and $L = EY \cap BX$. Then O, A and B are three distinct points on the line OE, and Y, K and L are three distinct points on the line EY, and each of these six points is different from $E = OE \cap EY$. Hence we can apply Pappus' theorem: the points $OK \cap BY$, $AK \cap LB$ and $AY \cap OL$ are collinear. But by definition $AY \cap OL = (a, ab)$, $OK \cap BY = (b, ba)$ and $AK \cap LB = X$, therefore $ab = ba$ holds. $\square$

The following statement shows that the Theorem of Pappus gives a stronger incidence condition than the Theorem of Desargues.

Theorem 3.34 (Hessenberg). *If the Theorem of Pappus holds in a projective plane Π, then the Theorem of Desargues also holds in Π.*

Proof. For $i = 1, 2$ let $A_i B_i C_i$ be two triangles which are in perspective from the point O and let t be the line joining the points $D = A_1 B_1 \cap A_2 B_2$ and $E = A_1 C_1 \cap A_2 C_2$. We show that the point $F = B_1 C_1 \cap B_2 C_2$ is also on t.

Let $K = A_1 C_1 \cap B_2 C_2$, $L = A_1 B_2 \cap OC_1$, $M = A_1 B_1 \cap OK$ and, finally, $N = A_2 B_2 \cap OK$ (see Figure 3.5). Applying Pappus' theorem for the collinear triples (O, B_1, B_2) and (A_1, K, C_1) we get that the points

$$OK \cap B_1 A_1 = M, \quad OC_1 \cap B_2 A_1 = L \quad \text{and} \quad B_1 C_1 \cap B_2 K = F$$

are also collinear. For the collinear triples (O, A_1, A_2) and (B_2, C_2, K) Pappus' theorem gives that the points

$$OC_2 \cap A_1 B_2 = L, \quad OK \cap A_2 B_2 = N \quad \text{and} \quad A_1 K \cap A_2 C_2 = E$$

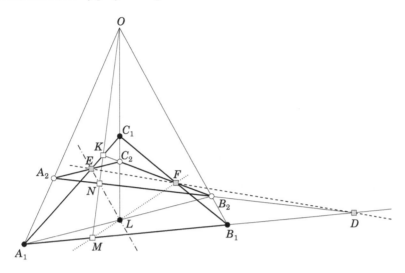

FIGURE 3.5
Hessenberg's theorem

are collinear. Finally, since (A_1, B_2, L) and (N, M, K) are collinear triples, we get that the points

$$A_1 M \cap B_2 N = D = A_1 B_1 \cap A_2 B_2,$$
$$A_1 K \cap LN = E = A_1 C_1 \cap A_2 C_2,$$
$$B_2 K \cap LM = F = B_1 C_1 \cap B_2 C_2$$

are collinear, which proves Desargues' theorem. □

Corollary 3.35. *If the Theorem of Pappus holds in a projective plane* Π, *then every ternary ring coordinatizing* Π *is a field.*

In the theory of abstract configurations a point-line incidence geometry $(\mathcal{P}, \mathcal{L}, I)$ is called a *symmetric n_k configuration* if $|\mathcal{P}| = |\mathcal{L}| = n$, there are k points on each line and there are k lines through each point. Thus the points and lines appearing in the Theorems of Pappus or Desargues form a symmetric 9_3 or 10_3 configuration, respectively. The Fano plane can be considered as a symmetric 7_3 configuration. So the following question is naturally raised: does there exist any characterization theorem for projective planes containing many Fano planes? Our goal is to answer this question. Let us start with a property of $PG(2, 2^r)$.

Let $\mathcal{S} = \{A, B, C, D\}$ be a set of four points in general position in the plane $PG(2, q)$. Then we can choose their representing vectors so that $\mathbf{a} + \mathbf{b} + \mathbf{c} = \mathbf{d}$. Suppose that $q = 2^r$. Then the characteristic of $GF(q)$ is 2, thus the previous equality implies $\mathbf{a}+\mathbf{b} = \mathbf{c}+\mathbf{d}$. The point with representing vector $\mathbf{a}+\mathbf{b}$ is on the line AB, while the point with representing vector $\mathbf{c}+\mathbf{d}$ is on the line CD, so $\mathbf{a}+\mathbf{b}$ is a representing vector of the point $AB \cap CD$. In the same way, we get that $\mathbf{a}+\mathbf{c}$ is a representing vector of the point $AC \cap BD$ and $\mathbf{b}+\mathbf{c}$ is a representing vector of the point $AD \cap BC$. Using $\mathrm{char}(\mathrm{GF}(q)) = 2$ again, we get

$$(\mathbf{a}+\mathbf{b}) + (\mathbf{a}+\mathbf{c}) + (\mathbf{b}+\mathbf{c}) = 2(\mathbf{a}+\mathbf{b}+\mathbf{c}) = \mathbf{0}.$$

So the three points, $AB \cap CD$, $AD \cap BD$ and $AD \cap BC$ are collinear. This means that $\mathcal{S}$ is contained in a subplane which is isomorphic to the Fano plane.

In the last part of this chapter projective planes having this property are characterized. The main result is the Theorm of Gleason (Corollary 3.45). Our approach is a mixture of the methods given in the original paper of *Gleason* [72] and the book of *Kárteszi* [105, Section 6.1].

Definition 3.36. *Let* Π *be a projective plane and* $\mathcal{S} = \{A, B, C, D\}$ *be a set of four points of* Π *in general position. We say that* $\mathcal{S}$ *is a* quadrangle *and the points* $AB \cap CD$, $AC \cap BD$ *and* $AD \cap BC$ *are called the* diagonal points *of* $\mathcal{S}$. *The quadrangle* $\mathcal{S}$ *has* Fano property *if its three diagonal points are collinear, and in this case the line containing the diagonal points is called the* diagonal line *of* $\mathcal{S}$. *The plane* Π *is called a* plane with Fano property *if all quadrangles of* Π *have Fano property.*

Theorem 3.37. *Let* Π *be a finite projective plane with Fano property and* $(\mathcal{R}, F)$ *be a ternary ring coordinatizing* Π. *Then* $(\mathcal{R}, +)$ *is an elementary abelian 2-group of order* 2^r.

Proof. Let X, Y, O and E be the base points of $(\mathcal{R}, F)$ and U denote the point $XY \cap OE$. By Lemma 3.10, $(\mathcal{R}, +)$ is a loop with unit element 0.

First, we show that $a + a = 0$ for all $0 \neq a \in \mathcal{R}$. Consider the points $A = (a, a)$, $A' = (0, a)$ and U. Then the point $P = AY \cap A'U$ has coordinates $(a, a + a)$. The diagonal points of the quadrangle $\{A, A', Y, U\}$ are X, P and O. As Π is a plane with Fano property, these three points are collinear, thus P is on the line OX, hence its second coordinate is 0 (see Figure 3.6).

Secondly, we prove that the addition is commutative. If at least one of a and b equals to 0, then $a + b = b + a$ obviously holds. If none of them is 0, then consider the three quadrangles and their diagonal points:

$$
\begin{array}{ll}
\{(a, a), (b, b), X, Y\} & \Rightarrow \{U, (b, a), (a, b)\} \\
\{(a, a+b), (0, b), X, Y\} & \Rightarrow \{U, (0, a+b), (a, b)\} \\
\{(b, b+a), (0, a), X, Y\} & \Rightarrow \{U, (0, b+a), (b, a)\}.
\end{array}
$$

The first diagonal line intersects both the second and the third in two points, hence the three diagonal lines must coincide. This means that both of the points $(0, a+b)$ and $(0, b+a)$ are the point of intersection of the diagonal line and the line OY, which implies the commutativity of the addition (see Figure 3.7).

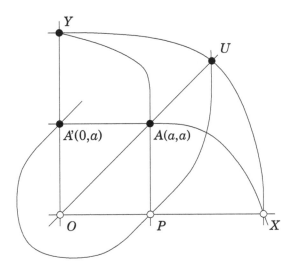

FIGURE 3.6
a+a=0

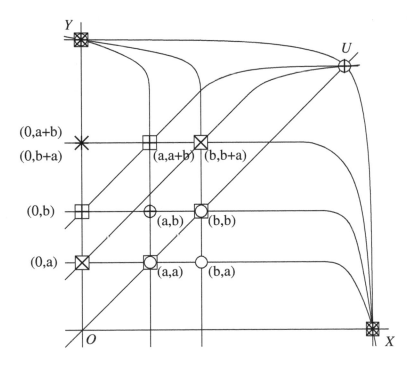

FIGURE 3.7
Addition is commutative

Now, we prove that the addition is associative. If at least one of a, b and c equals to 0, then $(a + b) + c = a + (b + c)$ obviously holds. If none of them is 0, then consider the three quadrangles and their diagonal points (when calculating the coordinates of the diagonal points, the commutativity of the addition is applied):

$$\{(a, a + b), (c, b + c), X, Y\} \quad \Rightarrow \{U, (c, a + b), (a, b + c)\}$$
$$\{(a, a + (b + c)), (0, b + c), X, Y\} \quad \Rightarrow \{U, (0, a + (b + c)), (a, b + c)\}$$
$$\{(c, c + (a + b)), (0, a + b), X, Y\} \quad \Rightarrow \{U, (0, c + (a + b)), (c, a + b)\}.$$

The first diagonal line intersects both the second and the third in two points, hence the three diagonal lines must coincide again. This means that both of the points $(0, a + (b + c))$ and $(0, (a + b) + c)$ are the point of intersection of the diagonal line and the line OY, which implies the associativity of the addition, so $(\mathcal{R}, +)$ is an abelian group.

Finally, as each element of the group has order 2, by the Theorem of Cauchy, the order of the group must be a power of 2.   □

As the cardinality of $\mathcal{R}$ equals to the order of Π, we immediately get the following.

Corollary 3.38. *The order of any finite projective plane with Fano property is 2^r with a suitable positive integer r.*

Let Π_q be a finite projective plane of order $q = 2^r$ with Fano property and $P I \ell$ be an incident point-line pair in Π_q. Our goal is to show that the group of elations $\Gamma_{P,\ell}$ is non-trivial. Our proof is based on the paper of *Gleason* [72].

Let $\{e_1, e_2, \ldots, e_q\}$ be the lines through P distinct from ℓ. For any point $A \neq P$ on ℓ, let $\pi_{i,j}^A$ denote the central projection of e_i to e_j from A. The following properties of these projections easily follow from Theorem 3.37.

Proposition 3.39. *For all $i, j, k \in \{1, 2, \ldots, q\}$*

- $(\pi_{i,j}^A)^{-1} = \pi_{j,i}^A$;

- $\pi_{i,j}^A \pi_{j,k}^A = \pi_{i,k}^A$;

- *if $B \neq P$ is also a point on ℓ, then $\pi_{i,j}^A \pi_{j,i}^B$ is a permutation of the points on e_i;*

- *the set*

$$J_i^{AB} = \{\pi_{i,j}^A \pi_{j,i}^B \colon j = 1, 2, \ldots, q\},$$

  with the composition as operation, is a sharply transitive permutation group on the set of points of $e_i \setminus \{P\}$.

Proof. The first three properties are obvious. For proving the fourth, choose the base points of a ternary ring $(\mathcal{R}, F)$ such that $Y = A$, $X = B$, $U = P$ and O is on the line e_i. Let the elements of $\mathcal{R}$ be $(\{1, 2, \ldots, q, \} \setminus \{i\}) \cup \{0\}$

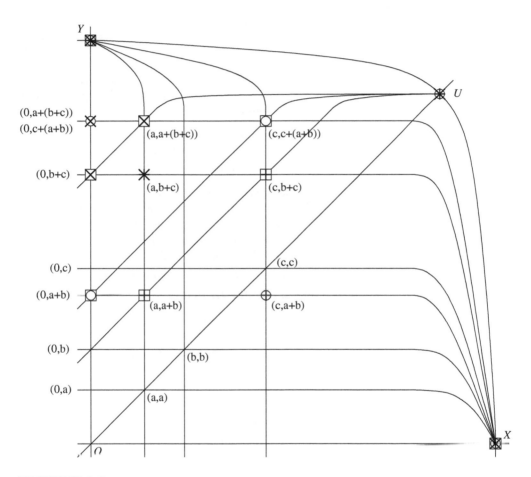

FIGURE 3.8
Addition is associative

and assign the coordinates such that the point $e_j \cap OY$ is $(0, j)$ if $j \neq i$, and $e_i \cap OY$ is $(0, 0)$. Let (k, k) be any point on the line e_i. Then, by the definition of addition in $(\mathcal{R}, F)$, we have

$$(k, k)^{\pi_{i,j}^A \pi_{j,i}^B} = (k, k + j)^{\pi_{j,i}^B} = (k + j, k + j).$$

Thus the elements of J^{AB} can be identified with the elements of $\mathcal{R}$ so that the operation corresponds to the addition of the ternary ring. By Theorem 3.37, $(\mathcal{R}, +)$ is an Abelian group on q elements which proves the last statement of the proposition. $\square$

Let $\overline{\varphi}$ be any element of $\Gamma_{P,\ell}$. As P is the center of $\overline{\varphi}$, the line e_i is fixed by $\overline{\varphi}$. Thus the restriction of $\overline{\varphi}$ on the line e_i, denoted by φ, is a permutation of the points of e_i. The next theorem characterizes these permutations.

Theorem 3.40. *There is a one-to-one correspondence between elements of $\Gamma_{P,\ell}$ and permutations of e_i which commute with every permutation of the form $\pi_{i,j}^C \pi_{j,i}^D$, where C and D are points on ℓ distinct from P.*

Proof. Let $\overline{\varphi} \in \Gamma_{P,\ell}$ and φ denote its restriction on the line e_i. We have already seen (see Figure 2.1) that the fixed points of any non-identity element of $\Gamma_{P,\ell}$ are exactly the points on the line ℓ. Hence any permutation on the points of the line e_i has at most one extension to an element of $\Gamma_{P,\ell}$.

First, we show that φ and $\pi_{i,j}^C \pi_{j,i}^D$ commute for all j, C and D. Let $R \neq P$ be any point on e_i. Then the points Y, R and $R^{\pi_{i,j}^C}$ are collinear, so the points $Y^{\overline{\varphi}} = Y$, $R^{\overline{\varphi}}$ and $(R^{\pi_{i,j}^C})^{\overline{\varphi}}$ are also collinear. Hence

$$(R^{\pi_{i,j}^C \overline{\varphi}})^{\pi_{j,i}^C} = R^{\overline{\varphi}}, \quad \text{so} \quad \pi_{i,j}^C \overline{\varphi} = \overline{\varphi} \pi_{i,j}^C.$$

We get in the same way that $\pi_{j,i}^D \overline{\varphi} = \overline{\varphi} \pi_{j,i}^D$, hence $\overline{\varphi}$ commutes with $\pi_{i,j}^C \pi_{j,i}^D$. It implies that φ and $\pi_{i,j}^C \pi_{j,i}^D$ commute on the line e_i.

Now, suppose that ψ is a permutation of the points of $e_i \setminus \{P\}$ which commutes with all permutations of the form $\pi_{i,j}^C \pi_{j,i}^D$. We extend ψ to a permutation $\overline{\psi}$ of the points of the plane and prove that $\overline{\psi} \in \Gamma_{P,\ell}$. Let us choose a point $C \neq P$ on the line ℓ. For any point R on the line ℓ let $R^{\overline{\psi}} = R$. If S is any point not on ℓ and $SP = e_j$, then let $S^{\overline{\psi}} = S^{\pi_{j,i}^C \psi \pi_{i,j}^C}$. As any line of the plane intersects ℓ and the points of ℓ are fixed, it is sufficient to prove that the points S, T and $ST \cap \ell = D$ are collinear if and only if the points $S^{\overline{\psi}}$, $T^{\overline{\psi}}$ and D are collinear. It obviously holds if $D = P$. If $D \neq P$, $SP = e_j$ and $ST = e_k$, then applying Proposition 3.39 and the commuting property of ψ we get

$$
\begin{aligned}
(S^{\overline{\psi}})^{\pi_{j,k}^D} &= S^{(\pi_{j,i}^C \psi \pi_{i,j}^C) \pi_{j,k}^D} = S^{\pi_{j,i}^C \psi \pi_{i,j}^C (\pi_{j,i}^D \pi_{i,k}^D)(\pi_{k,i}^C \pi_{i,k}^C)} \\
&= S^{\pi_{j,i}^C \psi (\pi_{i,j}^C \pi_{j,i}^D)(\pi_{i,k}^D \pi_{k,i}^C) \pi_{i,k}^C} = S^{\pi_{j,i}^C (\pi_{i,j}^C \pi_{j,i}^D)(\pi_{i,k}^D \pi_{k,i}^C) \psi \pi_{i,k}^C} \\
&= S^{(\pi_{j,i}^C \pi_{i,j}^C)(\pi_{j,i}^D \pi_{i,k}^D) \pi_{k,i}^C \psi \pi_{i,k}^C} = S^{\pi_{j,k}^D (\pi_{k,i}^C \psi \pi_{i,k}^C)} \\
&= T^{(\pi_{k,i}^C \psi \pi_{i,k}^C)} = T^{\overline{\psi}}.
\end{aligned}
$$

Hence the line joining D and $S^{\overline{\psi}}$ contains the point $T^{\overline{\psi}}$. This completes the proof. □

In order to move on, we need three lemmas from group theory. Their proofs are purely algebraic and elementary. We leave them as an exercise; the reader can find them in the paper by *Gleason* [72].

Lemma 3.41. *Let A, B and C be subgroups of a group and suppose that the elements of A and B can be so indexed that for each i $a_i b_i \in C$, and every element of C can be written in this form. Then A and B commute as complexes, so $AB = BA$.*

Lemma 3.42. *Let $A_1, A_2, \ldots, A_k$ be subgroups of a finite group and suppose that they mutually commute as complexes. Then $B = A_1 A_2 \ldots A_k$ is a group and every prime factor of the cardinality of B is a factor of the cardinality of A_i for some i.*

Lemma 3.43. *Let p be a prime and G be a group of permutations of a finite set $\mathcal{H}$. Suppose that for each element $s \in \mathcal{H}$ there exists an element $\alpha_s \in G$ of order p such that α_s fixes s but it has no other fixed point. Then G is transitive on $\mathcal{H}$.*

Theorem 3.44 (Gleason). *Let Π be a finite projective plane with Fano property. Then Π is a translation plane with respect to all of its lines.*

Proof. By Theorem 3.38, the order of Π is $q = 2^r$. We claim that the order of $\Gamma_{P,\ell}$ is q for all incident point-line pairs (P, ℓ). This implies that Π is (P, ℓ)-desarguesian for all incident point-line pairs (P, ℓ).

Let (P, ℓ) be an arbitrary incident point-line pair. Let $e_1, e_2, \ldots, e_q$ be the lines through P other than ℓ and $A, B_1, B_2, \ldots, B_{q-1}$ be the points on ℓ other than P. Pick a line e_i and for $j = 1, 2, \ldots, q-1$, consider the groups $J_i^{AB_j}$ defined in Proposition 3.39. We show that $J_i^{AB_j}$ and $J_i^{AB_k}$ commute as complexes for all j and k. As $J_i^{AB_j}$ is a group, its elements are the same as the elements of the group $J_i^{B_j A}$. Enumerate the elements of $J_i^{B_j A}$ in the order $\pi_{i,m}^{B_j} \pi_{m,i}^A$ and the elements of $J_i^{AB_k}$ in the order $\pi_{i,m}^A \pi_{m,i}^{B_k}$ for $m = 1, 2, \ldots, q$. Then

$$(\pi_{i,m}^{B_j} \pi_{m,i}^A)(\pi_{i,m}^A \pi_{m,i}^{B_k}) = \pi_{i,m}^{B_j} \pi_{m,i}^{B_k},$$

so the product is an element of $J_i^{B_j B_k}$. The cardinality of $J_i^{B_j B_k}$ is q, thus we can apply Lemma 3.41 and we get that $J_i^{AB_j} J_i^{AB_k} = J_i^{AB_k} J_i^{AB_j}$ for all j and k. Hence, by Lemma 3.42, $G = J_i^{AB_1} J_i^{AB_2} \ldots J_i^{AB_{q-1}}$ is a group. Each $J_i^{AB_j}$ has order q, so G is a 2-group, hence its center is non-trivial.

Let ψ be a non-identity element in the center of G. Then ψ commutes with all elements of type $\pi_{i,j}^C \pi_{j,i}^D$, because $J_i^{CD} \subset J_i^{AC} J_i^{AD} \subset G$. Hence, by Theorem 3.40, it can be extended to a non-identity (P, ℓ)-perspectivity, $\overline{\psi}$. This implies that $|\Gamma_{P,\ell}| > 1$.

We claim that the cardinality of $\Gamma_{P,\ell}$ is a constant $h > 1$. Any two points are collinear, thus it is enough to show that $|\Gamma_{P,\ell}| = |\Gamma_{R,\ell}|$ if P and R are distinct points on a line ℓ. Let $\overline{K}$ be the group of all collineations of Π which leave ℓ fixed. Then $\overline{K}$ induces a group K of permutations of the set of points on ℓ. For each point $T I \ell$, any non-identity element of $\Gamma_{T,\ell}$ corresponds to an element of K with the property that it has exactly one fixed point, T, and its order is 2. So we can apply Lemma 3.43 which gives that K is transitive on the points of ℓ. This means that $\overline{K}$ is also transitive on the points of ℓ. Let $\beta \in \overline{K}$ be a collineation mapping P to R. Then Lemma 2.24 implies that in the full collineation group of the plane $\beta^{-1}\Gamma_{P,\ell}\beta = \Gamma_{R,\ell}$.

Hence we can apply Lemma 2.35 with $h = 2$. This gives that Π is a translation plane with respect to ℓ. As ℓ was an arbitrary line, the theorem is proved. $\qquad\square$

By Theorem 2.40, the following corollary, called Gleason's theorem, is straightforward.

Corollary 3.45 (Gleason). *Let Π be a finite projective plane with Fano property. Then Π is isomorphic to $\mathrm{PG}(2, 2^r)$ with a suitable r.*

There have been many attempts to extend Gleason's theorem to odd primes and to prove the following, longstanding conjecture.

Conjecture 3.46. *Let Π be a finite projective plane and p be a prime. Suppose that any set of four points in general position in Π is contained in a subplane of order p. Then Π is isomorphic to $\mathrm{PG}(2, p^r)$ with a suitable r.*

The best known result is due to *Blokhuis* and *Sziklai* [28].

Theorem 3.47 (Blokhuis, Sziklai). *Let Π be a finite projective plane of order p^2, p prime. Suppose that any set of four points in general position in Π is contained in a Baer subplane of order p. Then Π is isomorphic to $\mathrm{PG}(2, p^2)$.*

Exercises

3.1. Determine the coordinate structure of the plane obtained by Hall's free extension process.

3.2. Determine the coordinate structure of the parabola plane.

3.3. Replace the parabola by the graph of $x \mapsto x^4$ over the reals. Determine the coordinate structure in this particular case. Show that the plane is not Desarguesian.

3.4. Generalize the result of the previous exercise for the graph of $x \mapsto x^{2k}$.

3.5. Classify those convex functions for which the translates of the graph together with the vertical lines form a Desarguesian affine plane (over the reals).

3.6. Try to find planar functions over finite fields.

3.7. Show that $x \mapsto x^{p+1}$ is a planar function over $\mathrm{GF}(p^m)$. Determine the coordinate structure of this plane when $m > 1$ is odd.

3.8. Complete the proof of Theorem 3.22 by verifying property (ii) in the definition of the kernel for k' (where $kk' = 1$).

3.9. Describe all quasifields of order 9. Deduce that there is a unique translation plane of order 9.

3.10. Complete the missing details about Hall quasifields.

3.11. Describe the equations of lines in case of translation planes constructed by Examples 3.29 and 3.30. What has to be changed if we consider their dual planes?

3.12. Prove that Π is (Y, OX)-transitive if and only if $(\mathcal{R}^*, \cdot)$ is a group (so multiplication is associative) and $F(a, b, cb) = (a + c)b$ for every $a, b, c \in \mathcal{R}$.

3.13. Check the algebraic properties of the quasifield defined in Example 3.29.

3.14. Determine for $r = 2$ when an André quasifield is associative (or distributive). What can be said in the general case?

3.15. Verify the properties mentioned right after Example 3.30 of the general André construction for a quasifield.

3.16. Show that on the plane coordinatized by quaternions, Pappus' theorem is not valid in general.

3.17. In our context Hessenberg's theorem means that various properties of the coordinate structure can also be deduced from Pappus' theorem. Try to do this for some properties (e.g., associativity of multiplication, distributivity).

4

Projective spaces of higher dimensions

This chapter is about combinatorial properties of projective spaces of higher dimensions, basic properties of collineations, polarities, quadrics and Hermitian varieties.

Let us first define the abstract notion of a (finite) projective space.

Definition 4.1. *Let S be a finite set together with some of its distinguished subset. To each of these distinguished subsets an integer number $-1 \leq d \leq n$ is associated. S is called a finite projective space of dimension n, the distinguished subsets are called subspaces of dimension d (where d is the number associated with the distinguished subset) if the following axioms are satisfied.*

1. *For every $-1 \leq d \leq n$ there is at least one subspace of dimension d, moreover*

    - *there is a unique subspace of dimension (-1), namely $\emptyset$;*

    - *there is a unique subspace of dimension n, namely S;*

    - *subspaces of dimension 0 are the one-element subsets of S.*

2. *If a subspace of dimension r is contained in a subspace of dimension s, then $r \leq s$ and, in case of $r = s$, the two subspaces coincide.*

3. *The intersection of subspaces is a subspace.*

4. *If the intersection of a subspace of dimension r and a subspace of dimension s is a subspace of dimension t, and the intersection of all subspaces containing both of the subspaces is a subspace of dimension u, then*

$$r + s = t + u.$$

5. *Each subspace of dimension 1 contains $q + 1 \geq 3$ elements.*

Subspaces of dimension 0, 1, 2 and $(n-1)$ are called points, lines, planes *and* hyperplanes, *respectively.*

Definition 4.2. *For any family of subspaces of a projective space,* $\mathcal{F} = \{\Sigma_1, \Sigma_2, \ldots, \Sigma_m\}$, *the intersection of all subspaces containing each element of* $\mathcal{F}$ *is called the* subspace generated by $\mathcal{F}$ *and it is denoted by* $\langle \Sigma_1, \Sigma_2, \ldots, \Sigma_m \rangle$.

In case of abstract projective space defined above we use geometric terminology, too. For example, when a subspace of dimension 0 is contained in a subspace of dimension 1, then we say that the point lies on the line.

In Chapter 1 we have already defined projective spaces of dimension 2. Definition 1.3 is somewhat different from the above definition. It is easy to see that replacing the elements e of $\mathcal{E}$ by the set of points $P \in \mathcal{P}$ that are incident with e, then we get the definition of a projective space of dimension 2.

Projective spaces can also be defined, similarly to Definition 1.3, by various sets of points and various incidence and containment relations between them. However, as we shall see, projective spaces of dimensions at least 3 are unique up to isomorphisms, so it does not really matter which of the different "good" definitions is used.

Linear algebra tells us that the following example gives a projective space.

Example 4.3. Let V_{n+1} be the vector space of dimension $(n + 1)$ over the finite field GF(q). Let $\mathcal{S}$ be the set of one-dimensional subspaces of V_{n+1} and the distinguished subsets be the subspaces of V_{n+1} and $\emptyset$. The dimension of a distinguished subset corresponding to a $(k + 1)$-dimensional subspace of V_{n+1} is defined as k, and that of $\emptyset$ as -1.

This projective space is called the Galois space of dimension n and it is denoted by PG(n, q).

The intersection and incidence properties, known from classical projective geometry, can easily be deduced from the axioms of higher dimensional projective spaces. As an example, we prove a property that will be used in the proof of Desargues' theorem (Theorem 4.5). To find and check other basic properties is left to the reader (see Exercises 4.1 and 4.2).

Lemma 4.4. *If the lines e and f of a projective space $\mathcal{S}$ intersect each other in a point M, $E_1 \neq M \neq E_2$ are different points of e, $F_1 \neq M \neq F_2$ are different points of f, then the lines E_1F_1 and E_2F_2 intersect each other.*

Proof. Lines are 1-dimensional, points are 0-dimensional subspaces. Therefore, axiom 4 of Definition 4.1 implies that there is a $1+1-0$, that is a 2-dimensional subspace that contains e and f. This subspace also contains the lines E_1F_1 and E_2F_2. Again, axiom 4 of Definition 4.1 gives that the dimension of their intersection is $2 - (1 + 1) = 0$; in other words they intersect in a point.   □

It is not difficult to prove the planar Desargues' theorem from the axioms of higher dimensional spaces.

Theorem 4.5. *If a projective plane* Π *can be embedded in a projective space* S *(of dimension three) and two triangles of* Π *are in perspective from a point, then they are also in perspective from a line.*

Proof. First we show that Desargues' theorem is true when the two triangles in S are not in a plane. So assume that the triangles ABC and $A'B'C'$ are in different planes and they are in perspective from the point O. We are going to show that they are in perspective from a line. Let Π and Π' be the two planes determined by the two triangles, respectively. As these planes are different, they meet in a unique line e. The points $AB \cap A'B', BC \cap B'C'$ and $CA \cap C'A'$ exist by Lemma 4.4, since any two of the lines OAA', OBB' and OCC' intersect each other in O. These points are contained in both planes Π and Π', hence they are on $e = \Pi \cap \Pi'$. This proves that the two triangles are in perspective from the line e.

Assume now that the projective plane Π can be embedded in a three-dimensional projective space S as a plane. Moreover, assume that the two triangles $A_1A_2A_3$ and $B_1B_2B_3$ of Π are in perspective from a point O. Let f be a line through O which is not contained in Π, let A and B be two distinct points on f different from O. The existence of these points follows from axiom 5 of Definition 4.1. Since the lines AA_i and BB_i $(i = 1, 2, 3)$ are contained in the plane generated by f and OA_iB_i, they intersect each other. Let $C_i = AA_i \cap BB_i$. Then the triangles $A_1A_2A_3$ and $C_1C_2C_3$ are in perspective from A, the triangles $B_1B_2B_3$ and $C_1C_2C_3$ are in perspective from B. The triangle $C_1C_2C_3$ is not contained in the plane Π. If it was, then A and B and hence the line f joining them would be in Π. By the first part of the proof these triangles are in perspective from a line. Therefore the triples of points $A_1A_2 \cap C_1C_2, A_2A_3 \cap C_2C_3, A_3A_1 \cap C_3C_1$ and $B_1B_2 \cap C_1C_2, B_2B_3 \cap C_2C_3, B_3B_1 \cap C_3C_1$ are collinear. Each side of the triangle $C_1C_2C_3$ intersects Π in exactly one point, hence $A_1A_2 \cap C_1C_2 = B_1B_2 \cap C_1C_2 = A_1A_2 \cap B_1B_2$. The analogous statement is also true for the other indices. The points $A_1A_2 \cap B_1B_2, A_2A_3 \cap B_2B_3$ and $A_3A_1 \cap B_3B_1$ are just the intersections of the corresponding sides of triangles $A_1A_2A_3$ and $B_1B_2B_3$. This implies that the two triangles are in perspective from a line. $\square$

In the previous chapter it was shown that Desargues' theorem implies that the plane can be coordinatized by a field. The coordinatization procedure can also be extended to a projective space. Using Theorems 3.31 and 4.5 a lengthy but not difficult computation shows the following theorem (the proof can be found in Chapter 5 of the book [174] by *Tsuzuku*).

Theorem 4.6. *Every finite projective space S of dimension $n \geq 3$ is isomorphic to the space* $\mathrm{PG}(n, q)$.

From now on we focus on the spaces $\mathrm{PG}(n, q)$, where $n \geq 1$. Some of the next propositions were proven in Chapter 1 for the particular case $n = 2$.

Let V_{n+1} be a vector space of dimension $(n+1)$ over the finite field $\mathrm{GF}(q)$, and let us denote the zero vector (origin) by $\mathbf{0}$. Define a relation on $V_{n+1} \setminus \{\mathbf{0}\}$ in the following way: $\mathbf{x} = (x_0, x_1, \ldots, x_n)$ and $\mathbf{y} = (y_0, y_1, \ldots, y_n)$ are in relation if and only if there is a $0 \neq \lambda \in \mathrm{GF}(q)$, for which $x_i = \lambda y_i$ for every $i = 0, 1, \ldots, n$. This is clearly an equivalence relation and the equivalence classes correspond to one-dimensional subspaces of V_{n+1}, in other words, to points of $\mathrm{PG}(n, q)$. The equivalence class of a vector $\mathbf{x}$ is denoted by $[\mathbf{x}]$. Any vector of an equivalence class is called a *representative vector* of the corresponding point. If $\mathbf{x} = (x_0, x_1, \ldots, x_n)$ is a representative vector of a point P, then we say that P has *homogeneous coordinates* $(x_0 : x_1 : \ldots : x_n)$. A set of points is called linearly dependent, if their representative vectors are dependent. A k-dimensional subspace of $\mathrm{PG}(n, q)$ consists of the points whose representative vectors satisfy $\mathbf{x}A = \mathbf{0}$, where A is a matrix of size $(n+1) \times (n-k)$ and rank $(n - k)$ that describes the subspace. In particular, hyperplanes can be described by vectors $\mathbf{u}^{\mathrm{T}} = (u_0, u_1, \ldots, u_n)$, where $(u_0, u_1, \ldots, u_n) \neq \mathbf{0}$. The vectors $(u_0, u_1, \ldots, u_n)$ and $(\lambda u_0, \lambda u_1, \ldots, \lambda u_n)$ define the same hyperplane if $0 \neq \lambda \in \mathrm{GF}(q)$. If $\mathbf{u}^{\mathrm{T}} = (u_0, u_1, \ldots, u_n)$ defines a hyperplane Σ, then we say that Σ has *homogeneous coordinates* $[u_0 : u_1 : \ldots : u_n]$. The point with homogeneous coordinates $(x_0 : x_1 : \ldots : x_n)$ is incident with the hyperplane with homogeneous coordinates $[u_0 : u_1 : \ldots : u_n]$ if and only if $\mathbf{x}\mathbf{u}^{\mathrm{T}} = \sum_{i=0}^{n} x_i u_i = 0$. The line joining two points consists of points represented by linear combinations of the representative vectors of the two points. In general, the subspace generated by a set of points consists of the points represented by linear combinations of the representative vectors of the points in the set.

The same model can be formulated over an arbitrary field $\mathbb{K}$, and defines a projective space (for $\mathbb{K} = \mathbb{R}$ and $n = 3$ it is the classical projective space); this projective space is denoted by $\mathrm{PG}(n, \mathbb{K})$.

Similarly to the construction of the affine plane $\mathrm{AG}(2, \mathbb{K})$ from the projective plane $\mathrm{PG}(2, \mathbb{K})$ in Example 1.22, we can construct the n-dimensional affine space $\mathrm{AG}(n, \mathbb{K})$ from the projective space $\mathrm{PG}(n, \mathbb{K})$ by deleting all points of a hyperplane. One could define affine spaces axiomatically but as the projective closure of an affine space of dimension at least three is a projective space $\mathrm{PG}(n, \mathbb{K})$, affine spaces will not be defined axiomatically. Affine spaces will be represented either as a projective space minus a hyperplane or directly by a vector space of dimension n. In the latter case a k-dimensional subspace of an affine space will be an additive coset (or translate) of a k-dimensional vector subspace of the vector space.

The algebraic description of projective spaces shows that the duality principle is valid for projective spaces. The *dual space* of an n-dimensional projective space $\mathcal{S}$ is the projective space $\mathcal{S}^*$ whose points are the hyperplanes of $\mathcal{S}$, lines are the $(n-2)$-dimensional subspaces of $\mathcal{S}$, and in general, k-dimensional subspaces of $\mathcal{S}^*$ are exactly the $(n - k - 1)$-dimensional subspaces of $\mathcal{S}$. The incidence relation is the reverse containment relation for subspaces of $\mathcal{S}$. If a theorem can be formulated by using the notion of subspaces, incidence and containment and it is true in a projective space; then it is also true in $\mathcal{S}^*$, so the dual of the theorem is also true in $\mathcal{S}$. Recall that the dual of a theorem is obtained by replacing subspaces by their duals, and incidence with containment.

If $\{P_0, P_1, \ldots, P_{n+1}\}$ is a set of $n + 2$ points in $\mathcal{S} = \mathrm{PG}(n, q)$ in general position (that is, no $n+1$ of them lie in a hyperplane), then we can associate a coordinate vector to any other point R of the space in the following way. The points P_i can be represented by $t_i \mathbf{x}_i$, where $0 \neq t_i \in \mathrm{GF}(q)$ and $\mathbf{x}_i \in V_{n+1}$. Since the vector space V_{n+1} has dimension $n + 1$, for every $t \in \mathrm{GF}(q)$ there exist $a_0, a_1, \ldots, a_n \in \mathrm{GF}(q)$, for which

$$t\mathbf{x}_{n+1} = \sum_{i=0}^{n} a_i \mathbf{x}_i.$$

If t is varied, then the a_i's change, however, the ratios a_i/a_j remain the same. Thus if $\mathbf{r}$ is a representative vector of the point R, then

$$\mathbf{r} = \sum_{i=0}^{n} t_i a_i \mathbf{x}_i,$$

where the elements $t_i \in \mathrm{GF}(q)$ are determined up to a constant factor. The $(n + 1)$-tuple $(t_0 : t_1 : \ldots : t_n)$ is called the *coordinate vector of R with respect to* $\{P_0, P_1, \ldots, P_n\}$. The points $P_0, P_1, \ldots, P_n$ are the *base simplex* of the coordinate system, P_{n+1} is the *unit point* of the coordinate system, finally the set of the above $n + 2$ points is called the *base points* of the coordinate system. Sometimes this set of points is also called the *frame of reference*.

For the study of combinatorial properties of the projective space $\mathrm{PG}(n, q)$ some notations are introduced. Let $\Theta(r) = (q^{r+1} - 1)/(q - 1)$ and for $r \leq s$ let $[r, s] = \prod_{i=r}^{s} (q^i - 1)$.

Theorem 4.7. *For subspaces of the projective space* $\mathrm{PG}(n, q)$ *the following properties hold:*

1. *the number of points is* $\Theta(n)$;

2. *for* $0 \leq k \leq n - 1$, *the number of k-dimensional subspaces is* $[n - k + 1, n + 1]/[1, k + 1]$;

3. *the number of m-dimensional subspaces containing a given subspace of dimension k is* $[m - k + 1, n - k]/[1, n - m]$, *if* $0 \leq k \leq m \leq n - 1$.

Proof. 1. The number of points of $\mathrm{PG}(n, q)$ is the number of 1-dimensional subspaces of an $(n + 1)$-dimensional vector space over $\mathrm{GF}(q)$. The vector space contains q^{n+1} elements, a 1-dimensional subspace contains q. Different 1-dimensional subspaces only meet in the origin; the vectors different from

0 are contained in a unique 1-dimensional subspace. Hence the number of 1-dimensional subspaces is

$$\frac{q^{n+1} - 1}{q - 1} = q^n + q^{n-1} + \ldots + q + 1 = \Theta(n).$$

2. A k-dimensional subspace is determined by $k + 1$ linearly independent points. Hence the number of different k-dimensional subspaces in an n-dimensional projective space is the total number of independent sets of points of size $k + 1$ divided by the number of independent sets of points of size $k + 1$ in a k-dimensional subspace. This gives

$$\frac{\Theta(n)(\Theta(n) - \Theta(0))(\Theta(n) - \Theta(1))\ldots(\Theta(n) - \Theta(k-1))}{\Theta(k)(\Theta(k) - \Theta(0))(\Theta(k) - \Theta(1))\ldots(\Theta(k) - \Theta(k-1))} =$$

$$[n - k + 1, n + 1]/[1, k + 1].$$

3. By the principle of duality, the number of m-dimensional subspaces of $\mathrm{PG}(n, q)$ containing a given k-dimensional subspace is the same as the number of $(n - m - 1)$-dimensional subspaces contained in an $(n - k - 1)$-dimensional subspace. This latter number is just the number of $(n - m - 1)$-dimensional subspaces in $\mathrm{PG}(n - k - 1, q)$ and, according to 2 above, it is

$$\frac{[(n - k - 1) - (n - m - 1) + 1, n - k]}{[1, n - m]} = \frac{[m - k + 1, n - k]}{[1, n - m]}.$$

$\square$

The number of d-dimensional subspaces of an n-dimensional vector space is often denoted as $\begin{bmatrix} n \\ d \end{bmatrix}_q$. These are somewhat similar to the ordinary binomial coefficients and called *Gaussian binomial* or *q-binomial coefficients*. With our notation they are

$$\begin{bmatrix} n \\ d \end{bmatrix}_q = \frac{[n - d + 1, n]}{[1, d]} = \frac{(q^n - 1)(q^{n-1} - 1)\ldots(q^{n-d+1} - 1)}{(q^d - 1)(q^{d-1} - 1)\ldots(q - 1)}.$$

Definition 4.8. *Let $\mathcal{S}$ and $\mathcal{S}'$ be two projective spaces. A collineation is a bijection $\phi : \mathcal{S} \to \mathcal{S}'$ which preserves the incidence of subspaces; that is when $\Pi_r \subset \Pi_s$ are two subspaces, then $\Pi_r^\phi \subset \Pi_s^\phi$.*
If $\mathcal{S} = \mathcal{S}'$, then we call ϕ a collineation of the projective space $\mathcal{S}$.

For $n = 2$ the definition above is seemingly different from Definition 2.1. However, it is not difficult to show that the two definitions are equivalent. If $\phi : \mathcal{S} \to \mathcal{S}$ is a bijection which preserves incidences of points and lines, then one can show that ϕ is a collineation. The proof is left to the reader. In Chapter 2 we have already seen two typical examples of collineations of $\mathrm{PG}(2, q)$. These examples can be extended to higher dimensional spaces over an arbitrary field $\mathbb{K}$ easily.

Example 4.9. Let A be an invertible $(n+1) \times (n+1)$ matrix with entries from $\mathbb{K}$. Then the mapping

$$\mathbf{x} \mapsto \mathbf{x}A$$

is a collineation of the projective space $\mathrm{PG}(n, \mathbb{K})$. These mappings are called *linear transformations*.

The mapping gives a permutation of the points, because of $\det A \neq 0$. It also preserves incidence since it preserves linear combinations.

If a linear transformation given by the matrix A transforms the points, then the action of this collineation on hyperplanes is described by the matrix $(A^{-1})^T$. Indeed, incidence of the point with representative vector $\mathbf{x}$ and a hyperplane defined by $\mathbf{u}$ is described by $\mathbf{x}\mathbf{u}^T = 0$, and if $\mathbf{x}$ is replaced by $\mathbf{x}A$ and $\mathbf{u}$ is replaced by $\mathbf{u}(A^{-1})^T$, then in the incidence equation we get $AA^{-1} = I$ in the middle.

As in classical projective geometry, also in $\mathrm{PG}(n, \mathbb{K})$, it can be seen that there is a unique linear transformation mapping an (ordered) set of $n+2$ points in general position to a given (ordered) set of $n+2$ points in general position. The detailed proof of this statement is given in Theorem 4.16; here we only sketch the main idea. We need to imagine the first $(n+1)$ points as a basis of the $(n+1)$ dimensional vector space, obtain the $(n+2)$-th point as a linear combination of the previous points and change the representative of the corresponding points in $\mathrm{PG}(n, \mathbb{K})$ in such a way that the first points are the base points and the last one is the unit point of the coordinate system. Do this for both the original set and the image set of points. Returning to the $(n+1)$-dimensional vector space there is a unique linear mapping that maps the representatives of the first $n+1$ points to the first $n+1$ representative vectors of the image set of points; the linear mapping will automatically map the last point to the last point of the image set.

This implies that when we have two coordinate systems in the projective space $\mathrm{PG}(n, \mathbb{K})$ and the base points are represented by the vectors $\mathbf{x}_i$ and $\mathbf{y}_i$ $(i = 0, \ldots, n)$, respectively, then there is a suitable linear transformation that can be used to change coordinates in the first system to the ones in the second system.

Example 4.10. Let σ be an automorphism of $\mathbb{K}$ and

$$\tilde{\sigma} \colon (x_0, x_1, \ldots, x_n) \mapsto (x_0^\sigma, x_1^\sigma, \ldots, x_n^\sigma).$$

If $\mathbf{x} = (x_0, x_1, \ldots, x_n)$, then we use the shorthand notation $\mathbf{x}^\sigma$ for $(x_0^\sigma, x_1^\sigma, \ldots, x_n^\sigma)$. Because of the identity

$$(\lambda \mathbf{x} + \mu \mathbf{y})^\sigma = \lambda^\sigma \mathbf{x}^\sigma + \mu^\sigma \mathbf{y}^\sigma$$

the mapping $\tilde{\sigma}$ takes any k-dimensional linear subspace of $\mathbb{K}^{n+1}$ into a k-dimensional linear subspace. Thus if we define $[\tilde{\sigma}]$ as

$$[\tilde{\sigma}][\mathbf{x}] = [\tilde{\sigma}(\mathbf{x})]$$

then we get a collineation of $PG(n, \mathbb{K})$. This is called the *automorphic collineation* induced from σ.

If σ_1 and σ_2 are automorphisms of $\mathbb{K}$, then $[\widetilde{\sigma_1 \circ \sigma_2}] = [\tilde{\sigma}_1] \circ [\tilde{\sigma}_2]$ and $[\tilde{\sigma}_1]^{-1} = [\widetilde{\sigma_1^{-1}}]$ obviously hold, hence the automorphic collineations form a subgroup of the full collineation group of $PG(n, \mathbb{K})$ and this subgroup is isomorphic to the group of automorphisms of the field $\mathbb{K}$.

We show that these are essentially the only types of collineations of $PG(n, q)$. This is called the *Fundamental Theorem of Projective Geometry*. Before the proof we recall some notions and theorems from linear algebra.

Let V_{n+1} be an $(n+1)$-dimensional vector space over the field $\mathbb{K}$. A mapping $\Phi: V_{n+1} \to V_{n+1}$ is called a *linear transformation*, if for all $\mathbf{v}_1, \mathbf{v}_2 \in V_{n+1}$ and for all $\alpha_1, \alpha_2 \in \mathbb{K}$

$$\Phi(\alpha_1 \mathbf{v}_1 + \alpha_2 \mathbf{v}_2) = \alpha_1 \Phi(\mathbf{v}_1) + \alpha_2 \Phi(\mathbf{v}_2).$$

The composition of linear transformations is also linear transformation and if Φ is bijective, then its inverse is a linear transformation, too. Hence the set of invertible linear transformations of a vector space with the composition as operation forms a group. This group is the *general linear group of V_{n+1}*, it is denoted by $GL(V_{n+1})$. If $V_{n+1} = \mathbb{K}^{n+1}$, then the group is denoted by $GL(n+1, \mathbb{K})$.

Let $\Phi \in GL(n+1, \mathbb{K})$ be an invertible linear transformation. Then Φ maps the linear subspaces of $\mathbb{K}^{n+1}$ into linear subspaces of the same dimension. In particular, Φ gives a bijection on the set of 1-dimensional subspaces, so Φ induces a mapping on the set of points of the projective space $PG(n, \mathbb{K})$. This mapping

$$[\Phi]: PG(n, \mathbb{K}) \to PG(n, \mathbb{K}), \qquad [\Phi]([\mathbf{x}]) = [\Phi(\mathbf{x})]$$

is a collineation, because it preserves linear combinations. If $\Phi, \Psi \in GL(n+1, \mathbb{K})$, then

$$[\Phi \circ \Psi] = [\Phi] \circ [\Psi] \text{ and } [\Phi^{-1}] = [\Phi]^{-1}$$

obviously hold. Hence the collineations obtained from invertible linear transformations form a subgroup of the whole collineation group of $PG(n, \mathbb{K})$. This subgroup is the *projective general linear group* and it is denoted by $PGL(n+1, \mathbb{K})$.

The mapping $\Phi \mapsto [\Phi]$ is a surjective group homomorphism. The kernel of this homomorphism contains the linear transformations $\Phi \in GL(n+1, \mathbb{K})$ for which $[\Phi(\mathbf{x})] = [\mathbf{x}]$ holds for all $\mathbf{0} \neq \mathbf{x} \in \mathbb{K}^{n+1}$. We show that these transformations are the multiplications with a scalar $\lambda \in \mathbb{K}$.

Definition 4.11. *Let V be a vector space over $\mathbb{K}$ and let $\Phi: V \to V$ be a linear transformation. A vector $\mathbf{x} \neq \mathbf{0}$ is an eigenvector of $\Phi: V \to V$ if there exists a scalar $\lambda \in \mathbb{K}$ such that $\Phi(\mathbf{x}) = \lambda \mathbf{x}$. In this case λ is the eigenvalue belonging to the eigenvector $\mathbf{x}$.*

Obviously, if $\mathbf{x}$ is an eigenvector of Φ with eigenvalue λ, then for all $0 \neq \mu \in \mathbb{K}$ the vector $\mu\mathbf{x}$ is also an eigenvector with eigenvalue λ.

The proof of the following proposition is straightforward.

Proposition 4.12. *A point $[\mathbf{x}] \in \mathrm{PG}(n, \mathbb{K})$ is a fixed point of the collineation $[\Phi] \in \mathrm{PGL}(n + 1, \mathbb{K})$ if and only if $\mathbf{x}$ is an eigenvector of Φ.*

Lemma 4.13. *Suppose that the vectors $\mathbf{v}_0, \mathbf{v}_1, \ldots, \mathbf{v}_n$ form a basis of $\mathbb{K}^{n+1}$. Let $\mathbf{v}_{n+1} = \sum_{i=0}^n \mathbf{v}_i$ and suppose that all of the vectors $\mathbf{v}_0, \mathbf{v}_1, \ldots, \mathbf{v}_{n+1}$ are eigenvectors of a linear transformation $\Phi \in \mathrm{GL}(n+1, \mathbb{K})$. Then the eigenvalues corresponding to these vectors are all equal to a non-zero element of $\mathbb{K}$.*

Proof. Let λ_i be the eigenvalue corresponding to $\mathbf{v}_i$. Then

$$\lambda_{n+1}\mathbf{v}_0 + \lambda_{n+1}\mathbf{v}_1 + \cdots + \lambda_{n+1}\mathbf{v}_n = \lambda_{n+1}\mathbf{v}_{n+1} = \Phi(\mathbf{v}_{n+1})$$
$$= \Phi(\mathbf{v}_0) + \cdots + \Phi(\mathbf{v}_n)$$
$$= \lambda_0\mathbf{v}_0 + \lambda_1\mathbf{v}_1 + \cdots + \lambda_n\mathbf{v}_n.$$

The decomposition of a vector as a linear combination of basis vectors is unique, so we get $\lambda_{n+1} = \lambda_0 = \lambda_1 = \cdots = \lambda_n = \lambda$. As $\Phi(\mathbf{v}_{n+1}) = \lambda\mathbf{v}_{n+1} \neq \mathbf{0}$, we get $\lambda \neq 0$. $\square$

Proposition 4.14. *Two linear transformations $\Phi_1, \Phi_2 \in \mathrm{GL}(n+1, \mathbb{K})$ induce the same collineation of $\mathrm{PG}(n, \mathbb{K})$ if and only if the linear transformation $\Phi_1 \circ \Phi_2^{-1}$ is a multiplication with a non-zero scalar of $\mathbb{K}$.*

Proof. By definition, $[\Phi_1] = [\Phi_2]$ if and only if $[\Phi_1 \circ \Phi_2^{-1}]$ fixes all points of $\mathrm{PG}(n, \mathbb{K})$. It happens if and only if all non-zero vectors are eigenvectors of the linear transformation $\Psi = \Phi_1 \circ \Phi_2^{-1}$.

If $\Phi_1 \circ \Phi_2^{-1}$ is the multiplication with a non-zero scalar, then all non-zero vectors are eigenvectors by definition. Now, suppose that all non-zero vectors are eigenvectors of Ψ. Let $\mathbf{v}_0, \mathbf{v}_1, \ldots, \mathbf{v}_n$ be a basis in $\mathbb{K}^{n+1}$. By Lemma 4.13, there exists an element λ in the multiplicative group $\mathbb{K}^*$ of $\mathbb{K}$ for which $\Psi(\mathbf{v}_i) = \lambda\mathbf{v}_i$ for all $i = 0, 1, \ldots, n$. Let $\mathbf{x}$ be an arbitrary vector and express it as a linear combination of the basis $\mathbf{v}_0, \mathbf{v}_1, \ldots, \mathbf{v}_n$. Applying Ψ we have

$$\Psi(\mathbf{x}) = \Psi(x_0\mathbf{v}_0 + x_1\mathbf{v}_1 + \cdots + x_n\mathbf{v}_n) = x_0\Psi(\mathbf{v}_0) + x_1\Psi(\mathbf{v}_1) + \cdots + x_n\Psi(\mathbf{v}_n)$$
$$= x_0\lambda\mathbf{v}_0 + x_1\lambda\mathbf{v}_1 + \cdots + x_n\lambda\mathbf{v}_n = \lambda x_0\mathbf{v}_0 + \lambda x_1\mathbf{v}_1 + \cdots + \lambda x_n\mathbf{v}_n = \lambda\mathbf{x},$$

so Ψ is the multiplication with λ. $\square$

Corollary 4.15. *The projective general group $\mathrm{PGL}(n + 1, \mathbb{K})$ is the factor group of $\mathrm{GL}(n + 1, \mathbb{K})$ with respect to the normal subgroup formed by the multiplications with non-zero scalars of $\mathbb{K}$. As the latter group is isomorphic to the multiplicative group $\mathbb{K}^*$ of $\mathbb{K}$, we get*

$$\mathrm{PGL}(n + 1, \mathbb{K}) \cong \mathrm{GL}(n + 1, \mathbb{K})/\mathbb{K}^*.$$

Theorem 4.16. *The group* $\mathrm{PGL}(n+1, \mathbb{K})$ *acts regularly on the set of ordered* $(n + 2)$*-tuples of points in general position.*

Proof. First we prove that the action is transitive. Let $A_0, \ldots, A_{n+1}$ and $B_0, \ldots, B_{n+1}$ be two ordered $(n + 2)$-tuples of points in general position. As any set of representative vectors for $n + 1$ points in general position forms a basis of $\mathbb{K}^{n+1}$, there exist representative vectors $\mathbf{a}_0, \ldots, \mathbf{a}_n$ for $A_0, \ldots, A_n$ and $\mathbf{b}_0, \ldots, \mathbf{b}_n$ for $B_0, \ldots, B_n$ such that

$$[\mathbf{a}_0 + \cdots + \mathbf{a}_n] = A_{n+1} \quad \text{and} \quad [\mathbf{b}_0 + \cdots + \mathbf{b}_n] = B_{n+1}.$$

As both sets of vectors $\{\mathbf{a}_0, \ldots, \mathbf{a}_n\}$ and $\{\mathbf{b}_0, \ldots, \mathbf{b}_n\}$ form a basis of $\mathbb{K}^{n+1}$, there exists a linear transformation $\Phi \in \mathrm{GL}(n + 1, \mathbb{K})$ for which $\Phi(\mathbf{a}_i) = \mathbf{b}_i$ holds if $0 \leq i \leq n$. Because of the linearity it means that

$$\Phi(\mathbf{a}_{n+1}) = \Phi(\mathbf{a}_0 + \cdots + \mathbf{a}_n) = \mathbf{b}_0 + \cdots + \mathbf{b}_n = \mathbf{b}_{n+1},$$

so the collineation $[\Phi] \in \mathrm{PGL}(n + 1, \mathbb{K})$ maps the point A_i into the point B_i for all $0 \leq i \leq n + 1$.

Now we prove that the action is sharply transitive. Suppose to the contrary that two linear transformations $[\Phi]$ and $[\Psi]$ take the point A_i to B_i for all $0 \leq i \leq n + 1$. Then $[\Phi^{-1} \circ \Psi] \in \mathrm{PGL}(n + 1, \mathbb{K})$ is different from the identity and it fixes all the points $A_0, \ldots, A_{n+1}$.

We can choose the representative vectors $\mathbf{a}_0, \mathbf{a}_1, \ldots, \mathbf{a}_n$ for the points $A_0, A_1, \ldots, A_n$ such that the sum $\mathbf{a}_0 + \mathbf{a}_1 + \cdots + \mathbf{a}_n$ represents the point A_{n+1}. Then each of the vectors $\mathbf{a}_0, \mathbf{a}_1, \ldots, \mathbf{a}_{n+1}$ is an eigenvector of $\Phi^{-1} \circ \Psi$, hence we can apply Lemma 4.13. This gives that all of the corresponding eigenvalues are equal to a fixed element $\lambda \in \mathbb{K}^*$. This means that $\Phi^{-1} \circ \Psi$ is the multiplication with λ, hence $[\Phi^{-1} \circ \Psi]$ is the identity. $\qquad\square$

The next theorem characterizes the automorphic collineations.

Theorem 4.17. *Let* $\mathbf{e}_0, \mathbf{e}_1, \ldots, \mathbf{e}_n$ *be the standard basis of* $\mathbb{K}^{n+1}$. *For* $i = 0, 1, \ldots, n$ *let* $E_i \in \mathrm{PG}(n, \mathbb{K})$ *be the point represented by* $\mathbf{e}_i$ *and let* E_{n+1} *be the point represented by* $\mathbf{e}_0 + \mathbf{e}_1 + \cdots + \mathbf{e}_n$. *Then a collineation* Ψ *of* $\mathrm{PG}(n, \mathbb{K})$ *is an automorphic collineation if and only if* Ψ *fixes each of the points* $E_0, E_1, \ldots, E_n$ *and* E_{n+1}.

Proof. If Ψ is an automorphic collineation induced from the automorphism σ, then Ψ fixes E_i for $i = 0, 1, \ldots, n + 1$, because $0^\sigma = 0$ and $1^\sigma = 1$.

Now suppose that Ψ fixes E_i for $i = 0, 1, \ldots, n + 1$. This means that if $0 \leq i \leq n - 1$, then Ψ maps the line $E_i E_n$ into itself. Thus for each of these i there exists a bijection $\phi_i \colon \mathbb{K} \to \mathbb{K}$ defined by the collineation Ψ :

$$\lambda^{\phi_i} = \mu \iff \Psi([\lambda \mathbf{e}_i + \mathbf{e}_n]) = [\mu \mathbf{e}_i + \mathbf{e}_n].$$

For $i = 0, 1, \ldots, n-1$ and $\lambda \in \mathbb{K}$ let $\mathcal{H}_\lambda^i$ denote the hyperplane with equation $X_i = \lambda X_n$. Then $\mathcal{H}_1^i$ is spanned by the set of points $\{E_j : 0 \leq j \leq n+1, j \neq i\}$, hence it is mapped by Ψ onto itself. As $E_i \notin \mathcal{H}_1^i$, this implies that the point

$$E_i E_n \cap \mathcal{H}_1^i = (0 : \cdots : 0 : \underbrace{1}_{i} : 0 : \cdots : 0 : 1)$$

is fixed by Ψ, which means that $1^{\phi_i} = 1$ for all $i = 0, 1, \ldots, n-1$.

Now, consider the point $P_\lambda^i = [\lambda e_i + e_n]$. For $i = 0, 1, \ldots, n-1$ we have $(P_\lambda^i)^\Psi = P_{\lambda^{\phi_i}}^i$ which implies

$$(\mathcal{H}_\lambda^i)^\Psi = \mathcal{H}_{\lambda^{\phi_i}}^i,$$

because $\mathcal{H}_\lambda^i$ is spanned by the set of points

$$\{E_j : 0 \leq j \leq n, \, j \neq i\} \cup \{P_{\lambda^{\phi_i}}^i\}.$$

This allows us to calculate the image of a point whose last coordinate is 1. Let $R = (\lambda_0 : \cdots : \lambda_{n-1} : 1)$. Then

$$R = \bigcap_{i=0}^{n-1} \mathcal{H}_{\lambda_i}^i,$$

hence

$$R^\Psi = \left(\bigcap_{i=0}^{n-1} \mathcal{H}_{\lambda_i}^i \right)^\Psi = \bigcap_{i=0}^{n-1} \mathcal{H}_{\lambda_i^{\phi_i}}^i = (\lambda_0^{\phi_0} : \lambda_1^{\phi_1} : \cdots : \lambda_{n-1}^{\phi_{n-1}} : 1).$$

We claim that the bijection $\phi_i \colon \mathbb{K} \to \mathbb{K}$ does not depend on i. Let $\gamma \in \mathbb{K}$ be an arbitrary element. Then the point $Q = (\gamma : \gamma : \ldots : \gamma : 1)$ is on the line $E_n E_{n+1}$. We know that

$$Q^\Psi = (\gamma^{\phi_0} : \gamma^{\phi_1} : \cdots : \gamma^{\phi_{n-1}} : 1).$$

As the line $E_n E_{n+1}$ is mapped into itself by Ψ, the hyperplane with equation $X_n = 0$ is also mapped into itself by Ψ, so the point $(1 : 1 : \cdots : 1 : 0)$ is fixed by Ψ. Hence

$$Q^\Psi = (\delta : \delta : \ldots : \delta : 1)$$

with a suitable $\delta \in \mathbb{K}$ which means that $\phi_0 = \phi_1 = \cdots = \phi_{n-1}$.

Replace the role of the first and the $(n+1)$-st homogeneous coordinate. We get that Ψ induces a permutation $\psi \colon \mathbb{K} \to \mathbb{K}$ such that if the first coordinate of a point T is 1 and $T = (1 : \lambda_1 : \cdots : \lambda_n)$, then

$$T^\Psi = (1 : \lambda_1^\psi : \cdots : \lambda_n^\psi).$$

This means that if we consider a point $U = (1 : \lambda_1 : \ldots : \lambda_{n-1} : 1)$, then

$$(1 : \lambda_1^\phi : \cdots : \lambda_{n-1}^\phi : 1) = U^\Psi = (1 : \lambda_1^\psi : \cdots : \lambda_{n-1}^\psi : 1),$$

hence the permutation ψ is equal to ϕ.

Finally we prove that $\phi_i \colon \mathbb{K} \to \mathbb{K}$ is an automorphism of $\mathbb{K}$. The property $0^\phi = 0$ comes from the definition, and we have already seen that $1^\phi = 1$. Let $0 \neq \lambda \in \mathbb{K}$ and $\mu \in \mathbb{K}$ arbitrary elements. Then the vectors $(1 : \lambda\mu : 0 : \cdots : 0 : \lambda)$ and $(1/\lambda : \mu : 0 : \cdots : 0 : 1)$ represent the same point, say P, of $\mathrm{PG}(n, \mathbb{K})$. Hence P^Ψ can be represented in two ways:

$$(1 : (\lambda\mu)^\phi : 0 : \cdots : 0 : \lambda^\phi) \quad \text{and} \quad ((1/\lambda)^\phi : \mu^\phi : 0 : \cdots : 0 : 1). \tag{4.1}$$

Putting $\mu = 1$ we get

$$\frac{1}{\lambda^\phi} = \frac{(1/\lambda)^\phi}{1}, \quad \text{so} \quad \frac{1}{\lambda^\phi} = \left(\frac{1}{\lambda}\right)^\phi.$$

Applying this for arbitrary μ, Equation (4.1) gives

$$\frac{1}{(\lambda\mu)^\phi} = \frac{(1/\lambda)^\phi}{\mu^\phi} = \frac{1}{\lambda^\phi}\frac{1}{\mu^\phi}, \quad \text{hence} \quad (\lambda\mu)^\phi = \lambda^\phi\mu^\phi.$$

We show that the permutation ϕ preserves the sum, too. Take the points L, M and N represented by the vectors $(1 : \lambda : 0 : 0 : \ldots : 0)$, $(0 : \mu : 0 : \ldots : 0 : 1)$ and $(1 : \lambda + \mu : 0 : 0 : \ldots : 1)$, respectively. These points are collinear, because

$$(1 : \lambda : 0 : 0 : \cdots : 0) + (0 : \mu : 0 : \cdots : 0 : 1) = (1 : \lambda + \mu : 0 : 0 : \cdots : 1).$$

Thus the points L^Φ, M^Φ and N^Φ are also collinear, so there exist $\alpha, \beta \in \mathbb{K}$ for which

$$\alpha(1 : \lambda^\phi : 0 : \cdots : 0) + \beta(0 : \mu^\phi : 0 : \cdots : 0 : 1) = (1 : (\lambda + \mu)^\phi : 0 : 0 : \cdots : 1).$$

From the equality of the first and of the last coordinates we get $\alpha = \beta = 1$. Hence the equality of the second coordinates gives

$$\lambda^\phi + \mu^\phi = (\lambda + \mu)^\phi,$$

so ϕ is an automorphism of $\mathbb{K}$. $\qquad\qquad\square$

We recall a definition from group theory. Let H_1 and H_2 be subgroups of the group G. Then G is the *semidirect product of H_1 and H_2* if the following hold:

- any element $g \in G$ is the product of an element from H_1 and an element from H_2,

- the intersection $H_1 \cap H_2$ contains only the identity of G,

- H_1 is a normal subgroup of G.

A semidirect product is denoted by either $G = H_1 \rtimes H_2$ or $G = H_2 \ltimes H_1$. Now we are ready to prove the Fundamental Theorem of Projective Geometry.

Theorem 4.18. *Let* $\mathbb{K}$ *be a field and* $n \geq 2$ *be an integer. Then the following hold:*

1. *Any element of the full collineation group* $\mathrm{Coll}(\mathrm{PG}(n, \mathbb{K}))$ *of the projective space* $\mathrm{PG}(n, \mathbb{K})$ *can be expressed as the composition of an automorphic collineation and a projective general linear transformation.*

   In analytic form any collineation of $\mathrm{PG}(n, \mathbb{K})$ *can be described as*

$$\mathbf{x} \mapsto \mathbf{x}^\sigma A,$$

   where σ *is an automorphism of* $\mathbb{K}$ *and* A *is a non-singular* $(n+1) \times (n+1)$ *matrix over* $\mathbb{K}$.

2. *The group* $\mathrm{PGL}(n+1, \mathbb{K})$ *is a normal subgroup in* $\mathrm{Coll}(\mathrm{PG}(n, \mathbb{K}))$.

3. $\mathrm{Coll}(\mathrm{PG}(n, \mathbb{K})) \cong \widetilde{\mathrm{Aut}(\mathbb{K})} \ltimes \mathrm{PGL}(n+1, \mathbb{K})$.

Proof. Let Φ be an arbitrary collineation of $\mathrm{PG}(n, \mathbb{K})$. For $i = 0, 1, \ldots, n+1$ let E_i denote the same point as in Theorem 4.17 and let $F_i = E_i^\Phi$. Then $\{F_0, F_1, \ldots, F_{n+1}\}$ is a set of $n+2$ points in general position, so by Theorem 4.16 there exists an element $\Psi \in \mathrm{PGL}(n+1, \mathbb{K})$ which maps F_i to E_i for all i. This means that the collineation $\Phi\Psi$ fixes all of the points $E_0, E_1, \ldots, E_{n+1}$. Hence, by Theorem 4.17, we get that $\Phi\Psi$ is an automorphic collineation, say $\tilde{\sigma}$, therefore $\Phi = \tilde{\sigma}\Psi^{-1}$.

The linear transformation Ψ can be described as $\mathbf{x} \mapsto \mathbf{x}A^{-1}$ with a suitable non-singular matrix A^{-1}, thus in analytic form Φ is given as $\mathbf{x} \mapsto \mathbf{x}^\sigma A$; this proves Part 1.

Let $\Phi \in \mathrm{PGL}(n+1, \mathbb{K})$ be a projective linear transformation given in analytic form as $\mathbf{x} \mapsto \mathbf{x}B$, and let $\Psi \in \mathrm{Coll}(\mathrm{PG}(n, \mathbb{K}))$ be an arbitrary collineation given in analytic form as $\mathbf{x} \mapsto \mathbf{x}^\sigma A$. If $C = (c_{ij})$ is an arbitrary matrix and τ is an automorphism, then let C^τ denote the matrix $C^\tau = (c_{ij}^\tau)$. With this notation $\Psi^{-1} \colon \mathbf{x} \mapsto \mathbf{x}^{\sigma^{-1}}(A^{-1})^{\sigma^{-1}}$, hence

$$\begin{aligned}
\mathbf{x}^{\Psi^{-1}\Phi\Psi} &= (\mathbf{x}^{\sigma^{-1}}(A^{-1})^{\sigma^{-1}})^{\Phi\Psi} \\
&= (\mathbf{x}^{\sigma^{-1}}(A^{-1})^{\sigma^{-1}}B)^\Psi = (\mathbf{x}^{\sigma^{-1}}(A^{-1})^{\sigma^{-1}}B)^\sigma A \\
&= \mathbf{x}A^{-1}B^\sigma A.
\end{aligned}$$

The matrices A and B are non-singular, so $A^{-1}B^\sigma A$ is a non-singular matrix, too. Thus $\Psi^{-1}\Phi\Psi \in \mathrm{PGL}(n+1, \mathbb{K})$ which proves Part 2.

Finally, by Theorem 4.17, the intersection of the set of automorphic collineations and the set of projective linear transformations consists of only the identity collineation. This property together with Parts 1 and 2 imply Part 3. $\square$

Corollary 4.19. *The full collineation group of* $\mathrm{PG}(n, q)$ *is isomorphic to the group* $\mathrm{P\Gamma L}(n+1, q)$.

If $q = p^r$, where p is prime, then the number of collineations of $\mathrm{PG}(n,q)$ is

$$|\mathrm{P\Gamma L}(n+1,q)| = r q^{n(n+1)/2} \prod_{i=2}^{n+1}(q^i - 1).$$

Polarities of finite projective spaces can be defined in the same way as in the case of classical projective spaces.

Definition 4.20. *Let S be a finite projective space and S^* be its dual space. A collineation $\alpha : S \to S^*$ is called a* correlation *of S. Every correlation α defines a collineation $S^* \to S$ by the action of α on hyperplanes of S. If α^2 is the identity on S, then α is called a* polarity.

If α is a polarity of S, P is a point and $\mathcal{H}$ is a hyperplane, then the hyperplane P^α is called the *polar* of P and the point $\mathcal{H}^\alpha$ is called the *pole* of $\mathcal{H}$. When a point Q is incident with the hyperplane P^α the point P is incident with the hyperplane Q^α. This follows immediately from the fact that α is an incidence-preserving mapping. In this case the points P and Q are *conjugate points*. If P is on the hyperplane P^α, then P is a *self-conjugate point*. The notions of polar subspace and self-conjugate subspace can be extended to arbitrary dimensions.

Definition 4.21. *Let α be a polarity of the projective space S. If Σ_k is a k-dimensional subspace ($k = 0, 1, \ldots, n-1$) of S, then the $(n-k-1)$-dimensional subspace Σ_k^α is called the* polar *of Σ_k.*

A k-dimensional subspace Σ_k is self-conjugate *if*

- $\Sigma_k \subseteq \Sigma_k^\pi$, if $k \leq (n-1)/2$,

- $\Sigma_k \supseteq \Sigma_k^\pi$, if $k \geq (n-1)/2$.

Let us remark that in the planar case the definitions above are valid for arbitrary projective planes, not only for Desarguesian planes. Polarities do not only play an important role in geometry, but also in extremal graph theory, as we will see later on (Theorems 12.14 and 12.15).

In what follows our aim is to describe polarities of $\mathrm{PG}(n,q)$. Of course, we shall use Theorem 4.18 for giving the classification.

Theorem 4.22. *Let α be a polarity of the projective space $\mathrm{PG}(n,\mathbb{K})$, $n \geq 2$. Then α^2 can be given analytically in the form*

$$\mathbf{x} \to \frac{1}{t}\mathbf{x}^{\sigma^2} A^\sigma (A^{-1})^\mathrm{T},$$

where σ is an automorphism of the field $\mathbb{K}$ for which σ^2 is the identity, $t \in \mathbb{K}$ is a fixed element such that $t \cdot t^\sigma = 1$ and A is an invertible matrix of size $(n+1) \times (n+1)$ satisfying $t A^\mathrm{T} = A^\sigma$.

Proof. If α is a polarity and $\mathbf{x}$ is the coordinate vector of a point, then, by Theorem 4.18,

$$(\mathbf{x}^\alpha)^\alpha = (\mathbf{x}^\sigma A)^\alpha = (\mathbf{x}^\sigma A)^\sigma (A^{-1})^\mathrm{T} = \mathbf{x}^{\sigma^2} A^\sigma (A^{-1})^\mathrm{T},$$

where σ is an automorphism of the field $\mathbb{K}$ and A is an $(n+1) \times (n+1)$ matrix giving a linear transformation. The coordinate vector on the right-hand side represents the same point as $\mathbf{x}$, hence there is an element $t_\mathbf{x} \in \mathbb{K}$ such that

$$t_\mathbf{x} \mathbf{x} = \mathbf{x}^{\sigma^2} A^\sigma (A^{-1})^\mathrm{T}. \tag{4.2}$$

First we show that the coefficient $t_\mathbf{x}$ does not depend on $\mathbf{x}$. Let X and Y be two distinct points and consider the vectors $\mathbf{x}, \mathbf{y}$ and $\mathbf{x} + \mathbf{y}$. Applying α we get

$$\begin{aligned} t_{\mathbf{x}+\mathbf{y}}(\mathbf{x} + \mathbf{y}) &= (\mathbf{x} + \mathbf{y})^{\sigma^2} A^\sigma (A^{-1})^\mathrm{T} = \mathbf{x}^{\sigma^2} A^\sigma (A^{-1})^\mathrm{T} + \mathbf{y}^{\sigma^2} A^\sigma (A^{-1})^\mathrm{T} \\ &= t_\mathbf{x} \mathbf{x} + t_\mathbf{y} \mathbf{y}, \end{aligned}$$

hence $(t_{\mathbf{x}+\mathbf{y}} - t_\mathbf{x})\mathbf{x} = (t_\mathbf{y} - t_{\mathbf{x}+\mathbf{y}})\mathbf{y}$. As $\mathbf{x}$ and $\mathbf{y}$ are linearly independent, this means that $t_{\mathbf{x}+\mathbf{y}} = t_\mathbf{x} = t_\mathbf{y}$. So the same coefficient belongs to any two distinct points. Let t denote this unique coefficient. Apply Equation (4.2) to the standard base vectors $\mathbf{e}_i$. As we have already seen in the proof of Theorem 4.17, $\mathbf{e}_i^\sigma = \mathbf{e}_i$, thus we get $tA^\mathrm{T} = A^\sigma$. So Equation (4.2) gives $\mathbf{x} = \mathbf{x}^{\sigma^2}$ for all $\mathbf{x}$, hence σ^2 must be the identity.

Now we distinguish two cases. If σ is the identity, then $tA^\mathrm{T} = A$, hence $t(tA^\mathrm{T})^\mathrm{T} = A$, so $t^2 = 1$. If σ is not the identity, then $tA^\mathrm{T} = A^\sigma$ which implies $t((tA^\mathrm{T})^\sigma)^\mathrm{T} = A^\sigma$, hence $t \cdot t^\sigma = 1$ because $(A^\mathrm{T})^\sigma = (A^\sigma)^\mathrm{T}$. $\qquad\square$

Let us look at the detailed description of the polarities of $\mathrm{PG}(n, q)$. We distinguish the two main cases again.

1. *σ is the identity.*

Then $t^2 = 1$ which implies $t = \pm 1$. If q is odd then we have to distinguish two subcases. If $t = 1$, then $A = A^\mathrm{T}$, hence the matrix describing the polarity is symmetric. In this case the polarity is called an *ordinary polarity*. Another terminology is to call it a *polarity defined by a quadric*, because the self-conjugate points of the polarity are the points of the quadric with equation $\mathbf{X} A \mathbf{X}^\mathrm{T} = 0$. In case of $t = -1$ we have $A = -A^\mathrm{T}$. As $\det A = (-1)^{n+1} \det A^\mathrm{T}$ and $\det A \neq 0$, such a polarity can only exist in odd dimensions. All points are self-conjugate; the polarity is called a *symplectic polarity* (or a *null polarity*).

For q even $1 = -1$ and $A = A^\mathrm{T}$. If each element in the main diagonal of A is 0, then all points are self-conjugate and the polarity is called a *null polarity* (or *symplectic polarity*) similarly to the odd case. If there is a non-zero element in the main diagonal of the matrix $A = (a_{ij})$, then the polarity is called a *pseudo polarity*. In this case the self-conjugate points form a hyperplane, since

$$\mathbf{x} A \mathbf{x}^\mathrm{T} = 0 \iff \sum_{i=0}^{n} a_{ii} x_i^2 = 0 \iff \left(\sum_{i=0}^{n} \sqrt{a_{ii}} x_i \right)^2 = 0.$$

Before considering the other case we prove a lemma about roots of a particular equation in $\mathrm{GF}(s^2)$.

Lemma 4.23. *Let σ be an automorphism of order 2 of the field $\mathbb{K}$ and suppose that $t \in \mathbb{K}$ satisfies the equation $t \cdot t^\sigma = 1$. Then there exists a $0 \neq \lambda \in \mathbb{K}$ such that $t\lambda^\sigma = \lambda$.*

Proof. Let $0 \neq \mu \in \mathbb{K}$ be an arbitrary element. If $t\mu^\sigma = \mu$, then we are done. If $t\mu^\sigma \neq \mu$, then let $t\mu^\sigma = \nu$. This implies

$$\nu^\sigma = (t\mu^\sigma)^\sigma = t^\sigma \mu = \frac{1}{t}\mu,$$

hence $t\nu^\sigma = \mu$. Thus $t(\mu + \nu)^\sigma = t\mu^\sigma + t\nu^\sigma = \nu + \mu = (\mu + \nu)$.

We claim that there exists a μ such that $\mu + \nu \neq 0$ also holds. If $\mu = -\nu$, then $\mu^\sigma = -\mu/t$. If it were true for all μ, then $\mu_1 \cdot \mu_2 \neq 0$ would imply that

$$\frac{-1}{t}(\mu_1 \cdot \mu_2) = (\mu_1 \cdot \mu_2)^\sigma = \mu_1^\sigma \cdot \mu_2^\sigma = \frac{-1}{t}\mu_1 \cdot \frac{-1}{t}\mu_2 = \frac{1}{t^2}\mu_1 \cdot \mu_2.$$

Thus $-t^2 = t$, hence $t = -1$ because $t \neq 0$. So $\mu^\sigma = \mu$ for all $\mu \in \mathbb{K}$. This would mean that σ is the identity, giving a contradiction. $\quad\square$

2. σ is not the identity.

In this case $q = s^2$ and σ is the involutory automorphism of the field $\mathrm{GF}(q)$. This automorphism maps $\mathbf{x}$ to $\mathbf{x}^s$. We know that $tA^{\mathrm{T}} = A^\sigma$. Instead of A consider the matrix λA, where λ is a non-zero root of the equation $t\lambda^\sigma = \lambda$. The existence of λ follows from Lemma 4.23. This matrix defines the same correlation as A, but $tA^{\mathrm{T}} = A^\sigma$ implies

$$(\lambda A)^{\mathrm{T}} = (t\lambda^\sigma A)^{\mathrm{T}} = \lambda^\sigma(tA^{\mathrm{T}}) = \lambda^\sigma A^\sigma = (\lambda A)^\sigma.$$

Hence it is sufficient to consider the case $A^{\mathrm{T}} = A^\sigma$.

This polarity is called a *Hermitian* (*or unitary*) *polarity*. The self-conjugate points of the polarity are defined by the equation $\mathbf{X}^\sigma A\mathbf{X}^{\mathrm{T}} = 0$, and the set of self-conjugate points forms a *Hermitian variety*.

Quadrics and Hermitian varieties can also be defined without referring to the polarities.

Definition 4.24. *Let*

$$Q(\mathbf{X}) = \sum_{i,j=0}^{n} a_{ij}X_iX_j$$

be a quadratic form. The set of points in $\mathrm{PG}(n,q)$, defined by $Q(\mathbf{X}) = 0$, is called the quadric (*or quadratic variety*) *belonging to Q.*

Let $q = s^2$, σ be the involutory automorphism $x \mapsto x^s$,

$$H(\mathbf{X}) = \sum_{i,j=0}^{n} t_{ij}X_i^\sigma X_j$$

be a Hermitian form, that is $t_{ij} = t_{ji}^\sigma$ for every pair (i, j). The set of points in $\mathrm{PG}(n, q)$, defined by $H(\mathbf{X}) = 0$ is called the Hermitian variety belonging to H.

A quadratic or Hermitian variety is singular if, by changing the coordinate system, one can achieve that the corresponding form contains fewer variables. If this cannot be achieved then the variety is non-singular.

We are going to show that in a suitable coordinate system the forms describing quadratic and Hermitian varieties can have simpler equations. At the beginning the cases of quadratic and Hermitian forms are treated simultaneously. In case of a quadratic form Q we assume q being odd and introduce $t_{ij} = (a_{ij} + a_{ji})/2$. $\overline{x}$ will denote x in case of Q and x^σ in case of H.

Lemma 4.25. *The form*

$$F(\mathbf{X}) = \sum_{i,j=0}^{n} t_{ij}\overline{X}_i X_j$$

can be transformed to

$$\sum_{i=1}^{r} a_i \overline{X}_i X_i$$

by appropriately changing the coordinate system, where $a_i = \overline{a}_i$ and $r \le n+1$.

Proof. If some $t_{ii} \ne 0$, then we may assume $t_{00} \ne 0$ by permuting the coordinates. If $t_{ii} = 0$ for all i, then we may assume $t_{01} \ne 0$. Let $X_0 = Y_0$, $X_1 = cY_0 + Y_1$ and $X_i = Y_i$ if $i > 1$. Then $F(\mathbf{X}) = \sum s_{ij}\overline{Y}_i Y_j$, where $s_{00} = t_{01}c + t_{10}\overline{c}$. Thus, if q is odd, then choosing $c = t_{01}^{-1}$ we get $s_{00} = 2 \ne 0$, while if q is even, then we can choose c such that $t_{01}c \notin \mathrm{GF}(\sqrt{q})$ and this implies $s_{00} \ne 0$.

Consider the form

$$F(\mathbf{X}) = \sum_{i,j=0}^{n} t_{ij}\overline{X}_i X_j$$

with $t_{00} \ne 0$. Then

$$F(\mathbf{X}) = t_{00}\overline{X}_0 X_0 + \overline{X}_0 \left(\sum_{j=1}^{n} t_{0j} X_j \right) + X_0 \left(\sum_{i=1}^{n} t_{i0}\overline{X}_i \right) + \sum_{i,j=1}^{n} t_{ij}\overline{X}_i X_j$$

$$= t_{00}^{-1} \left(\sum_{i=0}^{n} t_{i0}\overline{X}_i \right) \left(\sum_{j=0}^{n} t_{0j} X_j \right) + \sum_{i,j=1}^{n} t'_{ij}\overline{X}_i X_j.$$

Let $X_0 = Y_0 - t_{00}^{-1}\sum_{j=1}^{n} t_{1j}X_j$, and $X_i = Y_i$ for $i > 0$. Then

$$F(\mathbf{X}) = t_{00}\overline{Y}_0 Y_0 + \sum_{i,j=1}^{n} t'_{ij}\overline{Y}_i Y_j = t_{00}\overline{Y}_0 Y_0 + F'(\mathbf{Y}).$$

Applying this process for F', finally we conclude with

$$F(\mathbf{X}) = \sum_{i=0}^{r} a_i \overline{X}_i X_i. \tag{4.3}$$

The case $r < n$ appears if and only if all coefficients of the corresponding form G' are 0 after a step, which means that the variety is singular. □

Corollary 4.26. *A Hermitian variety, or a quadratic variety for q odd, is singular if and only if the matrix belonging to the variety is singular.*

It also follows from Lemma 4.25 that in $\mathrm{PG}(n, q)$ each singular variety is a cone whose base is a non-singular variety in a suitable r-dimensional subspace Π_r and whose vertex is an $(n-r-1)$-dimensional subspace Π_{n-r-1} such that $\Pi_r \cap \Pi_{n-r-1} = \emptyset$ holds. If the variety is given by Equation (4.3), then Π_{n-r-1} is the subspace defined by $X_0 = X_1 = \cdots = X_r = 0$. Thus the geometric properties of singular varieties follow inductively from the properties of non-singular ones in lower dimension. Because of this in the rest of the chapter only non-singular varieties are considered.

For finding simpler canonical forms for varieties we need some algebraic properties of finite fields.

Lemma 4.27. *If $q = s^2$, then each element of the subfield $\mathrm{GF}(s) \subset \mathrm{GF}(q)$ is an $(s+1)$-st power in $\mathrm{GF}(q)$.*

Proof. Let $\alpha \in \mathrm{GF}(q)$ be a primitive element. As $q - 1 = (s+1)(s-1)$, we get that α^{s+1} is a primitive element in $\mathrm{GF}(s)$, hence each non-zero element of $\mathrm{GF}(s)$ can be written as $(\alpha^{s+1})^k = (\alpha^k)^{s+1}$ with a suitable k. □

Corollary 4.28. *If $q = s^2$, then in $\mathrm{GF}(q)$ the number of roots of the equation $x^{s+1} = c$ is the following:*

- *0, if $c \notin \mathrm{GF}(s)$;*

- *1, if $c = 0$;*

- *$s + 1$, if $0 \neq c \in \mathrm{GF}(s)$.*

Proof. If $a \in \mathrm{GF}(q)$, then $a^{s+1} \in \mathrm{GF}(s)$ because $(a^{s+1})^s = a^{s^2+s} = a^{1+s}$. Thus $x^{s+1} = c$ has no solution if $c \notin \mathrm{GF}(s)$. The equation $x^{s+1} = 0$ has a unique root, $x = 0$. Hence all non-zero elements are roots of exactly one equation of type $x^{s+1} = c$ with $0 \neq c \in \mathrm{GF}(s)$. As a polynomial of degree $s + 1$ has at most $s + 1$ roots, by the pigeonhole principle, each of the $s - 1$ equations of type $x^{s+1} = c$ with $0 \neq c \in \mathrm{GF}(s)$ must have exactly $s + 1$ roots. □

Lemma 4.29. *If $q = s^2$, then in $\mathrm{GF}(q)$ the number of roots of the equation $x^s + x = c$ is the following:*

- 0, *if* $c \notin \mathrm{GF}(s)$;

- s, *if* $c \in \mathrm{GF}(s)$.

Proof. If $a \in \mathrm{GF}(q)$, then $a^s + a \in \mathrm{GF}(s)$ because $(a^s + a)^s = a^{s^2} + a^s = a + a^s$. Thus $x^s + x = c$ has no solution if $c \notin \mathrm{GF}(s)$. Hence each element of $\mathrm{GF}(q)$ is a root of exactly one equation of type $x^{s+1} = c$ with $c \in \mathrm{GF}(s)$. As a polynomial of degree s has at most s roots, by the pigeonhole principle, each of the s equations of type $x^s + x = c$ with $c \in \mathrm{GF}(s)$ must have exactly s roots. $\qquad\square$

Corollary 4.30. *If* $q = s^2$, *then there exists an* $\varepsilon \in \mathrm{GF}(q)$ *such that* $\varepsilon^{s+1} = -1$ *and* $\varepsilon^2 \neq 1$.

Lemma 4.31. *If* q *is odd, then each non-square element of* $\mathrm{GF}(q)$ *can be written as the sum of two square elements of* $\mathrm{GF}(q)$.

Proof. Let $\mathcal{S}$ and $\mathcal{N}$ denote the set of square elements and the set of non-square elements in $\mathrm{GF}(q)$, respectively. If there exists an element $e \in \mathcal{N}$ that can be written as $e = c^2 + d^2$ with $c, d \in \mathrm{GF}(q)$, then for an arbitrary $f \in \mathcal{N}$ we have

$$f = e\frac{f}{e} = (c^2 + d^2)\frac{f}{e} = \left(c^2\frac{f}{e}\right) + \left(d^2\frac{f}{e}\right).$$

As $e, f \in \mathcal{N}$ implies $f/e \in \mathcal{S}$, this yields that f can also be written as the sum of two squares.

Suppose to the contrary that the statement of the lemma is false. This implies that the sum of any two elements of the set $\bar{\mathcal{S}} = \mathcal{S} \cup \{0\}$ is also in $\bar{\mathcal{S}}$. Let p denote the characteristic of $\mathrm{GF}(q)$. Then the additive inverse of an element $g \in \mathrm{GF}(q)$ is $(p-1)g = g + g + \cdots + g$, where the number of summands in the sum is $p - 1$. Hence $\bar{\mathcal{S}}$ contains the additive inverse of any of its elements. This means that $(\bar{\mathcal{S}}, +)$ is a subgroup of the additive group of $\mathrm{GF}(q)$. The cardinality of $\bar{\mathcal{S}}$ is $\frac{q-1}{2} + 1 = \frac{q+1}{2}$, thus $|\bar{\mathcal{S}}|$ cannot divide q. This contradiction proves the statement. $\qquad\square$

Theorem 4.32. *In* $\Gamma\mathrm{G}(n, s^2)$ *the form belonging to a non-singular Hermitian variety can be transformed to*

$$H_n(\mathbf{X}) = \sum_{i=0}^n X_i^{s+1}$$

by a suitable changing of coordinates.

Proof. By Lemma 4.25, the form belonging to a non-singular Hermitian variety can be transformed to

$$\sum_{i=0}^n a_i \overline{X}_i X_i$$

where $a_i = \bar{a}_i$, hence $a_i \in \mathrm{GF}(s)$ for all i. By Lemma 4.27, for all a_i there exists $b_i \in \mathrm{GF}(s^2)$ such that $a_i = b_i^{s+1}$. Let $Y_i = b_i X_i$, then

$$\sum_{i=0}^{n} a_i \overline{X}_i X_i = \sum_{i=0}^{n} b_i^{s+1} X_i^s X_i = \sum_{i=0}^{n} Y_i^{s+1},$$

which is the required form. $\square$

Using the canonical form we can determine the number of points of a Hermitian variety, which is the same as the number of self-conjugate points of a Hermitian (unitary) polarity.

Theorem 4.33. *In* $\mathrm{PG}(n, s^2)$ *a non-singular Hermitian variety contains*

$$u(n, s^2) = \frac{\left(s^{n+1} + (-1)^n\right)\left(s^n - (-1)^n\right)}{s^2 - 1}$$

points.

Proof. We prove it by induction on the dimension. The 1-dimensional non-singular Hermitian variety has canonical form $X_0^{s+1} + X_1^{s+1} = 0$. Hence it does not contain the point $(0 : 1)$ and it contains the point $(1 : t)$ if and only if $t^{s+1} = -1$. By Corollary 4.28, this equation has exactly $s + 1$ roots. As $u(1, s^2) = s + 1$, this proves the theorem for $n = 1$.

Suppose that the statement is true for all $k < n$ and consider the n-dimensional non-singular Hermitian variety $\mathcal{U}_n$ with canonical equation

$$X_0^{s+1} + \sum_{i=1}^{n} X_i^{s+1} = 0.$$

Take the hyperplane Σ_∞ with equation $X_0 = 0$. Then the coordinates of any point in $\mathcal{U}_n \cap \Sigma_\infty$ satisfy the equation $\sum_{i=1}^{n} X_i^{s+1} = 0$, hence the intersection is an $(n - 1)$-dimensional non-singular Hermitian variety $\mathcal{U}_{n-1}$. Let E be the point with coordinate-vector $\mathbf{e} = (1 : 0 : 0 : \ldots : 0)$. Then E is not on $\mathcal{U}_n$. Consider the lines joining E and the points of Σ_∞. These lines give a partition of the set of points $\mathrm{PG}(n, s^2) \setminus \{E\}$. Let A be a point in Σ_∞ with coordinate-vector $\mathbf{a} = (0 : a_1 : a_2 : \ldots : a_n)$. Now, if A_t is any point on the line EA different from E, then its coordinate-vector can be written as

$$\mathbf{a_t} = t\mathbf{e} + \mathbf{a} = (t : a_1 : a_2 : \ldots : a_n)$$

where $t \in \mathrm{GF}(s^2)$.

First, suppose that $A \in \mathcal{U}_{n-1}$. Then

$$A_t \in \mathcal{U}_n \iff t^{s+1} + \sum_{i=1}^{n} a_i^{s+1} = t^{s+1} = 0 \iff t = 0,$$

so the line EA_t contains only one point, namely A, of $\mathcal{U}_n$. On the other hand, if $A \notin \mathcal{U}_{n-1}$, then $\sum_{i=1}^{n} a_i^{s+1} = b \neq 0$. In this case

$$A_t \in \mathcal{H}_n \iff t^{s+1} + \sum_{i=1}^{n} a_i^{s+1} = t^{s+1} + b = 0 \iff t^{s+1} = -b.$$

The equality

$$b^s = \left(\sum_{i=1}^{n} a_i^{s+1} \right)^s = \sum_{i=1}^{n} (a_i^{s+1})^s = \sum_{i=1}^{n} a_i^{s+1} = b$$

implies that $0 \neq b \in \mathrm{GF}(s)$. Thus, by Corollary 4.28, the equation $t^{s+1} = -b$ has $s+1$ distinct roots in $\mathrm{GF}(s^2)$, so the line EA_t contains $s+1$ points of $\mathcal{H}_n$. The two cases together show that $u(n, s^2)$ satisfies the recurrence relation

$$u(n, s^2) = u(n-1, s^2) + (s+1) \left(\frac{s^{2n} - 1}{s^2 - 1} - u(n-1, s^2) \right),$$

which gives the required result after elementary manipulations. $\qquad \square$

In the planar case a Hermitian variety is called a *Hermitian curve*.

Corollary 4.34. *In the plane* $\mathrm{PG}(2, s^2)$ *any non-singular Hermitian curve contains* $s^3 + 1$ *points.*

Let $\mathcal{U}_n$ be a non-singular Hermitian variety in $\mathrm{PG}(n, s^2)$ and $H_n(\mathbf{X}) = \mathbf{X}H(\mathbf{X}^s)^{\mathrm{T}}$ be the corresponding form. Let us define the sesquilinear form $G(\mathbf{p}, \mathbf{r})$ belonging to $\mathcal{U}_n$ as $G(\mathbf{p}, \mathbf{r}) = \mathbf{p}H(\mathbf{r}^s)^{\mathrm{T}}$. Then elementary calculation shows that for all $t \in \mathrm{GF}(s^2)$

$$H_n(\mathbf{p} + t\mathbf{r}) = H_n(\mathbf{p}) + tG(\mathbf{p}, \mathbf{r}) + (tG(\mathbf{p}, \mathbf{r}))^s + t^{s+1}H_n(\mathbf{r}).$$

The sesquilinear form $G(\mathbf{p}, \mathbf{r})$ helps us to determine the number of common points of a line and $\mathcal{U}_n$.

Definition 4.35. *Let* $\mathcal{U}_n$ *be a non-singular Hermitian variety. The line* ℓ *is a tangent to* $\mathcal{U}_n$ *if* $|\ell \cap \mathcal{U}_n| = 1$, *and* ℓ *is an* $(s+1)$-secant *of* $\mathcal{U}_n$ *if* $|\ell \cap \mathcal{U}_n| = s+1$.

Lemma 4.36. *Let* P *and* R *be two distinct points and* $\mathcal{U}_n$ *be a non-singular Hermitian variety. Then the following hold:*

- *If* $H_n(\mathbf{p})H_n(\mathbf{r}) \neq (G(\mathbf{p}, \mathbf{r}))^{s+1}$, *then the line* PR *is an* $(s+1)$-secant *of* $\mathcal{U}_n$.

- *If* $H_n(\mathbf{p})H_n(\mathbf{r}) = (G(\mathbf{p}, \mathbf{r}))^{s+1}$, *then*

  - *if* $H_n(\mathbf{p}) = H_n(\mathbf{r})G(\mathbf{p}, \mathbf{r}) = 0$, *then the line* PR *is contained in* $\mathcal{U}_n$,
  - *otherwise the line* PR *is a tangent to* $\mathcal{U}_n$.

Proof. Let R_t be the point with coordinate-vector $\mathbf{r} + t\mathbf{p}$. Then R_t is on the line PR and $R_t \neq P$. Let $H_n(\mathbf{r}) = a$, $H_n(\mathbf{p}) = b$ and $G(\mathbf{p}, \mathbf{r}) = c$ be shorthand notations. Then $a^s = a$, $b^s = b$, hence $a, b \in \mathrm{GF}(s)$. The point R_t is on $\mathcal{U}_n$ if and only if

$$H_n(\mathbf{r} + t\mathbf{p}) = a + tc + (tc)^s + t^{s+1}b = 0. \tag{4.4}$$

First, suppose that P is on $\mathcal{U}_n$, hence $b = 0$. If $c \neq 0$, then Equation (4.4) can be written as $x^s + x = -a$ where $x = tc$. By Lemma 4.29, it has s distinct roots, because $-a \in \mathrm{GF}(s)$. Each root corresponds to a point $P \neq R_t \in \mathcal{U}_n$, so the line PR contains $1 + s$ points of $\mathcal{U}_n$ which proves the first part of the statement in this case. If $c = 0$ also holds, then Equation (4.4) becomes $a = 0$. This is an identity if R is on $\mathcal{U}_n$, and a contradiction otherwise, which proves the first part of the second statement in this case.

Now, suppose that $b \neq 0$. Then Equation (4.4) can be written as

$$\left(t^s + \frac{c}{b}\right)\left(t + \frac{c^s}{b}\right) = \frac{c^{s+1} - ab}{b^2}.$$

According to Corollary 4.28, it has one root if $ab = c^{s+1}$, which proves the second part of the second statement in this case, and it has $s + 1$ roots if $ab \neq c^{s+1}$, which proves the first statement of the lemma in this case. $\square$

Corollary 4.37. *Let P and R be two distinct points, $\mathcal{U}_n$ be a non-singular Hermitian variety in $\mathrm{PG}(n, s^2)$. If $P \in \mathcal{U}_n$, then*

- *if $R \in \mathcal{U}_n$ then the line PR is completely contained in $\mathcal{U}_n$ if and only if $G(\mathbf{p}, \mathbf{r}) = 0$,*

- *if $R \notin \mathcal{U}_n$ then the line PR is a tangent to $\mathcal{U}_n$ if and only if $G(\mathbf{p}, \mathbf{r}) = 0$.*

Corollary 4.38. *Let $\mathcal{U}_n$ be a non-singular Hermitian variety and ℓ be a line in $\mathrm{PG}(n, s^2)$. Then the intersection $\mathcal{U}_n \cap \ell$ consists of 0, 1, $s + 1$ or $s^2 + 1$ points.*

Definition 4.39. *The points P and R are* conjugates *with respect to the non-singular Hermitian variety $\mathcal{U}_n$ if $G(\mathbf{p}, \mathbf{r}) = 0$. The equation $G(\mathbf{X}, \mathbf{p}) = 0$ defines a hyperplane H_P. This hyperplane is called the* polar hyperplane *of P. If $P \in \mathcal{U}_n$, then H_P is called the* tangent hyperplane *to $\mathcal{U}_n$ at P.*

Lemma 4.40. *Let $\mathcal{U}_n$ be a non-singular Hermitian variety in $\mathrm{PG}(n, s^2)$. Then the mapping φ from $\mathrm{PG}(n, s^2)$ to its dual space defined as*

$$\varphi \colon P \mapsto H_P$$

is a Hermitian polarity and its self-conjugate points are the points of $\mathcal{U}_n$.

Proof. We may assume without loss of generality that $\mathcal{U}_n$ is given by its canonical equation $\sum_{i=0}^{n} X_i^{s+1} = 0$. Let I be the $(n+1) \times (n+1)$ identity matrix. Then φ acts as the multiplication by I. As $I^s = I^{\mathrm{T}}$ is obvious, by Theorem 4.22, φ is an Hermitian polarity. $\square$

The next statement immediately follows from Lemma 4.36.

Corollary 4.41. *Let $\mathcal{U}_n$ be a non-singular Hermitian variety in $\mathrm{PG}(n, s^2)$, $\Sigma_k \subset \mathcal{U}_n$ be a k-dimensional subspace, $P \in \Sigma_k$ be a point and H_P be the tangent hyperplane to $\mathcal{U}_n$ at P. Then*

- *H_P contains Σ_k;*

- *a line PR is an $(s+1)$-secant of $\mathcal{U}_n$ if and only if $R \notin H_P$.*

In particular, if $n = 2$, then Corollaries 4.34 and 4.41, and Lemma 4.40 imply the next statement.

Corollary 4.42. *Let $\mathcal{U}_2$ be a non-singular Hermitian curve and $P \in \mathcal{U}_2$ be a point in the plane $\mathrm{PG}(2, s^2)$. Then there is a unique tangent line to $\mathcal{U}_2$ at P and there are s^2 secant lines to $\mathcal{U}_2$ through P. Any line of the plane is either a tangent or a secant to $\mathcal{U}_2$. There are altogether $s^3 + 1$ tangent lines and $s^4 - s^3 + s^2$ secant lines in the plane.*

Definition 4.43. *The group of linear transformations fixing a non-singular Hermitian variety $\mathcal{U}_n$ in $\mathrm{PG}(n, s^2)$, $n \geq 2$, is called the* unitary group *and it is denoted by $\mathrm{PGU}(n+1, s^2)$.*

Theorem 4.44. *The unitary group acts transitively on the set of points of the corresponding non-singular Hermitian variety $\mathcal{U}_n$.*

Proof. Let $H_n(\mathbf{X})$ be the Hermitian form and $G(\mathbf{p}, \mathbf{r})$ be the sesquilinear form belonging to $\mathcal{U}_n$. Let $C \in \mathcal{U}_n$ be any point and $\lambda \in \mathrm{GF}(s^2)$ be any root of the equation $\lambda^s + \lambda = 0$. Define the mapping $\varphi_{C,\lambda}$ on the set of points of $\mathrm{PG}(n, s^2)$ as

$$\varphi_{C,\lambda} \colon \mathbf{x} \mapsto \mathbf{x} + \lambda G(\mathbf{x}, \mathbf{c})\mathbf{c}.$$

We claim that $\varphi_{C,\lambda}$ is a well-defined, non-singular linear transformation with center C which preserves $\mathcal{U}_n$ setwise. The sesquilinearity of G implies that $(\alpha \mathbf{x})^{\varphi_{C,\lambda}} = \alpha(\mathbf{x}^{\varphi_{C,\lambda}})$ and

$$G(\alpha_1 \mathbf{x}_1 + \alpha_2 \mathbf{x}_2, \mathbf{c}) = \alpha_1 G(\mathbf{x}_1, \mathbf{c}) + \alpha_2 G(\mathbf{x}_2, \mathbf{c}),$$

thus $\varphi_{C,\lambda}$ is well-defined and linear. It is non-singular, because

$$
\begin{aligned}
(\mathbf{x})^{\varphi_{C,\lambda}} = (\mathbf{y})^{\varphi_{C,\lambda}} &\iff \mathbf{x} + \lambda G(\mathbf{x}, \mathbf{c})\mathbf{c} = \mathbf{y} + \lambda G(\mathbf{y}, \mathbf{c})\mathbf{c} \\
&\iff \mathbf{x} - \mathbf{y} = \lambda G(\mathbf{y} - \mathbf{x}, \mathbf{c})\mathbf{c} \\
&\implies \mathbf{x} - \mathbf{y} = \beta \mathbf{c} \implies G(\mathbf{y} - \mathbf{x}, \mathbf{c}) = -\beta G(\mathbf{c}, \mathbf{c}) = \mathbf{0} \\
&\implies \mathbf{x} = \mathbf{y}.
\end{aligned}
$$

If $P \in \mathcal{U}_n$ is any point, then its image has coordinate vector $\mathbf{p} + \lambda G(\mathbf{p}, \mathbf{c})\mathbf{c}$. On the one hand, this means that $C^{\varphi_{C,\lambda}} = C$ for all λ, and for any point $P \neq C$, its image $P^{\varphi_{C,\lambda}}$ is on the line joining C and P, thus C is a center of $\varphi_{C,\lambda}$. On the other hand, using $H_n(\mathbf{p}) = H_n(\mathbf{c}) = 0$ and $(G(\mathbf{p}, \mathbf{c}))^s = G(\mathbf{c}, \mathbf{p})$, we get

$$
\begin{aligned}
H_n(\mathbf{p} + \lambda G(\mathbf{p}, \mathbf{c})\mathbf{c}) &= H_n(\mathbf{p}) + (\lambda G(\mathbf{p}, \mathbf{c})) G(\mathbf{c}, \mathbf{p}) \\
&\quad + (\lambda G(\mathbf{p}, \mathbf{c}))^s G(\mathbf{p}, \mathbf{c}) + (\lambda G(\mathbf{p}, \mathbf{c}))^{s+1} H_n(\mathbf{c}) \\
&= (\lambda + \lambda^s) \cdot G(\mathbf{p}, \mathbf{c}) G(\mathbf{c}, \mathbf{p}) \\
&= 0,
\end{aligned}
$$

so $P^{\varphi_{C,\lambda}} \in \mathcal{U}_n$.

Suppose that the line ℓ is an $(s+1)$-secant of $\mathcal{U}_n$ through C. Let $C \neq P \in (\mathcal{U}_n \cap \ell)$ be any point. Then, by Lemma 4.36, $G(\mathbf{p}, \mathbf{c}) \neq 0$, so if $\lambda_1 \neq \lambda_2$ distinct roots of the equation $\lambda^s + \lambda = 0$, then, by definition, $P^{\varphi_{C,\lambda_1}} \neq P^{\varphi_{C,\lambda_2}}$. It follows from the previous observations that $P^{\varphi_{C,\lambda}} \in (\mathcal{U}_n \cap \ell)$ also holds for all roots. As the equation has s distinct roots, the set of points

$$
\{P^{\varphi_{C,\lambda}} : \lambda^s + \lambda = 0\}
$$

consists of s distinct points and this is a subset of $(\mathcal{U}_n \cap \ell) \setminus \{C\}$, whose cardinality is also s. Hence the two sets must be the same. In other words, if P_1 and P_2 are points of the set $(\mathcal{U}_n \cap \ell) \setminus \{C\}$, then we can choose λ such that $P_1^{\varphi_{C,\lambda}} = P_2$.

Now we are ready to prove the transitivity of the unitary group. Let P and R be two distinct points of $\mathcal{U}_n$. If the line PR is an $(s+1)$-secant of $\mathcal{U}_n$ then take a point C on the line PR different from both P and R. Then we have already seen that there exists λ such that $\varphi_{C,\lambda}$ maps P to R. If the line PR lies on $\mathcal{U}_n$, then let $S \in \mathcal{U}_n$ be a point such that both PS and RS are $(s+1)$-secants to $\mathcal{U}_n$. The point S exists because $\mathcal{U}_n$ is non-singular. Take the points $P \neq C \neq S$ on the line PS and $R \neq C' \neq S$ on the line RS. Then we can choose λ and λ' such that $P^{\varphi_{C,\lambda}} = S$ and $S^{\varphi_{C',\lambda'}} = R$, hence the linear transformation $\varphi_{C,\lambda}\varphi_{C',\lambda'}$ maps P to R, proving the theorem. $\qquad\square$

Theorem 4.45. *Let $\mathcal{U}_n$ be a non-singular Hermitian variety in $\mathrm{PG}(n, s^2)$, $P \in \mathcal{U}_n$ be a point and H_P be the tangent hyperplane to $\mathcal{U}_n$ at P. If $n \geq 3$, then $\mathcal{U}_n \cap H_P$ is a cone $\mathcal{K}$. The vertex of $\mathcal{K}$ is P and its base is a non-singular Hermitian variety $\mathcal{U}_{n-2}$.*

Proof. If a linear transformation φ fixes $\mathcal{U}_n$ and maps a point $P \in \mathcal{U}_n$ to $P' \in \mathcal{U}_n$, then φ maps H_P to $H_{P'}$. Thus, by Theorems 4.32 and 4.44, we may assume without loss of generality that $\mathcal{U}_n$ is given by its canonical equation, $\sum_{i=0}^n X_i^{s+1} = 0$, and P is the point $(1 : \epsilon : 0 : \cdots : 0)$ where ϵ is a fixed root of the equation $x^{s+1} = -1$ (the existence of ϵ follows from Corollary 4.30). Then $\epsilon^s = -\frac{1}{\epsilon}$, and elementary calculation gives that H_P has equation $\epsilon X_0 = X_1$.

Thus

$$\mathcal{U}_n \cap H_P = \{(1 : \epsilon : tr_2 : tr_3 : \cdots : tr_n) \colon t \in \mathrm{GF}(s^2) \text{ and } \sum_{i=2}^{n} r_i^{s+1} = 0\}$$

$$\cup \{(0 : 0 : r_2 : r_3 : \cdots : r_n) \colon \sum_{i=2}^{n} r_i^{s+1} = 0\}$$

$$= \{\mathbf{p} + t\mathbf{r} \colon t \in \mathrm{GF}(s^2) \text{ and } R \in \mathcal{U}_{n-2}\} \cup \mathcal{U}_{n-2}.$$

This set of points is a cone with vertex P and base $\mathcal{U}_{n-2}$ by definition. $\qquad \square$

In the next part of the chapter we prove the analogous theorems for non-singular quadrics.

Theorem 4.46. *In* $\mathrm{PG}(n,q)$ *the form* $Q_n(\mathbf{X})$ *belonging to a non-singular quadratic variety* $\mathcal{Q}_n$ *can be transformed to one of the following canonical forms:*

- *If n is even, then* $Q_n(\mathbf{X}) = X_0 X_1 + X_2 X_3 + \cdots + X_{n-2} X_{n-1} + X_n^2.$

- *If n is odd, then either*

  - $Q_n(\mathbf{X}) = X_0 X_1 + X_2 X_3 + \cdots + X_{n-1} X_n$, *or*
  - $Q_n(\mathbf{X}) = X_0 X_1 + X_2 X_3 + \cdots + X_{n-3} X_{n-2} + f(X_{n-1}, X_n),$
    where f is an irreducible homogeneous quadratic polynomial over $\mathrm{GF}(q).$

Proof. We distinguish two cases according to the parity of q.

First, suppose that q is odd. In this case, by Lemma 4.25, the form can be transformed to $Q_n(\mathbf{X}) = \sum_{i=0}^{n} a_i X_i^2$. We may assume that a_i is a square if $0 \le i \le m$ and a_i is a non-square if $m < i \le n$. This means that we can fix an arbitrary non-square element e and find suitable elements b_i such that $a_i = b_i^2$ if $0 \le i \le m$ and $a_i = e b_i^2$ if $m < i \le n$. Thus

$$Q_n(\mathbf{X}) = \sum_{i=0}^{n} a_i X_i^2 = \sum_{i=0}^{m} b_i^2 X_i^2 + \sum_{i=m+1}^{n} e b_i^2 X_i^2.$$

Substituting $Y_i = b_i X_i$ we get

$$Q_n(\mathbf{X}) = \sum_{i=0}^{m} Y_i^2 + e \left(\sum_{i=m+1}^{n+1} Y_i^2 \right).$$

Let us call the pair $(m+1; n-m)$ the *quadratic pair* of the form. The quadratic pair can be changed by transformations. As

$$e Q_n(\mathbf{X}) = e \left(\sum_{i=0}^{m} Y_i^2 \right) + \sum_{i=m+1}^{n+1} (e Y_i)^2,$$

the quadratic pair $(k; l)$ of a form can be changed to $(l; k)$. It follows from Lemma 4.31 that there exist $c, d \in \mathrm{GF}(q)$ for which $e = c^2 + d^2$ holds. For a fixed pair of indices i and j let $Z_i = cY_i - dY_j$ and $Z_j = dY_i + cY_j$. Then

$$Z_i^2 + Z_j^2 = (c^2 + d^2)(Y_i^2 + Y_j^2) = e(Y_i^2 + Y_j^2).$$

Thus if a form has quadratic pair $(k; l)$, then it can be transformed to a form with quadratic pair $(k + 2; l - 2)$. It is easy to see that with the repeated applications of these two transformations the following quadratic pairs of the form are available:

- if n is even, then both of $(n; 1)$ and $(n/2; n/2 + 1)$;

- if n is odd, then either $(n+1; 0)$ or $(n; 1)$, and either $((n+1)/2; (n+1)/2)$ or $((n - 1)/2; (n + 3)/2)$, depending on the parities of the entries of the starting quadratic pair $(m + 1; n - m)$.

First, consider the n even case. If $q \equiv 3 \pmod 4$, then -1 is a non-square, hence we can choose $e = -1$ and take the form with quadratic pair $(n/2; n/2 + 1)$. Then

$$Q_n(\mathbf{X}) = \sum_{i=0}^{\frac{n}{2}-1} Z_i^2 - \sum_{i=\frac{n}{2}}^{n} Z_i^2.$$

For $i = 0, 1, \ldots, \frac{n}{2} - 1$ let $W_i = Z_i + Z_{i+\frac{n}{2}}$ and $W_{i+\frac{n}{2}} = Z_i - Z_{i+\frac{n}{2}}$. Then $W_i W_{i+\frac{n}{2}} = Z_i^2 - Z_{i+\frac{n}{2}}^2$, hence the form can be written as

$$Q_n(\mathbf{X}) = \sum_{i=0}^{\frac{n}{2}-1} Z_i^2 - \sum_{i=\frac{n}{2}}^{n} Z_i^2 = \sum_{i=0}^{\frac{n}{2}-1} W_i W_{i+\frac{n}{2}} - Z_n^2,$$

which gives the required form after multiplying by -1. If $q \equiv 1 \pmod 4$, then there exists $r \in \mathrm{GF}(q)$ with $r^2 = -1$. Take the form with quadratic pair $(n; 1)$ and for $i = 0, 1, \ldots, n/2 - 1$ let $W_i = Z_i + rZ_{i+\frac{n}{2}}$ and $W_{i+\frac{n}{2}} = Z_i - rZ_{i+\frac{n}{2}}$. Then $W_i W_{i+\frac{n}{2}} = Z_i^2 + Z_{i+\frac{n}{2}}^2$, hence the form can be written as

$$Q_n(\mathbf{X}) = \sum_{i=0}^{n-1} Z_i^2 + eZ_n^2 = \sum_{i=0}^{\frac{n}{2}-1} W_i W_{i+\frac{n}{2}} + eZ_n^2.$$

After multiplying with e and introducing the new variables $V_i = eW_i$ for $i \geq n/2$, we get the required form.

If n is odd, then the situation is more complicated, but the basic ideas are the same as for n even. In the case $q \equiv 3 \pmod 4$ choose $e = -1$. Then the form can be transformed to one of the forms with quadratic pair $((n + 1)/2; (n + 1)/2)$ or $((n - 1)/2; (n + 3)/2)$. After introducing the new variables

$W_i = Z_i + Z_{i+\frac{n-1}{2}}$ and $W_{i+\frac{n+1}{2}} = Z_i - Z_{i+\frac{n-1}{2}}$ in the first case for $i = 0, 1, \ldots, \frac{n-1}{2}$ and in the second case for $i = 0, 1, \ldots, \frac{n-3}{2}$, the form becomes

$$Q_n(\mathbf{X}) = \begin{cases} \sum_{i=0}^{\frac{n-1}{2}} Z_i^2 - \sum_{i=\frac{n+1}{2}}^{n} Z_i^2 = \sum_{i=0}^{\frac{n-1}{2}} W_i W_{i+\frac{n+1}{2}}, \\ \sum_{i=0}^{\frac{n-3}{2}} Z_i^2 - \sum_{i=\frac{n-1}{2}}^{n} Z_i^2 = \sum_{i=0}^{\frac{n-3}{2}} W_i W_{i+\frac{n+1}{2}} - (Z_{n-1}^2 + Z_n^2). \end{cases}$$

As $q \equiv 3 \pmod 4$ implies that the polynomial $f(z_{n-1}, z_n) = z_{n-1}^2 + z_n^2$ is irreducible, this is the required form.

If $q \equiv 1 \pmod 4$, then let $r^2 = -1$ again. Then the form can be transformed to one of the forms with quadratic pair $(n+1; 0)$ or $(n; 1)$. After introducing the new variables $W_i = Z_i + rZ_{i+\frac{n-1}{2}}$ and $W_{i+\frac{n+1}{2}} = Z_i - rZ_{i+\frac{n-1}{2}}$ in the first case for $i = 0, 1, \ldots, \frac{n-1}{2}$ and in the second case for $i = 0, 1, \ldots, \frac{n-3}{2}$, the form becomes

$$Q_n(\mathbf{X}) = \begin{cases} \sum_{i=0}^{n} Z_i^2 = \sum_{i=0}^{\frac{n-1}{2}} W_i W_{i+\frac{n+1}{2}}, \\ \sum_{i=0}^{n-1} Z_i^2 + eZ_n^2 = \sum_{i=0}^{\frac{n-3}{2}} W_i W_{i+\frac{n+1}{2}} + (Z_{n-1}^2 + eZ_n^2). \end{cases}$$

Now $q \equiv 1 \pmod 4$ implies that $-e$ is a non-square element, hence the polynomial $f(z_{n-1}, z_n) = z_{n-1}^2 + ez_n^2$ is irreducible. Thus this is the required form.

Now consider the q even case.

Step 1. First, we show that if a non-singular form $Q_k(\mathbf{X}) = \sum_{0 \le i \le j \le k} a_{ij} X_i X_j$ contains at least three variables, then we can eliminate X_0^2. If $a_{ii} = 0$ for some i, then we can do it by a permutation of the variables. Suppose that $a_{ii} \ne 0$ for all i. In $GF(2^r)$ the mapping $x \mapsto x^2$ is an automorphism, hence each element has a unique square root. If a_{ij} were 0 for all $i \ne j$, then

$$Q_k(\mathbf{X}) = \sum_{i=0}^{k} a_{ii} X_i^2 = \left(\sum_{i=0}^{k} \sqrt{a_{ii}} X_i \right)^2$$

would be singular. As there are at least three variables, we may assume without loss of generality that $a_{12} \ne 0$ also holds. Then

$$Q_k(\mathbf{X}) = a_{11} X_1^2 + X_1 (a_{01} X_0 + a_{12} X_2 + \cdots + a_{1k} X_k) + \sum_{i \le j,\, i \ne 1 \ne j} a_{ij} X_i X_j.$$

Let $Y_2 = a_{01} X_0 + a_{12} X_2 + \cdots + a_{1k} X_k$ and $Y_i = X_i$ for $i \ne 2$. As $a_{12} \ne 0$, the variable X_2 can be expressed as $X_2 = b_0 Y_0 + b_2 Y_2 + \cdots + b_k Y_k$, a linear combination of the Y_i's that does not contain Y_1. So we get

$$\begin{aligned} Q_k(\mathbf{X}) &= a_{11} Y_1^2 + Y_1 Y_2 + \sum_{i \le j,\, i \ne 1 \ne j} c_{ij} Y_i Y_j \\ &= c_{00} Y_0^2 + a_{11} Y_1^2 + Y_1 Y_2 + \sum_{i \le j,\, i \ne 1,\, 2 \le j} c_{ij} Y_i Y_j \end{aligned}$$

with suitable coefficients c_{ij}. The coefficient $a_{11} \neq 0$, hence there exists a $d \in \mathrm{GF}(2^r)$ such that $d^2 = c_{00}/a_{11}$. Now define $Z_1 = Y_1 + dY_0$ and $Z_i = Y_i$ for $i \neq 1$. Then $Y_1 = Z_1 + dZ_0$, hence

$$Q_k(\mathbf{X}) = c_{00}Z_0^2 + a_{11}(Z_1 + dZ_0)^2 + (Z_1 + dZ_0)Z_2 + \sum_{i \leq j,\, i \neq 1 \neq j} c_{ij}Z_iZ_j$$

$$= a_{11}Z_1^2 + dZ_0Z_2 + Z_1Z_2 + \sum_{i \leq j,\, i \neq 1 \neq j} c_{ij}Z_iZ_j.$$

Step 2. Thus we may assume that the original form does not contain X_0^2. The form is non-singular, so there exists a $k > 0$ for which $a_{0k} \neq 0$. By a suitable permutation of the variables we get $a_{01} \neq 0$, hence the form can be written as

$$Q_n(\mathbf{X}) = a_{01}X_0X_1 + \sum_{i \leq j,\, i,j=1}^{n} a_{ij}X_iX_j.$$

Now, put $Y_1 = a_{01}X_1 + a_{11}X_1 + a_{12}X_2 + \cdots + a_{1n}X_n$ and $Y_i = X_i$ for $i \neq 1$. Then

$$Q_n(\mathbf{X}) = Y_0Y_1 + \sum_{i \leq j,\, i,j=1}^{n} b_{ij}Y_iY_j$$

$$= (Y_0 + b_{11}Y_1 + b_{12}Y_2 + \cdots + b_{1n}Y_n)Y_1 + \sum_{i \leq j,\, i,j=2}^{n} b_{ij}Y_iY_j.$$

Finally, let $Z_0 = Y_0 + b_{11}Y_1 + b_{12}Y_2 + \cdots + b_{1n}Y_n$ and $Z_i = Y_i$ for $i > 0$. Then

$$Q_n(\mathbf{X}) = Z_0Z_1 + \sum_{i \leq j,\, i,j=2}^{n} b_{ij}Z_iZ_j.$$

We can repeat this process if the number of variables in the last sum is at least three, so finally we obtain

$$Q_n(\mathbf{X}) = \begin{cases} X_0X_1 + X_2X_3 + \cdots + X_{n-2}X_{n-1} + eX_n^2, & \text{if } n \text{ is even}, \\ X_0X_1 + X_2X_3 + \cdots + X_{n-3}X_{n-2} + f(X_{n-1}, X_n), & \text{if } n \text{ is odd}, \end{cases}$$

where f is a homogeneous quadratic polynomial. If n is even, then we get the required form after introducing the new variable $X_n' = \sqrt{e}X_n$. If n is odd and f is irreducible, this is the required form. If f is the product of two linear factors, say $(rx_{n-1} + sx_n)(tx_{n-1} + ux_n)$, then we get the required form after introducing the new variables $Y_{n-1} = rX_{n-1} + sX_n$ and $Y_n = tX_{n-1} + uX_n$. This proves the theorem. $\square$

It follows from the previous theorem that the number of projectively distinct non-singular quadrics in $\mathrm{PG}(n, q)$ is one if n is even and two if n is odd.

Definition 4.47. *A non-singular quadric in* $\mathrm{PG}(n, q)$ *is called a* parabolic *quadric if* $n > 2$ *even and a* conic *if* $n = 2$. *If* n *is odd and the corresponding canonical quadratic form can be written as*

$$X_0 X_1 + X_2 X_3 + \cdots + X_{n-3} X_{n-2} + X_{n-1} X_n,$$

or

$$X_0 X_1 + X_2 X_3 + \cdots + X_{n-3} X_{n-2} + f(X_{n-1}, X_n),$$

where f *is an irreducible polynomial, then the quadric is called* hyperbolic *or* elliptic, *respectively.*

Quadratic surfaces are interesting objects in their own right; they have many applications in other areas of mathematics. For example, we will see in Theorem 12.27 how one can construct extremal graphs from elliptic quadrics of $\mathrm{PG}(3, q)$.

Using the canonical forms one can calculate the number of k-dimensional subspaces lying on a non-singular quadric. We do it only for $k = 0$; for the general proof we refer to the book [88].

Theorem 4.48. *In* $\mathrm{PG}(n, q)$ *the number of points on a non-singular quadric* $\mathcal{Q}_n$ *is the following:*

- *If* n *is even, then the parabolic quadric contains*

  - $p(n, q) = \frac{q^n - 1}{q - 1}$

  points.

- *If* n *is odd, then the elliptic and the hyperbolic quadric contains*

  - $e(n, q) = \frac{(q^{(n+1)/2} + 1)(q^{(n-1)/2} - 1)}{q - 1},$
  - $h(n, q) = \frac{(q^{(n-1)/2} + 1)(q^{(n+1)/2} - 1)}{q - 1}$

  points, respectively.

Proof. We prove it by induction on the dimension. If $n = 1$, then the canonical equation of the hyperbolic quadric is $X_0 X_1 = 0$ and it contains two points, $(0 : 1)$ and $(1 : 0)$, while the canonical equation of the elliptic quadric is $f(X_0, X_1) = 0$ and it contains no points because f is irreducible. So $h(1, q) = 2$ and $e(1, q) = 0$; the statement is true in this case. If $n = 2$, then the conic with equation $X_0 X_1 + X_2^2 = 0$ contains $p(2, q) = q + 1$ points, because if $x_0 = 0$, then the only point is $(0 : 1 : 0)$, while if $x_0 \neq 0$, then the points of the conic can be coordinatized as $(1 : -t^2 : t)$ where t is an arbitrary element of $\mathrm{GF}(q)$.

We claim that all functions $p(n, q)$, $e(n, q)$ and $h(n, q)$ satisfy the recurrence relation

$$a(n + 2, q) = q a(n, q) + 1 + q^{n+1}.$$

The canonical equations in $\mathrm{PG}(n+2,q)$ can be written as

$$X_0 X_1 + F(X_2, X_3, \ldots, X_{n+2}) = 0, \tag{4.5}$$

where $F(X_2, X_3, \ldots, X_{n+2}) = 0$ is the canonical equation of the corresponding quadric in $\mathrm{PG}(n,q)$. First, consider the points of $\mathcal{Q}_{n+2}$ in the hyperplane Σ_∞ with equation $X_0 = 0$. In this hyperplane there are two types of points. As Equation (4.5) becomes $F(X_2, X_3, \ldots, X_{n+2}) = 0$, the coordinate-vector of a point of $\mathcal{Q}_{n+2} \cap \Sigma_\infty$ either has the form $(0 : t : x_2 : x_3 : \cdots : x_{n+2})$ with $F(x_2, x_3, \ldots, x_{n+2}) = 0$ and $t \in \mathrm{GF}(q)$, or $(0 : t : 0 : 0 : \cdots : 0)$. The number of points of the former type is $qa(n,q)$ because we have q choices for t, while there is a unique point of the latter type because of the homogeneity of the coordinates. The coordinate-vector of a point of $\mathrm{PG}(n+2,q) \setminus \Sigma_\infty$ can be written as $(1 : x_1 : x_2 : x_3 : \cdots : x_{n+2})$. This point is on $\mathcal{Q}_{n+2}$ if and only if $x_1 = -F(x_2, x_3, \ldots, x_{n+2})$ holds. We can choose $x_2, x_3, \ldots, x_{n+1}$ independently from $\mathrm{GF}(q)$, thus there are q^{n+1} points of this type.

Thus, by the induction hypothesis, it is enough to show the following:

- $\dfrac{q^{n+2}-1}{q-1} = q\dfrac{q^n-1}{q-1} + 1 + q^{n+1}$;

- $\dfrac{(q^{(n+3)/2}+1)(q^{(n+1)/2}-1)}{q-1} = q\dfrac{(q^{(n+1)/2}+1)(q^{(n-1)/2}-1)}{q-1} + 1 + q^{n+1}$;

- $\dfrac{(q^{(n-1)/2}+1)(q^{(n+1)/2}-1)}{q-1} = q\dfrac{(q^{(n+1)/2}+1)(q^{(n+3)/2}-1)}{q-1} + 1 + q^{n+1}$.

Elementary calculations prove these identities. $\square$

Let $\mathcal{Q}_n$ be a non-singular quadric and $Q_n(\mathbf{X})$ be the corresponding quadratic form. Let us define the bilinear form $G(\mathbf{p}, \mathbf{r})$ belonging to $\mathcal{Q}_n$ as

$$G(\mathbf{p}, \mathbf{r}) = Q_n(\mathbf{p} + \mathbf{r}) - Q_n(\mathbf{p}) - Q_n(\mathbf{r}).$$

Then elementary calculation shows that for all $t \in \mathrm{GF}(q)$

$$Q_n(\mathbf{p} + t\mathbf{r}) = Q_n(\mathbf{p}) + tG(\mathbf{p}, \mathbf{r}) + t^2 Q_n(\mathbf{r}).$$

The bilinear form $G(\mathbf{p}, \mathbf{r})$ helps us to determine the number of common points of a line and $\mathcal{Q}_n$.

Definition 4.49. *Let $\mathcal{Q}_n$ be a non-singular quadric. The line ℓ is a* tangent *to $\mathcal{Q}_n$ if $|\ell \cap \mathcal{Q}_n| = 1$, and ℓ is a* secant *of $\mathcal{Q}_n$ if $|\ell \cap \mathcal{Q}_n| = 2$.*

Lemma 4.50. *Let P be a point on the non-singular quadric $\mathcal{Q}_n$. Then the following hold:*

- *If $R \neq P$ is an arbitrary point and $G(\mathbf{p}, \mathbf{r}) \neq 0$, then the line PR is a secant of $\mathcal{Q}_n$.*

- *If $R \notin \mathcal{Q}_n$ is a point and $G(\mathbf{p}, \mathbf{r}) = 0$, then the line PR is a tangent to $\mathcal{Q}_n$.*

- If $P \neq R \in \mathcal{Q}$ is a point and $G(\mathbf{p}, \mathbf{r}) = 0$ then the line PR is completely contained in $\mathcal{Q}_n$.

Proof. Let R_t be the point with coordinate-vector $\mathbf{r} + t\mathbf{p}$. Then R_t is on the line PR and $R_t \neq P$. As

$$Q_n(\mathbf{r} + t\mathbf{p}) = Q_n(\mathbf{r}) + tG(\mathbf{r}, \mathbf{p}) + t^2 Q_n(\mathbf{p}) = Q_n(\mathbf{r}) + tG(\mathbf{r}, \mathbf{p}),$$

R_t is on $\mathcal{Q}_n$ if and only if

$$Q_n(\mathbf{r}) + tG(\mathbf{r}, \mathbf{p}) = 0. \tag{4.6}$$

If $G(\mathbf{r}, \mathbf{p}) \neq 0$, then Equation (4.6) has a unique solution, $t = -Q_n(\mathbf{r})/G(\mathbf{r}, \mathbf{p})$. This proves the first statement of the theorem. If $G(\mathbf{r}, \mathbf{p}) = 0$ and $Q_n(\mathbf{r}) \neq 0$, then Equation (4.6) has no solution, while if $G(\mathbf{r}, \mathbf{p}) = Q_n(\mathbf{r}) = 0$, then all $t \in \mathrm{GF}(q)$ satisfy Equation (4.6). These prove the second and third statements of the theorem. $\square$

Corollary 4.51. *Let $\mathcal{Q}_n$ be a quadric and ℓ be a line in $\mathrm{PG}(n, q)$. Then the intersection $\mathcal{Q}_n \cap \ell$ consists of 0, 1, 2 or $q + 1$ points.*

Proof. If $\mathcal{Q}_n$ is non-singular, then the statement follows immediately from the previous lemma. If $\mathcal{Q}_n$ is singular whose base is a non-singular quadric $\mathcal{Q}_r \subset \Sigma_r$ and vertex is an $(n - r - 1)$-dimensional subspace Σ_{n-r-1}, then there are four types of lines. Lines joining two points of Σ_{n-r-1} meet $\mathcal{Q}_n$ in $q + 1$ points. If $P \in \Sigma_{n-r-1}$ and $R \in \mathcal{Q}_r$, then the line PR lies on $\mathcal{Q}_n$, while if $R \in \Sigma_r \setminus \mathcal{Q}_r$, then the line PR contains only one point of $\mathcal{Q}_n$. Finally, if a line ℓ has empty intersection with Σ_{n-r-1}, then $\langle \ell, \Sigma_{n-r-1} \rangle \cap \Sigma_r$ is a line ℓ', and $|\ell \cap \mathcal{Q}_n| = |\ell' \cap \mathcal{Q}_n|$, so the statement follows from Lemma 4.50. $\square$

Definition 4.52. *A point P is called a* nucleus *of the non-singular quadric $\mathcal{Q}_n$ if $G(\mathbf{p}, \mathbf{r}) = 0$ holds for all points $R \in \mathrm{PG}(n, q)$.*

Theorem 4.53. *Let $\mathcal{Q}_n$ be a non-singular quadric in $\mathrm{PG}(n, q)$. Then $\mathcal{Q}_n$ has a nucleus if and only if both n and q are even. In this case, if $\mathcal{Q}_n$ is given with its canonical equation, as in Theorem 4.46, then the nucleus is the point $(0 : 0 : \dots, ; 0 : 1)$.*

Proof. Suppose that the point $A = (a_0 : a_1 : \dots : a_n)$ is a nucleus of $\mathcal{Q}_n$. We may assume that $\mathcal{Q}_n$ is given with its canonical equation, as in Theorem 4.46. Then the condition $G(\mathbf{a}, \mathbf{x}) = 0$ with $\mathbf{x} = (x_0 : x_1 : \dots : x_n)$ gives the following:

- If $\mathcal{Q}_n = \mathcal{P}_n$: $(a_0 x_1 + a_1 x_0) + \dots + (a_{n-2} x_{n-1} + a_{n-1} x_{n-2}) + 2 a_n x_n = 0.$

- If $\mathcal{Q}_n = \mathcal{H}_n$: $(a_0 x_1 + a_1 x_0) + \dots + (a_{n-1} x_n + a_n x_{n-1}) = 0.$

- If $\mathcal{Q}_n = \mathcal{E}_n$ and $f(x_{n-1}, x_n) = x_{n-1}^2 + x_{n-1} x_n + d x_n^2$:
  $(a_0 x_1 + a_1 x_0) + \dots + (a_{n-3} x_{n-2} + a_{n-2} x_{n-3}) + (2 a_{n-1} + a_n) x_{n-1} + (a_{n-1} + 2 d a_n) x_n = 0.$

Substituting the base points

$$E_i = (0 : 0 : \ldots : 0 : \underbrace{1}_{i} : 0 : \ldots : 0) \qquad (i = 0, 1, \ldots, n)$$

in these equations, we get $a_i = 0$ for all i in the cases of the hyperbolic and elliptic quadrics, while in the case of the parabolic quadric $a_i = 0$ if $i < n$ and $2a_n = 0$. Hence $\mathcal{Q}_n$ has no nucleus if $\mathcal{Q}_n$ is either hyperbolic or elliptic, and in the parabolic case the point $(0 : 0 : \ldots : 0 : 1)$ is a nucleus if and only if $2 = 0$ in $\mathrm{GF}(q)$, thus if and only if q is a power of 2. $\qquad \square$

Definition 4.54. *The points P and R are* conjugates *with respect to the non-singular quadric $\mathcal{Q}_n$ if $G(\mathbf{p}, \mathbf{r}) = 0$.*

If P is not a nucleus of $\mathcal{Q}_n$, then the equation $G(\mathbf{p}, \mathbf{X}) = 0$ defines a hyperplane H_P. This hyperplane is called the polar hyperplane *of P.*

If $P \in \mathcal{Q}_n$, then H_P is called the tangent hyperplane *to $\mathcal{Q}_n$ at P.*

The next statement immediately follows from Lemma 4.50.

Corollary 4.55. *Let $\mathcal{Q}_n$ be a non-singular quadric, $\Sigma_k \subset \mathcal{Q}_n$ be a k-dimensional subspace, $P \in \Sigma_k$ be a point and H_P be the tangent hyperplane to $\mathcal{Q}_n$ at P in $\mathrm{PG}(n, q)$. Then*

- *H_P contains Σ_k;*

- *a line PR is a secant of $\mathcal{Q}_n$ if and only if $R \notin H_P$.*

Lemma 4.56. *Let $\mathcal{Q}_n$ be a non-singular quadric. Then the following hold:*

- *If at least one of q and n is odd, then the mapping φ from $\mathrm{PG}(n, q)$ to its dual space defined as*

$$\varphi \colon P \mapsto H_P$$

is a polarity. If q is odd, then φ is an ordinary polarity and its self-conjugate points are the points of $\mathcal{Q}_n$. If q is even, then φ is a null polarity.

- *If both q and n are even, then the nucleus of $\mathcal{Q}_n$ is incident with all tangent hyperplanes of $\mathcal{Q}_n$.*

Proof. Define the $(n+1) \times (n+1)$ matrices $A_\mathcal{P}$, $A_\mathcal{H}$ and $A_\mathcal{E}$, according to the canonical equations of $\mathcal{Q}_n$, in the following way. First, let B be the matrix

$$B = \begin{pmatrix} 0 & 1 & 0 & \ldots & \ldots & \ldots & 0 \\ 1 & 0 & 0 & \ldots & \ldots & \ldots & 0 \\ 0 & 0 & 0 & 1 & 0 & \ldots & 0 \\ 0 & 0 & 1 & 0 & 0 & \ldots & \vdots \\ \vdots & & & \ddots & \ddots & \ddots & \vdots \\ 0 & 0 & \ldots & \ldots & 0 & 0 & 1 \\ 0 & 0 & \ldots & \ldots & 0 & 1 & 0 \end{pmatrix}$$

of size $n \times n$ if n is even and of size $(n-1) \times (n-1)$ if n is odd. Secondly, extend B by one row and one column if n is even as

$$A_{\mathcal{P}} = \begin{pmatrix} & & & 0 \\ & B & & \vdots \\ & & & 0 \\ 0 & \cdots & 0 & 2a_n \end{pmatrix},$$

and by two rows and two columns if n is odd as

$$A_{\mathcal{H}} = \begin{pmatrix} & & & 0 & 0 \\ & B & & \vdots & \vdots \\ & & & 0 & 0 \\ 0 & \cdots & 0 & 0 & 1 \\ 0 & \cdots & 0 & 1 & 0 \end{pmatrix} \quad \text{and} \quad A_{\mathcal{E}} = \begin{pmatrix} & & & 0 & 0 \\ & B & & \vdots & \vdots \\ & & & 0 & 0 \\ 0 & \cdots & 0 & 2 & 1 \\ 0 & \cdots & 0 & 1 & 2d \end{pmatrix}$$

where $f(x_{n-1}, x_n) = x_{n-1}^2 + x_{n-1}x_n + dx_n^2$ is an irreducible polynomial. Then φ acts as the multiplication by the corresponding matrix $A_{\mathcal{P}}$, $A_{\mathcal{H}}$ or $A_{\mathcal{E}}$. Elementary calculation shows that $\det B = \pm 1 \neq 0$, hence $\det A_{\mathcal{P}} = 2a_n$. Thus $\det A_{\mathcal{P}} = 0$ if and only if $2 = 0$, that is when q is even. If n is odd, then $\det A_{\mathcal{H}} = \pm 1 \neq 0$ and $\det A_{\mathcal{E}} = \pm 1(4d - 1) \neq 0$ because the polynomial f is irreducible. Each of the matrices $A_{\mathcal{P}}$, $A_{\mathcal{H}}$ and $A_{\mathcal{E}}$ is symmetric, and hence, by Theorem 4.22, defines a polarity if its determinant is non-zero. The polarity is ordinary if q is odd and it is a null polarity when q is even and n is odd. If q is odd, then a point P having representative vector $\mathbf{p}$ is self-conjugate if and only if $P \in H_P$, which happens if and only if $(\mathbf{p}A_{(.)})\mathbf{p}^{\mathrm{T}} = 0 \iff P \in \mathcal{Q}_n$.

If both q and n are even, then the nucleus of $\mathcal{Q}_n$ is the point $N = (0 : 0 : \cdots : 0 : 1)$. If P is a point on $\mathcal{Q}_n$ with representative vector $\mathbf{p}$ then the hyperplane H_P is defined by the vector $\mathbf{p}A_{\mathcal{P}}$ whose last coordinate is 0, because $2a_n = 0$. Hence $N \in H_P$ which proves the last statement of the theorem. $\qquad\square$

Definition 4.57. *The group of linear transformations fixing a non-singular quadric $\mathcal{Q}_n$ in $\mathrm{PG}(n, q)$, $n \geq 2$, is called the orthogonal group and it is denoted by $\mathrm{PGO}(n+1, q)$, $\mathrm{PGO}_+(n+1, q)$ or $\mathrm{PGO}_-(n+1, q)$ according to whether $\mathcal{Q}_n$ is parabolic, hyperbolic or elliptic.*

Theorem 4.58. *The orthogonal group acts transitively on the set of points of the corresponding non-singular quadric $\mathcal{Q}_n$.*

Proof. Let C be any point in $\mathrm{PG}(n, q) \setminus \mathcal{Q}_n$ which is different from the nucleus of $\mathcal{Q}_n$ (if it exists). Let $Q_n(\mathbf{X})$ be the quadratic form and $G(\mathbf{p}, \mathbf{r})$ be the bilinear form belonging to $\mathcal{Q}_n$. Define the mapping φ_C on the set of points of $\mathrm{PG}(n, q)$ as

$$\varphi_C : \mathbf{x} \mapsto Q_n(\mathbf{c})\mathbf{x} - G(\mathbf{x}, \mathbf{c})\mathbf{c}.$$

The mapping is well-defined because the bilinearity of G implies that $(\alpha \mathbf{x})^{\varphi_C} = \alpha(\mathbf{x}^{\varphi_C})$. No line through C is on $\mathcal{Q}_n$, hence if a line ℓ through C meets $\mathcal{Q}_n$, then, by Corollary 4.51, it contains 1 or 2 points of $\mathcal{Q}_n$. If ℓ is a tangent to $\mathcal{Q}_n$ at the point P, then, by Lemma 4.50, $G(\mathbf{p}, \mathbf{c}) = 0$. Thus P is a fixed point of φ_C. If ℓ is a secant of $\mathcal{Q}_n$ and $\ell \cap \mathcal{Q}_n = \{A, B\}$, then, again by Lemma 4.50, $G(\mathbf{a}, \mathbf{b}) \neq 0$. The points A, B and C are collinear and distinct, hence there is a unique $0 \neq t \in \mathrm{GF}(q)$ such that $\mathbf{b} = \mathbf{a} + t\mathbf{c}$. This means that

$$Q_n(\mathbf{b}) = Q_n(\mathbf{a} + t\mathbf{c}) = Q_n(\mathbf{a}) + tG(\mathbf{a}, \mathbf{c}) + t^2 Q_n(\mathbf{c}).$$

As A and B are on $\mathcal{Q}_n$ we get

$$0 = Q_n(\mathbf{b}) = tG(\mathbf{a}, \mathbf{c}) + t^2 Q_n(\mathbf{c}),$$

hence $t \neq 0$ implies

$$t = -\frac{G(\mathbf{a}, \mathbf{c})}{Q_n(\mathbf{c})}, \quad \text{so } \mathbf{b} = \mathbf{a} - \frac{G(\mathbf{a}, \mathbf{c})}{Q_n(\mathbf{c})}\mathbf{c}.$$

Thus, by definition and by the bilinearity of G, we have

$$\mathbf{b}^{\varphi_C} = Q_n(\mathbf{c})\mathbf{b} - G(\mathbf{b}, \mathbf{c})\mathbf{c}$$

$$= Q_n(\mathbf{c})\mathbf{a} - G(\mathbf{a}, \mathbf{c})\mathbf{c} - G\left(\mathbf{a} - \frac{G(\mathbf{a}, \mathbf{c})}{Q_n(\mathbf{c})}\mathbf{c}, \mathbf{c}\right)\mathbf{c}$$

$$= Q_n(\mathbf{c})\mathbf{a} - 2G(\mathbf{a}, \mathbf{c})\mathbf{c} + \frac{G(\mathbf{a}, \mathbf{c})}{Q_n(\mathbf{c})}G(\mathbf{c}, \mathbf{c})\mathbf{c}.$$

If q is even, then $G(\mathbf{c}, \mathbf{c}) = 0$, while if q is odd, then $G(\mathbf{c}, \mathbf{c}) = 4Q_n(\mathbf{c})$, so in both cases $\mathbf{b}^{\varphi_C} = Q_n(\mathbf{c})\mathbf{a}$; thus φ_C maps B to A. The role of A and B is symmetric, hence φ_C maps B to A, too. This means that φ_C changes A and B, and so fixes $\mathcal{Q}_n$ setwise. The mapping φ_C is a linear transformation by definition. As its image contains all points of $\mathcal{Q}_n$, the dimension of its image is n, so φ is non-singular; it is an element of the orthogonal group.

Finally, let P and R be two distinct points of $\mathcal{Q}_n$. If the line PR is a secant of $\mathcal{Q}_n$, then take a point C on the line PR different from both P and R. Then φ_C maps P to R. If the line PR lies on $\mathcal{Q}_n$, then let $S \in \mathcal{Q}_n$ be a point such that both PS and RS are secants to $\mathcal{Q}_n$. The point S exists because $\mathcal{Q}_n$ is non-singular. Take the points $P \neq C \neq S$ on the line PS and $R \neq C' \neq S$ on the line RS. Then the linear transformation $\varphi_C \varphi_{C'}$ maps P to R, which proves the theorem. $\qquad\square$

We say that two quadrics have the *same character* if they are both parabolic, both hyperbolic or both elliptic.

Theorem 4.59. *Let $\mathcal{Q}_n$ be a non-singular quadric, $P \in \mathcal{Q}_n$ be a point and H_P be the tangent hyperplane to $\mathcal{Q}_n$ at P in $\mathrm{PG}(n, q)$. If $n \geq 2$, then $\mathcal{Q}_n \cap H_P$ is a cone $\mathcal{K}$. The vertex of $\mathcal{K}$ is P, its base is a non-singular quadric $\mathcal{Q}_{n-2}$, and the quadrics $\mathcal{Q}_n$ and $\mathcal{Q}_{n-2}$ have the same character.*

Proof. If a linear transformation φ fixes $\mathcal{Q}_n$ and maps a point $P \in \mathcal{Q}_n$ to $P' \in \mathcal{Q}_n$, then φ maps H_P to $H_{P'}$. Thus, by Theorems 4.46 and 4.58, we may assume without loss of generality that $\mathcal{Q}_n$ is given by its canonical equation and P is the point $(1 : 0 : 0 : \cdots : 0)$. Let $X_0 X_1 + F(X_2, \ldots, X_n) = 0$ be a shorthand notation of the canonical equation. Then elementary calculation gives that H_P has equation $X_1 = 0$, thus

$$
\begin{aligned}
\mathcal{Q}_n \cap H_P =& \{(1 : 0 : tr_2 : tr_3 : \cdots : tr_n) \colon t \in \mathrm{GF}(q) \text{ and } F(r_2, r_3, \ldots, r_n) = 0\} \\
& \cup \{(0 : 0 : r_2 : \cdots : r_n)\} \\
=& \{\mathbf{p} + t\mathbf{r} \colon t \in \mathrm{GF}(q) \text{ and } R \in \mathcal{Q}_{n-2}\} \cup \mathcal{Q}_{n-2},
\end{aligned}
$$

where the non-singular quadric $\mathcal{Q}_{n-2}$ is defined by the equation $F(X_2, \ldots, X_n) = 0$. This set of points is a cone with vertex P and base $\mathcal{Q}_{n-2}$ by definition. As, by Theorem 4.46, $\mathcal{Q}_{n-2}$ and $\mathcal{Q}_n$ have the same character, the theorem is proved. $\qquad\square$

Definition 4.60. *A subspace of maximum dimension on a quadric $\mathcal{Q}_n$ is called a* generator *and its dimension g is called the* projective index of $\mathcal{Q}_n$. *If $\mathcal{Q}_n$ is non-singular, then its* character w *is defined by $w = 2g - n + 3$.*

Theorem 4.61. *Let $n > 2$, $\mathcal{Q}_n$ and $\mathcal{Q}_{n-2}$ be non-singular quadrics of the same character and suppose that $\mathcal{Q}_n$ has projective index g and character w. Then*

- *any point of $\mathcal{Q}_n$ lies in a generator;*

- *the projective index of $\mathcal{Q}_{n-2}$ is $g - 1$;*

- *$w = 1$, 2 or 0, according as $\mathcal{Q}_n$ is parabolic, hyperbolic or elliptic quadric.*

Proof. First we prove Part 1. Let $P \in \mathcal{Q}_n$ be any point and Σ_g be a generator of $\mathcal{Q}_n$. If $P \in \Sigma_g$, then there is nothing to prove. If $P \notin \Sigma_g$, then $H_P \cap \Sigma_g$ is a subspace Σ_{g-1}. By Theorem 4.59, H_P contains the cone with vertex P and base Σ_{g-1}; as this cone is a subspace of dimension $(g - 1) + 1 = g$, the point P lies in a generator.

The previous argument also shows that the projective index of $\mathcal{Q}_{n-2}$ is at least $g - 1$. If its projective index were greater than $g - 1$, then there would be a subspace $\Sigma_{g'} \subset \mathcal{Q}_{n-2}$ with $g' > g$. Hence the join of P and $\Sigma_{g'}$ would be a subspace on $\mathcal{Q}_n$ having dimension at least $g + 1$, a contradiction.

Finally, Part 3 follows from the knowledge of the projective indices in low dimensions. If $n = 1$, then the hyperbolic quadric contains 2 points, thus its projective index is 0, while the elliptic quadric has no point, so its projective index is (-1). This gives that the projective index of $\mathcal{H}_{2k+1}$ is k and the projective index of $\mathcal{E}_{2k+1}$ is $k - 1$ for all $k > 0$. If $n = 2$, then a conic contains $q + 1$ points, but no lines, hence its projective index is 0. This implies that the projective index of $\mathcal{P}_{2k}$ is $k - 1$ for all $k \geq 1$. Putting these values to the formula $w = 2g - n + 3$ we get the last statement of the theorem. $\qquad\square$

In particular, if $n = 3$, then an elliptic quadric $\mathcal{E}_3$ does not contain any line, hence, by Corollary 4.51, no three of its points are collinear. A hyperbolic quadric $\mathcal{H}_3$ has projective index 1, hence, by Part 1 of Theorem 4.61, there are some generator lines through any of its points. In the next theorem we give a geometric description of the generators on $\mathcal{H}_3$.

Theorem 4.62. *Let $\mathcal{H}_3$ be a hyperbolic quadric in $\mathrm{PG}(3, q)$. Then $\mathcal{H}_3$ contains $2(q+1)$ generators. These lines can be divided into two sets, $\mathcal{R}_1$ and $\mathcal{R}_2$, such that each of them contains $q + 1$ pairwise skew generators and any line from one set intersects every line from the other set. If $P \in \mathcal{H}_3$ is any point, then exactly one element of $\mathcal{R}_1$ and exactly one element of $\mathcal{R}_2$ contains P.*

Proof. Let ℓ be a generator and $P_i \in \ell$ be a point of $\mathcal{H}_3$. Let H_{P_i} be the tangent plane to $\mathcal{H}_3$ at P_i. As $\mathcal{H}_1$ consists of two distinct points, Theorem 4.59 gives that $H_{P_i} \cap \mathcal{H}_3$ is a pair of two lines meeting in P_i. One of these lines is ℓ; let us denote the other one by e_i.

We claim that if $P_i \neq P_j$ are distinct points on ℓ, then e_i and e_j are skew lines. Suppose to the contrary that e_i and e_j are coplanar; let $R = e_i \cap e_j$ be their point of intersection. Let T be an arbitrary point in the plane $\Sigma_2 = \langle P_i, P_j, R \rangle$ which is not contained in the set $\ell \cup e_i \cup e_j$. Let g be a line in Σ_2 through T and distinct from each of the lines ℓ, e_i and e_j. Then $g \cap \mathcal{H}_3$ contains the distinct points $g \cap \ell$, $g \cap e_i$ and $g \cap e_j$, hence, by Corollary 4.51, g is contained in $\mathcal{H}_3$, so the whole plane Σ_2 is contained in $\mathcal{H}_3$, a contradiction. (If $q = 2$, then this argument does not work. In this case Σ_2 consists of seven points, six of them contained in $\ell \cup e_i \cup e_j$. We leave it as an exercise to prove that the seventh point must belong to $\mathcal{H}_3$, too.)

There are $q+1$ points on the line ℓ, hence, in this way, we have found a set of $q + 1$ pairwise skew generators, $\mathcal{R}_1 = \{e_1, e_2, \ldots, e_{q+1}\}$, of $\mathcal{H}_3$. These lines altogether contain $(q + 1)^2$ points which is the total number of points on $\mathcal{H}_3$ by Theorem 4.48. Thus there is a unique element of $\mathcal{R}_1$ through each point of $\mathcal{H}_3$. If $R_i \in \mathcal{H}_3$ is any point, then let f_{R_i} be the other line of $H_{R_i} \cap \mathcal{H}_3$, that is the line not contained in $\mathcal{R}_1$. As f_{R_i} is contained in $\mathcal{H}_3$, each element of $\mathcal{R}_1$ intersects f_{R_i} in a unique point. This implies that if f_{R_i} and f_{R_j} are distinct lines, then they are skew, because if they were coplanar, then any two elements of $\mathcal{R}_1$ would also be coplanar. Hence the lines of type f_{R_i} are pairwise skew and their union covers $\mathcal{H}_3$. This means that there are $(q + 1)^2/(q + 1)$ lines of this type. Let $\mathcal{R}_2$ denote this set of lines. Then $\mathcal{R}_1$ and $\mathcal{R}_2$ obviously satisfy the requirements of the theorem. $\square$

Corollary 4.63. *Let $\mathcal{H}_3$ be a hyperbolic quadric in $\mathrm{PG}(3, q)$, $\mathcal{R}_1$ and $\mathcal{R}_2$ be its two sets of generators and Π be any plane. Then $\mathcal{H}_3 \cap \Pi$ either contains two lines, one from $\mathcal{R}_1$ and one from $\mathcal{R}_2$, or it is a conic.*

Proof. There are $q+1$ planes through any element of $\mathcal{R}_1$, hence each of these planes contains a unique element of $\mathcal{R}_2$ and vice versa. So the total number of planes containing two generators is $(q + 1)^2$.

For a point $P \in \mathcal{H}_3$ let Σ_P denote the polar plane of P. By Corollary 4.55, Σ_P contains two generators and $q - 1$ tangent lines to $\mathcal{H}_3$; all other lines through P are secants to $\mathcal{H}_3$. Hence all planes intersecting Σ_P in a tangent line must intersect $\mathcal{H}_3$ in a conic. The number of this type of planes is $(q-1)q = q^2 - q$. A conic consists of $q+1$ points, thus the total number of planes intersecting $\mathcal{H}_3$ in a conic is $(q+1)^2(q^2-q)/(q+1) = q^3 - q$.

The sum $(q+1)^2 + (q^3-q)$ equals to the number of planes, which completes the proof. □

The two sets of generators of $\mathcal{H}_3$ are closely connected to the classical Theorem of Gallucci about the 16 points of intersections of skew lines in a 3-dimensional projective space. This theorem plays an important role in the theory of coordinatization of projective planes. Recently, using this theorem, *Bamberg* and *Penttila* [14] gave a geometric proof of Wedderburn's little theorem which states that every finite division ring is commutative. As a consequence, they obtained a geometric proof that a finite Desarguesian projective space is Pappian. For the sake of completeness we present a proof of Gallucci's theorem.

Theorem 4.64 (Gallucci). *Let* $\{e_1, e_2, e_3, e_4\}$ *and* $\{f_1, f_2, f_3, f_4\}$ *be two sets of pairwise skew lines in* $\mathrm{PG}(3,\mathbb{K})$. *Consider the possible 16 points of intersection of type* $e_i \cap f_j$. *If 15 of these points exist, then the 16th point also exists.*

Proof. We may assume without loss of generality that $P_{ij} = e_i \cap f_j$ exists if at least one of i and j is less than 4. As e_1, e_2, e_3 and e_4 are pairwise skew, the points $P_{11}, P_{12}, P_{21}, P_{22}$ and P_{33} are in general position, hence we may also assume that $P_{11} = (1:0:0:0)$, $P_{12} = (0:1:0:0)$, $P_{21} = (0:0:1:0)$, $P_{22} = (0:0:0:1)$ and $P_{33} = (1:1:1:1)$ (see Figure 4.1).

As the point P_{13} is on the line e_1 and the point P_{23} is on the line e_2, there exist $\alpha \neq 0 \neq \beta$ elements of $\mathbb{K}$ such that $P_{13} = (1:\alpha:0:0)$ and $P_{23} = (0:0:1:\beta)$. The point P_{33} is on the line $P_{13}P_{23}$, hence there exist $\gamma, \delta \in \mathbb{K}$ such that

$$(1:1:1:1) = \gamma \cdot (1:\alpha:0:0) + \delta \cdot (0:0:1:\beta) = (\gamma : \gamma\alpha : \delta : \delta\beta).$$

From this equation we get $\alpha = \beta = 1$, thus $P_{13} = (1:1:0:0)$ and $P_{23} = (0:0:1:1)$. In the same way, using the points on the line f_3, we get $P_{31} = (1:0:1:0)$ and $P_{32} = (0:1:0:1)$.

The points on the line f_4 can be expressed as linear combinations of points on the lines f_1 and f_2, thus there are non-zero elements κ, λ and μ such that $P_{14} = (1:\kappa:0,:0)$, $P_{24} = (0:0:1:\lambda)$ and $P_{34} = (1:1:\mu:\mu)$. As these three points are collinear, we get

$$(1:1:\mu:\mu) = \nu \cdot (1:\kappa:0:0) + \rho \cdot (0:0:1:\lambda) - (\nu : \nu\kappa : \rho : \rho\lambda)$$

with suitable $\nu, \rho \in \mathbb{K}$. Hence $\kappa = \lambda$, which implies $P_{14} = (1:\kappa:0:0)$, $P_{24} = (0:0:1:\kappa)$ and $P_{34} = (1:\kappa:1:\kappa)$. In the same way, there

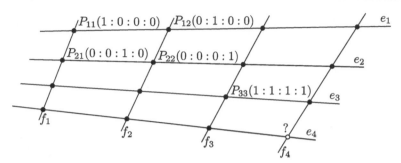

FIGURE 4.1
Gallucci's theorem

exists $0 \neq \tau \in \mathbb{K}$ such that $P_{41} = (1 : 0 : \tau : 0)$, $P_{42} = (0 : 1 : 0 : \tau)$ and $P_{43} = (1 : 1 : \tau : \tau)$. As

$$(1 : \kappa : \tau : \kappa\tau) = (1 : \kappa : 0 : 0) + \tau \cdot (0 : 0 : 1 : \kappa) = (1 : 0 : \tau : 0) + \kappa \cdot (0 : 1 : 0 : \tau),$$

this means that the point $(1 : \kappa : \tau : \kappa\tau)$ lies on both lines e_4 and f_4, hence the point of intersection $e_4 \cap f_4$ exists. □

Lemma 4.65. *Let e and f be two skew lines and P be a point in $\mathrm{PG}(3, \mathbb{K})$. If $P \notin e \cup f$ then there is a unique line through P which meets both e and f.*

Proof. As e and f are skew lines and $P \notin e \cup f$, the line f meets the plane $\langle P, e \rangle$ in a unique point, say R. Then the distinct lines PR and e are coplanar, hence they have a unique point of intersection. If $e \cap PR = S$, then the line $g = PRS$ satisfies the conditions of the lemma. If another line through P were also to meet both e and f, then, by Lemma 4.4, e and f would be coplanar, giving a contradiction. □

Theorem 4.66. *In $\mathrm{PG}(3, \mathbb{K})$ for any three pairwise skew lines there exists a unique hyperbolic quadric containing each of them.*

Proof. Let e_1, e_2 and e_3 be three pairwise skew lines. Take three distinct points, F_1, F_2 and P_{33} on e_3. By Lemma 4.65, there exist lines f_1, f_2 and f_3 such that for $i = 1, 2$ and $j = 1, 2, 3$ the points $P_{ij} = e_i \cap f_j$ exist, f_i contains the point F_i and f_3 contains the point P_{33}. After a suitable linear transformation we may assume without loss of generality that $P_{11} = (1 : 0 : 0 : 0)$, $P_{12} = (0 : 1 : 0 : 0)$, $P_{21} = (0 : 0 : 1 : 0)$, $P_{22} = (0 : 0 : 0 : 1)$ and $P_{33} = (1 : 1 : 1 : 1)$. Then the same calculation as in the proof of Gallucci's theorem gives that $P_{13} = (1 : 1 : 0 : 0)$, $P_{23} = (0 : 0 : 1 : 1)$, $F_1 = (1 : 0 : 1 : 0)$ and $F_2 = (0 : 1 : 0 : 1)$. This means that the equations of the lines are the following:

$$e_1 : X_2 = X_3 = 0, \quad e_2 : X_0 = X_1 = 0, \quad e_3 : X_0 = X_2 \text{ and } X_1 = X_3. \quad (4.7)$$

Suppose that a quadric $\mathcal{H}$ with equation

$$\sum_{i=0}^{3} a_i X_i^2 + \sum_{i<j,\,i,j=0}^{3} b_{ij} X_i X_j = 0 \qquad (4.8)$$

contains the lines e_1, e_2 and e_3. As for $i,j = 1,2$ the points P_{ij} are on $\mathcal{H}$; we get $a_0 = a_1 = a_2 = a_3 = 0$. Substituting the coordinates of the points $P_{13} = (1:1:0:0)$, $P_{23} = (0:0:1:1)$, $F_1 = (1:0:1:0)$ and $F_2 = (0:1:0:1)$ to Equation (4.8) we get $b_{01} = b_{23} = b_{02} = b_{13} = 0$. So the equation of $\mathcal{H}$ can be written as $b_{03} X_0 X_3 + b_{12} X_1 X_2 = 0$. Finally, as $P_{33} \in \mathcal{H}$, we get $b_{03} + b_{12} = 0$.

Hence only the hyperbolic quadric with equation $X_0 X_3 - X_1 X_2 = 0$ could contain the lines e_1, e_2 and e_3. Using Equations (4.7), elementary calculation shows that this quadric contains the three lines which proves the theorem. $\square$

The set of $q + 1$ lines meeting each of three pairwise skew lines is called a *regulus*. Generators of higher dimensional quadrics and reguli will also be considered in Chapter 5.

The detailed description of unitary and orthogonal groups is beyond the scope of this book. We only mention their basic properties. For a detailed study of them we refer the reader to Sections 22, 23 and 26 of the book [93].

Theorem 4.67. *Let $\mathcal{U}_n$ be a non-singular Hermitian variety in $\mathrm{PG}(n, s^2)$. Then the order of the unitary group fixing $\mathcal{U}_n$ is*

$$|\mathrm{PGU}(n+1, s^2)| = s^{n(n+1)/2} \prod_{i=2}^{n+1} (s^i - (-1)^i).$$

Corollary 4.68. *There are*

$$\frac{|\mathrm{PGL}(n+1, s^2)|}{|\mathrm{PGU}(n+1, s^2)|} = s^{n(n+1)/2} \prod_{i=2}^{n+1} (s^i + (-1)^i)$$

non-singular Hermitian varieties in $\mathrm{PG}(n, s^2)$.

Theorem 4.69. *Let $n > 1$ be even and $\mathcal{P}_n$ be a non-singular parabolic quadric in $\mathrm{PG}(n, q)$. Then the order of the orthogonal group fixing $\mathcal{P}_n$ is*

$$|\mathrm{PGO}(n+1, q)| = q^{n^2/4} \prod_{i=1}^{n/2} (q^{2i} - 1).$$

Corollary 4.70. *If $n > 1$ is even, then there are*

$$\frac{|\mathrm{PGL}(n+1, q)|}{|\mathrm{PGO}(n+1, q)|} = q^{n(n+2)/4} \prod_{i=1}^{n/2} (q^{2i+1} - 1)$$

non-singular parabolic quadrics in $\mathrm{PG}(n, q)$.

Theorem 4.71. *Let $n > 1$ be odd and $\mathcal{H}_n$ be a non-singular hyperbolic quadric in* $\mathrm{PG}(n, q)$. *Then the order of the orthogonal group fixing $\mathcal{H}_n$ is*

$$|\mathrm{PGO}_+(n+1, q)| = 2q^{(n^2-1)/4}(q^{(n+1)/2} - 1) \prod_{i=1}^{(n-1)/2} (q^{2i} - 1).$$

Corollary 4.72. *If $n > 1$ is odd, then there are*

$$\frac{|\mathrm{PGL}(n+1, q)|}{|\mathrm{PGO}_+(n+1, q)|} = \frac{q^{(n+1)^2/4}(q^{(n+1)/2} + 1)}{2} \prod_{i=1}^{(n-1)/2} (q^{2i+1} - 1)$$

non-singular hyperbolic quadrics in $\mathrm{PG}(n, q)$.

Theorem 4.73. *Let $n > 1$ be odd and $\mathcal{H}_n$ be a non-singular elliptic quadric in* $\mathrm{PG}(n, q)$. *Then the order of the orthogonal group fixing $\mathcal{H}_n$ is*

$$|\mathrm{PGO}_-(n+1, q)| = 2q^{(n^2-1)/4}(q^{(n+1)/2} + 1) \prod_{i=1}^{(n-1)/2} (q^{2i} - 1).$$

Corollary 4.74. *If $n > 1$ is odd, then there are*

$$\frac{|\mathrm{PGL}(n+1, q)|}{|\mathrm{PGO}_-(n+1, q)|} = \frac{q^{(n+1)^2/4}(q^{(n+1)/2} - 1)}{2} \prod_{i=1}^{(n-1)/2} (q^{2i+1} - 1)$$

non-singular elliptic quadrics in $\mathrm{PG}(n, q)$.

For the sake of completeness we give the canonical form of a null polarity in $\mathrm{PG}(n, q)$. The proof of the following theorem is similar to the proof of Theorem 4.25, so we skip it.

Theorem 4.75. *The matrix of a null polarity in* $\mathrm{PG}(n, q)$ *with n odd has canonical form*

$$B = \begin{pmatrix} 0 & 1 & 0 & \cdots & \cdots & \cdots & 0 \\ -1 & 0 & 0 & \cdots & \cdots & \cdots & 0 \\ 0 & 0 & 0 & 1 & 0 & \cdots & 0 \\ 0 & 0 & -1 & 0 & 0 & \cdots & \vdots \\ \vdots & & & \ddots & \ddots & \ddots & \vdots \\ 0 & 0 & \cdots & \cdots & 0 & 0 & 1 \\ 0 & 0 & \cdots & \cdots & 0 & -1 & 0 \end{pmatrix}.$$

Let us remark that if q is even, then $-1 = 1$, hence the canonical matrix of a null polarity is the same as the canonical matrix of a hyperbolic quadric, which is consistent with the fact that a hyperbolic quadric defines a null polarity in this case.

Exercises

4.1. Let Σ_{k-1}^1 and Σ_{k-1}^2 be two skew $(k-1)$-dimensional subspaces and P be a point in $\mathrm{PG}(2k-1, \mathbb{K})$. Show that if $P \notin \Sigma_{k-1}^1 \cup \Sigma_{k-1}^2$ then there is a unique line through P which meets both Σ_{k-1}^1 and Σ_{k-1}^2.

4.2. Let $\{\ell_1, \ell_2, \ldots, \ell_m\}$ be a set of lines in $\mathrm{PG}(n, \mathbb{K})$. Determine the possible values of the dimension of the subspace $\langle \ell_1, \ell_2, \ldots, \ell_m \rangle$.

4.3. Prove that any singular quadric in $\mathrm{PG}(2, q)$, q odd, is one of the following:

- a line,
- a single point,
- two intersecting lines.

4.4. Prove that any singular quadric in $\mathrm{PG}(3, q)$, q odd, is one of the following:

- a plane,
- a line,
- two intersecting planes,
- a quadratic cone whose vertex is a point and whose base is a conic.

4.5. Show that if a quadric $\mathcal{Q}$ has equation $\mathbf{X}A\mathbf{X}^{\mathrm{T}} = 0$ and B is the matrix of a non-singular linear transformation ϕ, then the equation of $\mathcal{Q}^\phi$ is $\mathbf{X}B^{-1}A(B^{-1})^{\mathrm{T}}\mathbf{X}^{\mathrm{T}} = 0$.

4.6. Show that in $\mathrm{PG}(n, q)$, q and n odd, if a non-singular quadric $\mathcal{Q}$ has equation $\mathbf{X}A\mathbf{X}^{\mathrm{T}} = 0$, then $\mathcal{Q}$ is hyperbolic or elliptic according as $(-1)^{(n+1)/2} \det A$ is a square or a non-square element in $\mathrm{GF}(q)$.

4.7. Let $\mathcal{Q}_3$ be a non-singular quadric in $\mathrm{PG}(3, q)$, q odd, and α denote the ordinary polarity defined by $\mathcal{Q}_3$. Show that if the line ℓ is a secant to $\mathcal{Q}_3$, then ℓ^α is

- a secant to $\mathcal{Q}_3$ if $\mathcal{Q}_3$ is a hyperbolic quadric;
- an external line to $\mathcal{Q}_3$ if $\mathcal{Q}_3$ is an elliptic quadric.

4.8. Suppose that a non-singular quadric $\mathcal{Q}_3$ in $\mathrm{PG}(3, q)$, q odd, contains the base points $E_1 = (1 : 0 : 0 : 0)$ and $E_2 = (0 : 1 : 0 : 0)$, and the polar line of $E_1 E_2$ in the ordinary polarity defined by $\mathcal{Q}_3$ is the line $E_3 E_4$. Show that the equation of $\mathcal{Q}_3$ can be written as

$$X_0 X_1 + a_{22}X_2^2 + 2a_{23}X_2 X_3 + a_{33}X_3^2 = 0.$$

4.9. Give an alternative proof of Theorem 4.46 in the case q odd and $n = 3$.

4.10. Let $\mathcal{Q}_3$ be an elliptic quadric in $\mathrm{PG}(3,q)$, q odd. Show that the equation of $\mathcal{E}_3$ can be written as

$$X_0 X_1 + a X_2^2 - (a+1) X_3^2 = 0$$

where $a \in \mathrm{GF}(q)$ and $\frac{a}{a+1}$ is a non-square element.

5

Higher dimensional representations

There are several structures whose geometric and combinatorial properties are easier to study if they are embedded into higher dimensional spaces. In this chapter some of these objects are investigated.

In $\mathrm{PG}(3, \mathbb{K})$ not only the points and planes but also the lines can be represented by homogeneous coordinates. These coordinates were introduced by *Julius Plücker* in 1846 and they are named after him. These are predecessors and special cases of *Grassmann coordinates* which describe k-dimensional subspaces in an n-dimensional projective space. The method works for an arbitrary field $\mathbb{K}$, but we consider only the case $\mathbb{K} = \mathrm{GF}(q)$.

Definition 5.1. *Let ℓ be a line in $\mathrm{PG}(3, q)$ through the distinct points X and Y. If $\mathbf{x} = (x_0 : x_1 : x_2 : x_3)$ and $\mathbf{y} = (y_0 : y_1 : y_2 : y_3)$ are homogeneous coordinates of X and Y, respectively, then the Plücker coordinates of the line ℓ are*

$$\mathbf{p} = (p_{01} : p_{02} : p_{03} : p_{23} : p_{31} : p_{12}),$$

where $p_{ij} = x_i y_j - x_j y_i$ for all $0 \leq i, j \leq 3$.

Let us notice that the Plücker coordinates correspond to 2×2 determinants, because

$$p_{ij} = \begin{vmatrix} x_i & x_j \\ y_i & y_j \end{vmatrix}.$$

The next proposition summarizes the most important properties of these coordinates.

Proposition 5.2.

1. *The Plücker coordinates are homogeneous coordinates; they are determined up to a scalar multiple.*

2. *The Plücker coordinates of a line do not depend on the choice of the two points on the line.*

3. *If $(p_{01} : p_{02} : p_{03} : p_{23} : p_{31} : p_{12})$ are Plücker coordinates of a line, then*

$$p_{01}p_{23} + p_{02}p_{31} + p_{03}p_{12} = 0.$$

4. *If at least one of the field elements p_{01}, p_{02}, p_{03}, p_{23}, p_{31} and p_{12} is non-zero and $p_{01}p_{23} + p_{02}p_{31} + p_{03}p_{12} = 0$, then there is a line whose Plücker coordinates are $(p_{01} : p_{02} : p_{03} : p_{23} : p_{31} : p_{12})$.*

Proof. We prove the first and second statements simultaneously. If X and Y are two distinct points in $PG(3, q)$ having coordinate-vectors $\mathbf{x}$ and $\mathbf{y}$, then the coordinate-vector of an arbitrary point on the line XY can be written as $\alpha\mathbf{x} + \beta\mathbf{y}$ where $\alpha, \beta \in GF(q)$. Thus we have to show that the Plücker coordinates defined by the pairs $(\mathbf{x}, \mathbf{y})$ and $(\alpha_1\mathbf{x} + \beta_1\mathbf{y}, \alpha_2\mathbf{x} + \beta_2\mathbf{y})$ (where $\alpha_1\beta_2 - \alpha_2\beta_1 \neq 0$ because the points are distinct) are scalar multiples of each other. This follows from the multiplicative property of determinants at once:

$$\begin{vmatrix} \alpha_1 x_i + \beta_1 y_i & \alpha_1 x_j + \beta_1 y_j \\ \alpha_2 x_i + \beta_2 y_i & \alpha_2 x_j + \beta_2 y_j \end{vmatrix} = \begin{vmatrix} \alpha_1 & \beta_1 \\ \alpha_2 & \beta_2 \end{vmatrix} \cdot \begin{vmatrix} x_i & x_j \\ y_i & y_j \end{vmatrix}.$$

A straightforward calculation proves the third statement:

$$\begin{aligned} p_{01}p_{23} + p_{02}p_{31} + p_{03}p_{12} &= (x_0y_1 - x_1y_0)(x_2y_3 - x_3y_2) \\ &\quad + (x_0y_2 - x_2y_0)(x_3y_1 - x_1y_3) \\ &\quad + (x_0y_3 - x_3y_0)(x_1y_2 - x_2y_1) \\ &= x_0x_2y_1y_3 - x_0x_3y_1y_2 - x_1x_2y_0y_3 + x_1x_3y_0y_2 \\ &\quad + x_0x_3y_1y_2 - x_0x_1y_2y_3 - x_2x_3y_0y_1 + x_1x_2y_0y_3 \\ &\quad + x_0x_1y_2y_3 - x_0x_2y_1y_3 - x_1x_3y_0y_2 + x_2x_3y_1y_1 \\ &= 0. \end{aligned}$$

For proving the fourth statement we may assume without loss of generality that $p_{01} \neq 0$. Let

$$X = (0 : 1 : p_{02}/p_{01} : p_{03}/p_{01}) \text{ and } Y = (-p_{01} : 0 : p_{12} : -p_{31}).$$

Then applying the equality $p_{23} = -(p_{02}p_{31} + p_{03}p_{12})/p_{01}$, an obvious calculation gives that the Plücker coordinates of the line XY are $(p_{01} : p_{02} : p_{03} : p_{23} : p_{31} : p_{12})$. □

Corollary 5.3. *Suppose that at least one of the field elements p_{01}, p_{02}, p_{03}, p_{23}, p_{31} and p_{12} is not equal to 0. Then there exists a line in $PG(3, q)$ whose Plücker coordinates are p_{ij} if and only if*

$$p_{01}p_{23} + p_{02}p_{31} + p_{03}p_{12} = 0.$$

The equation $X_0X_3 + X_1X_4 + X_2X_5 = 0$ defines a hyperbolic quadric $\mathcal{H}_5$ in $PG(5, q)$. Thus there exists a bijection between the lines of $PG(3, q)$ and the points of $\mathcal{H}_5$. The line with Plücker coordinates $(p_{01} : p_{02} : p_{03} : p_{23} : p_{31} : p_{12})$ corresponds to the point of $\mathcal{H}_5$ having the same homogeneous coordinates. This is the *Klein correspondence*.

Lemma 5.4. *In* $\mathrm{PG}(3, q)$ *two lines* e *and* f *with Plücker coordinates* $\mathbf{e} = (p_{01}^e : p_{02}^e : p_{03}^e : p_{23}^e : p_{31}^e : p_{12}^e)$ *and* $\mathbf{f} = (p_{01}^f : p_{02}^f : p_{03}^f : p_{23}^f : p_{31}^f : p_{12}^f)$ *have a point in common if and only if*

$$p_{01}^e p_{23}^f + p_{02}^e p_{31}^f + p_{03}^e p_{12}^f + p_{01}^f p_{23}^e + p_{02}^f p_{31}^e + p_{03}^f p_{12}^e = 0.$$

Proof. Let $X = (x_0 : x_1 : x_2 : x_3)$ and $Y = (y_0 : y_1 : y_2 : y_3)$ be two points on the line e, and $S = (s_0 : s_1 : s_2 : s_3)$ and $T = (t_0 : t_1 : t_2 : t_3)$ be two points on the line f. The two lines have a point in common if and only if there exists a point U whose homogeneous coordinates can be expressed as a linear combination of the coordinates of both pairs (X, Y) and (S, T). This happens if and only if the coordinate-vectors of X, Y, S and T are linearly dependent, which is equivalent to

$$\begin{vmatrix} x_0 & x_1 & x_2 & x_3 \\ y_0 & y_1 & y_2 & y_3 \\ s_0 & s_1 & s_2 & s_3 \\ t_0 & t_1 & t_2 & t_3 \end{vmatrix} = 0.$$

If we apply the Laplacian determinant expansion by minors in the first two rows, we get the statement. $\square$

A line in $\mathrm{PG}(3, q)$ can be represented as the intersection of two planes, too. This leads to the following definition.

Definition 5.5. *Let* ℓ *be a line in* $\mathrm{PG}(3, q)$ *contained in the distinct planes* U *and* V. *If* $\mathbf{u} = [u_0 : u_1 : u_2 : u_3]$ *and* $\mathbf{v} = [v_0 : v_1 : v_2 : v_3]$ *are homogeneous coordinates of* U *and* V, *respectively, then the* dual Plücker coordinates *of the line* ℓ *are*

$$\hat{\mathbf{p}} = (\widehat{p_{01}} : \widehat{p_{02}} : \widehat{p_{03}} : \widehat{p_{23}} : \widehat{p_{31}} : \widehat{p_{12}})$$

where $\widehat{p_{ij}} = u_i v_j - u_j v_i$ *for all* $0 \le i, j \le 3$.

By the Principle of Duality, the dual Plücker coordinates satisfy the four statements of Proposition 5.2, Corollary 5.3 and Lemma 5.4. As any line can be described in two ways, on the one hand, as joining two points and on the other hand, as an intersection of two planes, there must be some relations between Plücker coordinates and dual Plücker coordinates. The next theorem gives this connection.

Theorem 5.6. *Suppose that a line* ℓ *in* $\mathrm{PG}(3, q)$ *has Plücker coordinates*

$$\mathbf{p} = (p_{01} : p_{02} : p_{03} : p_{23} : p_{31} : p_{12})$$

and dual Plücker coordinates

$$\hat{\mathbf{p}} = (\widehat{p_{01}} : \widehat{p_{02}} : \widehat{p_{03}} : \widehat{p_{23}} : \widehat{p_{31}} : \widehat{p_{12}}).$$

Then

$$\hat{\mathbf{p}} = (p_{23} : p_{31} : p_{12} : p_{01} : p_{02} : p_{03}).$$

Proof. Suppose that both planes U and V contain the points X and Y. Then $\mathbf{u}\mathbf{x}^{\mathrm{T}} = \mathbf{u}\mathbf{y}^{\mathrm{T}} = \mathbf{v}\mathbf{x}^{\mathrm{T}} = \mathbf{v}\mathbf{y}^{\mathrm{T}} = 0$. We may assume without loss of generality that $p_{01} \neq 0$. Consider the planes U' and V' with coordinate-vectors $\mathbf{u}' = [p_{12} : -p_{02} : p_{01} : 0]$ and $\mathbf{v}' = [p_{13} : -p_{03} : 0 : p_{01}]$, respectively. We claim that $U' \cap V' = \ell$. As the distinct points X and Y uniquely determine ℓ, it is enough to prove that both points are incident with both planes, that is $\mathbf{u}'\mathbf{x}^{\mathrm{T}} = \mathbf{u}'\mathbf{y}^{\mathrm{T}} = \mathbf{v}'\mathbf{x}^{\mathrm{T}} = \mathbf{v}'\mathbf{y}^{\mathrm{T}} = 0$.

Consider the determinant

$$\begin{vmatrix} x_0 & x_1 & x_2 \\ x_0 & x_1 & x_2 \\ y_0 & y_1 & y_2 \end{vmatrix}.$$

Its value is obviously 0, hence if we expand it according to the first row, then we get $0 = x_0 p_{12} - x_1 p_{02} + x_2 p_{01}$, thus $\mathbf{u}'\mathbf{x}^{\mathrm{T}} = 0$. In the same way, if we expand the determinants

$$\begin{vmatrix} x_0 & x_1 & x_3 \\ x_0 & x_1 & x_3 \\ y_0 & y_1 & y_3 \end{vmatrix}, \quad \begin{vmatrix} y_0 & y_1 & y_2 \\ x_0 & x_1 & x_2 \\ y_0 & y_1 & y_2 \end{vmatrix} \text{ and } \begin{vmatrix} y_0 & y_1 & y_3 \\ x_0 & x_1 & x_3 \\ y_0 & y_1 & y_3 \end{vmatrix},$$

then we get the equations $\mathbf{v}'\mathbf{x}^{\mathrm{T}} = 0$, $\mathbf{u}'\mathbf{y}^{\mathrm{T}} = 0$ and $\mathbf{v}'\mathbf{y}^{\mathrm{T}} = 0$, respectively. So by the dual of Proposition 5.2, Part 2, the dual Plücker coordinates of ℓ can be calculated from the coordinates of U' and V'. The result of this calculation (after applying Corollary 5.3 in the case of $\widehat{p_{01}}$) gives

$$\begin{aligned} \widehat{p_{01}} &= -p_{12}p_{03} - p_{02}p_{31} \\ &= p_{01}p_{23}, \\ \widehat{p_{02}} &= p_{01}p_{31}, \\ \widehat{p_{03}} &= p_{01}p_{12}, \\ \widehat{p_{23}} &= p_{01}^2, \\ \widehat{p_{31}} &= p_{01}p_{02}, \\ \widehat{p_{12}} &= p_{01}p_{03}, \end{aligned}$$

which proves the theorem. $\square$

The point-line incidence and the plane-line containment can also be described with Plücker coordinates.

Lemma 5.7. *Let $\ell = (p_{01} : p_{02} : p_{03} : p_{23} : p_{31} : p_{12})$ be a line in $\mathrm{PG}(3, q)$. Consider the matrices*

$$\Lambda = \begin{pmatrix} 0 & -p_{23} & -p_{31} & -p_{12} \\ p_{23} & 0 & -p_{03} & p_{02} \\ p_{31} & p_{03} & 0 & -p_{01} \\ p_{12} & -p_{02} & p_{01} & 0 \end{pmatrix} \text{ and } \hat{\Lambda} = \begin{pmatrix} 0 & -p_{01} & -p_{02} & -p_{03} \\ p_{01} & 0 & -p_{12} & p_{31} \\ p_{02} & p_{12} & 0 & -p_{23} \\ p_{03} & -p_{31} & p_{23} & 0 \end{pmatrix}$$

whose entries come from the Plücker coordinates of ℓ. Then

- *a point $X = (x_0 : x_1 : x_2 : x_3)$ is on ℓ if and only if $\mathbf{x}\Lambda = \mathbf{0}$,*

- *the plane $U = [u_0 : u_1 : u_2 : u_3]$ contains the line ℓ if and only if $\mathbf{u}\hat{\Lambda} = \mathbf{0}$.*

Proof. Let S and T be two distinct points on ℓ. For $i = 0, 1, 2, 3$ let E_i be the base point of the coordinate system whose $(i+1)$-st coordinate is 1 and each of the other three coordinates is 0. Then the points E_0, E_1, E_2 and E_3 form a tetrahedron $\mathcal{T}$. If X is not on ℓ, then the plane $\langle X, \ell \rangle$ contains at most three vertices of $\mathcal{T}$. Hence the point X is on ℓ if and only if the line XE_i and ℓ meet for all $i = 0, 1, 2, 3$. By Lemma 5.4, it happens if and only if

$$0 = \begin{vmatrix} 1 & 0 & 0 & 0 \\ x_0 & x_1 & x_2 & x_3 \\ s_0 & s_1 & s_2 & s_3 \\ t_0 & t_1 & t_2 & t_3 \end{vmatrix} = \begin{vmatrix} 0 & 1 & 0 & 0 \\ x_0 & x_1 & x_2 & x_3 \\ s_0 & s_1 & s_2 & s_3 \\ t_0 & t_1 & t_2 & t_3 \end{vmatrix}$$

$$= \begin{vmatrix} 0 & 0 & 1 & 0 \\ x_0 & x_1 & x_2 & x_3 \\ s_0 & s_1 & s_2 & s_3 \\ t_0 & t_1 & t_2 & t_3 \end{vmatrix} = \begin{vmatrix} 0 & 0 & 0 & 1 \\ x_0 & x_1 & x_2 & x_3 \\ s_0 & s_1 & s_2 & s_3 \\ t_0 & t_1 & t_2 & t_3 \end{vmatrix}.$$

Expanding these determinants by the first row and using $p_{ij} = s_i t_j - s_j t_i$ we get the four coordinates of the vector $\mathbf{x}\Lambda$.

If we define $\hat{0} := 0$ and apply the operation $\hat{\ }$ to Λ entrywise, then, by Theorem 5.6, we get $\hat{\Lambda}$. Thus the second part of the lemma follows from the first one by duality. □

Let

$$A = \begin{pmatrix} 0 & -a_{23} & -a_{31} & -a_{12} \\ a_{23} & 0 & -a_{03} & a_{02} \\ a_{31} & a_{03} & 0 & -a_{01} \\ a_{12} & -a_{02} & a_{01} & 0 \end{pmatrix} \tag{5.1}$$

be a skew symmetric matrix with at least one non-zero entry. Elementary calculation shows that $\det A = (a_{01}a_{23} + a_{02}a_{31} + a_{03}a_{12})^2$. If $\det A = 0$, then, by Corollary 5.3, there exists a line f whose Plücker coordinates are $(a_{01} : a_{02} : a_{03} : a_{23} : a_{31} : a_{12})$. In the other case, if $\det A \neq 0$, then A defines a null polarity.

Definition 5.8. *In $\mathrm{PG}(3, q)$ the* linear complex *defined by the skew symmetric matrix A is the set of lines $\ell = (p_{01} : p_{02} : p_{03} : p_{23} : p_{31} : p_{12})$ whose Plücker coordinates satysfy the equation*

$$a_{23}p_{01} + a_{31}p_{02} + a_{12}p_{03} + a_{01}p_{23} + a_{02}p_{31} + a_{03}p_{12} = 0.$$

The linear complex is called special *if $\det A = 0$, and it is called* general *if $\det A \neq 0$.*

According to Corollary 5.3 and Lemma 5.4, a special linear complex is nothing else than the set of all lines intersecting a distinguished line ℓ. The Plücker coordinates of ℓ are determined by the entries of A. The line ℓ is called the *axis of the linear complex*. As a line contains $q+1$ points and there are q^2+q+1 lines through each point, each special linear complex consists of $1+(q+1)(q^2+q)$ lines.

Proposition 5.9. *The elements of a general linear complex are the self-conjugate lines of a null polarity.*

Proof. Let the general linear complex $\mathcal{G}$ be defined by the skew symmetric matrix A whose entries are the same as in (5.1). Then A can be considered as a matrix of a null polarity, too. The line joining the points $X = (x_0 : x_1 : x_2 : x_3)$ and $Y = (y_0 : y_1 : y_2 : y_3)$ is self-conjugate in the null polarity if and only if the point Y is on the polar plane of the point X. It happens if and only if $\mathbf{x}A\mathbf{y}^\mathrm{T} = 0$, which can be written in detail with coordinates as

$$
\begin{aligned}
0 &= (x_1a_{23} + x_2a_{31} + x_3a_{12})y_0 + (x_0(-a_{23}) + x_2a_{03} + x_3(-a_{02}))y_1 \\
&\quad + (x_0(-a_{31}) + x_1(-a_{03}) + x_3a_{01})y_2 + (x_0(-a_{12}) + x_1a_{02} + x_2(-a_{01}))y_3 \\
&= a_{23}(x_1y_0 - x_0y_1) + a_{31}(x_2y_0 - x_0y_2) + a_{12}(x_3y_0 - x_0y_3) \\
&\quad + a_{01}(x_3y_2 - x_2y_3) + a_{02}(x_1y_3 - y_1x_3) + a_{03}(x_2y_1 - x_1y_2).
\end{aligned}
$$

By definition, this equality holds if and only if the line XY belongs to $\mathcal{G}$. $\square$

In a null polarity each point of $\mathrm{PG}(3,q)$ is self-conjugate. Thus the set of self-conjugate lines through a given point P is the pencil of lines with carrier P in the polar plane of P. Hence if $\mathcal{G}$ is a general linear complex, then there are $q+1$ lines of $\mathcal{G}$ through each point of $\mathrm{PG}(3,q)$, and there are $q+1$ lines of $\mathcal{G}$ in each plane of $\mathrm{PG}(3,q)$. As there are $q+1$ points on each line, counting incident pairs (P,ℓ) with $\ell \in \mathcal{G}$ in two different ways gives that the number of lines of $\mathcal{G}$ is q^3+q^2+q+1. In Chapter 10 we will see that $\mathcal{G}$ is the set of lines of the classical generalized quadrangle $W(q)$.

Using Lemmas 5.4 and 5.7 we can translate the incidence properties of points, lines and planes of $\mathrm{PG}(3,q)$, via the Klein correspondence, to the geometric properties of a hyperbolic quadric in $\mathrm{PG}(5,q)$ and vice versa. The most important connections are listed in Table 5.1. The proofs of these correspondences are straightforward calculations, so we leave most of them as exercises. For the sake of delight, we only sketch two proofs.

Theorem 5.10. *Let e and f be two lines in $\mathrm{PG}(3,q)$, and P_e and P_f be the corresponding points of the Klein quadric $\mathcal{H}_5$. Then e and f are coplanar if and only if the line P_eP_f is contained in $\mathcal{H}_5$.*

Proof. According to Lemma 5.4, e and f have a point in common if and only if

$$
p_{01}^e p_{23}^f + p_{02}^e p_{31}^f + p_{03}^e p_{12}^f + p_{01}^f p_{23}^e + p_{02}^f p_{31}^e + p_{03}^f p_{12}^e = 0.
$$

By Lemma 4.50, the line $P_e P_f$ is contained in $\mathcal{H}_5$ if and only if $G(P_e, P_f) = 0$, where in case of the Klein quadric

$$
\begin{aligned}
G(P_e, P_f) &= (p_{01}^e + p_{01}^f)(p_{23}^e + p_{23}^f) + (p_{02}^e + p_{02}^f)(p_{31}^e + p_{31}^f) \\
&\quad + (p_{03}^e + p_{03}^f)(p_{12}^e + p_{12}^f) \\
&\quad - (p_{01}^e p_{23}^e + p_{02}^e p_{31}^e + p_{03}^e p_{12}^e) - (p_{01}^f p_{23}^f + p_{02}^f p_{31}^f + p_{03}^f p_{12}^f) \\
&= p_{01}^e p_{23}^f + p_{01}^f p_{23}^e + p_{02}^e p_{31}^f + p_{02}^f p_{31}^e + p_{03}^e p_{12}^f + p_{03}^f p_{12}^e.
\end{aligned}
$$

Thus the two conditions coincide, which proves the theorem. □

Theorem 5.11. *Let $\mathcal{L}_U$ be the set of $q^2 + q + 1$ lines contained in the plane U of $\mathrm{PG}(3, q)$, and $\mathcal{P}_U$ be the set of the corresponding $q^2 + q + 1$ points of the Klein quadric $\mathcal{H}_5$. Then the points of $\mathcal{P}_U$ form a plane in $\mathrm{PG}(5, q)$ which is one of the generators of $\mathcal{H}_5$.*

Proof. Let $\mathbf{u} = [u_0 : u_1 : u_2 : u_3]$ be a coordinate vector of U. By Theorem 5.7, a line $e = (p_{01} : p_{02} : p_{03} : p_{23} : p_{31} : p_{12})$ is contained in U if and only if

$$
\begin{aligned}
-u_1 p_{01} - u_2 p_{02} - u_3 p_{03} &= 0, \\
u_0 p_{01} - u_2 p_{12} - u_3 p_{31} &= 0, \\
u_0 p_{02} + u_1 p_{12} - u_3 p_{23} &= 0, \\
u_0 p_{03} - u_1 p_{31} + u_2 p_{23} &= 0.
\end{aligned} \tag{5.2}
$$

This can be written as $A_{\mathbf{u}} e^{\mathrm{T}} = \mathbf{0}$, where the matrix $A_{\mathbf{u}}$ is defined by U as

$$
A_{\mathbf{u}} = \begin{pmatrix}
-u_1 & -u_2 & -u_3 & 0 & 0 & 0 \\
u_0 & 0 & 0 & 0 & -u_3 & -u_2 \\
0 & u_0 & 0 & -u_3 & 0 & u_1 \\
0 & 0 & u_0 & u_2 & -u_1 & 0
\end{pmatrix}.
$$

We claim that $\operatorname{rank} A_{\mathbf{u}} = 3$. If $A_{\mathbf{u}}^i$ denotes the i-th row vector of $A_{\mathbf{u}}$ for $i = 1, 2, 3, 4$, then

$$
u_0 A_{\mathbf{u}}^1 + u_1 A_{\mathbf{u}}^2 + u_2 A_{\mathbf{u}}^3 + u_3 A_{\mathbf{u}}^4 = \mathbf{0},
$$

hence $\operatorname{rank} A_{\mathbf{u}} \leq 3$. On the other hand, at least one coordinate of $\mathbf{u}$ is non-zero. If $u_j \neq 0$ for some $j \in \{0, 1, 2, 3\}$, then elementary calculation shows that after deleting the $(j + 1)$-st row of $A_{\mathbf{u}}$, the remaining three rows are linearly independent.

This means that in $\mathrm{PG}(5, q)$ the set of equations $A_{\mathbf{u}} \mathbf{x}^{\mathrm{T}} = \mathbf{0}$ defines a subspace of dimension $5 - 3 = 2$, that is a plane. This plane contains $q^2 + q + 1$ points. As the coordinates of each of the $q^2 + q + 1$ lines of $\mathrm{PG}(3, q)$ lying in the plane U satisfy Equations (5.2), the set of points of this plane must be $\mathcal{P}_U$, which proves the statement. □

By duality, we get the following corollary at once.

Corollary 5.12. *Let $\mathcal{L}_R$ be the set of $q^2 + q + 1$ lines through the point R in* PG$(3, q)$, *and $\mathcal{P}_R$ be the set of the corresponding $q^2 + q + 1$ points of the Klein quadric $\mathcal{H}_5$. Then the points of $\mathcal{P}_R$ form a plane in* PG$(5, q)$ *which is one of the generators of $\mathcal{H}_5$.*

In the next part of the chapter partitions of projective spaces into their subspaces are considered. We will show a surprising connection between these structures and translation planes.

Definition 5.13. *An r-spread $\mathcal{S}^r$ of the projective space* PG(n, q) *is a set of r-dimensional subspaces of* PG(n, q) *with the property that each point of* PG(n, q) *is incident with exactly one element of $\mathcal{S}^r$.*

It follows from the definition at once that if PG(n, q) has an r-spread, then $\Theta(r)|\Theta(n)$. This necessary condition is also sufficient. First, we give an algebraic proof.

Theorem 5.14. *The projective space* PG(n, q) *has an r-spread if and only if $(r + 1) \mid (n + 1)$.*

Proof. If there exists an r-spread $\mathcal{F}^r$ of PG(n, q), then each point of the space is contained in exactly one r-dimensional subspace belonging to $\mathcal{F}^r$, hence $\Theta(n)/\Theta(r)$ is a positive integer. By Theorem 4.7, this implies $q^{r+1} - 1 \mid q^{n+1} - 1$, which is equivalent to $(r + 1) \mid (n + 1)$ (the proof of this number theoretic statement can be found for example in Chapter 1 of [134]).

Now, suppose that there exists a positive integer s such that $n + 1 = (s + 1)(r + 1)$. Let f be an irreducible polynomial over GF(q) with degree $(r + 1)$ and let α be a root of f in the field extension GF(q^{r+1}). Then each element χ of GF(q^{r+1}) can be uniquely written as

$$\chi = x_0 + x_1\alpha + \ldots + x_r\alpha^r,$$

where $x_i \in$ GF(q) for all $0 \le i \le r$.

If we take $s + 1$ elements χ_i of GF(q^{r+1}), then for $0 \le i \le s$ we get

$$\chi_i = x_{i0} + x_{i1}\alpha + \ldots + x_{ir}\alpha^r,$$

where $x_{ij} \in$ GF(q) for all $0 \le i \le r$ and $0 \le j \le s$. The lexicographically ordered $n + 1$ elements $x_{ij} \in$ GF(q) can be considered as the homogeneous coordinate-vector of a point in PG(n, q). In this way we assign to each point of PG(n, q) an ordered $(s + 1)$-tuple $(\chi_0, \chi_1, \ldots, \chi_s)$ whose elements come from GF(q^{r+1}).

Let $\tau_0, \tau_1, \ldots, \tau_s$ be elements of GF(q^{r+1}) and assume that at least one of them is not equal to 0. Consider the equations

$$\tau_i\chi_0 = \tau_0\chi_i \qquad (i = 1, 2, \ldots, s).$$

This is a set of $s(r + 1)$ linearly independent equations in variables x_{ij}, hence they define an $n - s(r + 1) = r$ dimensional subspace Π_r in PG(n, q).

TABLE 5.1
Klein correspondence

PG$(3, q)$	PG$(5, q)$
$(q^2 + 1)(q^2 + q + 1)$ lines in the space	— $(q^2 + 1)(q^2 + q + 1)$ points on $\mathcal{H}_5$
two skew lines	— two points on $\mathcal{H}_5$ whose joining line is a secant
two intersecting lines	— two points on $\mathcal{H}_5$ whose joining line is on $\mathcal{H}_5$
a pencil of lines	— points on a line on $\mathcal{H}_5$
the $q^2 + q + 1$ lines on a plane	— points on a generator plane of $\mathcal{H}_5$ (these are called *Latin planes*)
the set of $q^3 + q^2 + q + 1$ planes of the space	— the set of $q^3 + q^2 + q + 1$ Latin planes of $\mathcal{H}_5$
the $q^2 + q + 1$ lines through a point	— points on a generator plane of $\mathcal{H}_5$ (these are called *Greek planes*)
the set of $q^3 + q^2 + q + 1$ points of the space	— the set of $q^3 + q^2 + q + 1$ Greek planes of $\mathcal{H}_5$
there is a unique line joining two points	— two Latin planes meet in a unique point
two planes meet in a line	— two Greek planes meet in a unique point
there are $q + 1$ points on any line	— there are $q + 1$ Latin planes through any point on $\mathcal{H}_5$
any line is contained in $q + 1$ planes	— there are $q + 1$ Greek planes through any point on $\mathcal{H}_5$
the tangent lines to a conic	— conic in a Greek plane
the generators of a quadratic cone	— conic in a Latin plane
a regulus	— a conic, the intersection of $\mathcal{H}_5$ and a plane that does not contain any line of $\mathcal{H}_5$
the $q^2 + 1$ lines of a regular spread	— an ellipti quadric $\mathcal{E}_3$, the intersection of $\mathcal{H}_5$ and a 3-dimensional subspace whose polar line is skew to $\mathcal{H}_5$
a spread of $q^2 + 1$ lines	— $q^2 + 1$ points of $\mathcal{H}_5$, one in each Latin plane and one in each Greek plane
a packing by regular spreads	— a partition of $\mathcal{H}_5$ by elliptic quadrics
a special linear complex	— the intersection of $\mathcal{H}_5$ and one of its tangent hyperplanes
a general linear complex	— a parabolic quadric $\mathcal{P}_4$, the intersection of $\mathcal{H}_5$ and a non-tangent hyperplane

Each ordered $(s+1)$-tuple $\tau = (\tau_0, \tau_1, \ldots, \tau_s)$ corresponds to a point $P(\tau)$ in $\mathrm{PG}(s, q^{r+1})$; distinct $(s+1)$-tuples correspond to distinct points. The space $\mathrm{PG}(s, q^{r+1})$ contains $(q^{(r+1)(s+1)} - 1)/(q^{r+1} - 1)$ points, thus the subspaces Π_r contain altogether

$$\frac{q^{(r+1)(s+1)} - 1}{q^{r+1} - 1} \cdot \frac{q^{r+1} - 1}{q - 1} = \frac{q^{n+1} - 1}{q - 1}$$

points of $\mathrm{PG}(n, q)$. This number equals the total number of points in $\mathrm{PG}(n, q)$; hence the union of subspaces Π_r gives an r-spread of $\mathrm{PG}(n, q)$. □

The smallest dimension in which a spread exists is $n = 3$. In this case $r = 1$, thus we are looking for a line spread. The set of points of $\mathrm{PG}(3, q)$ must be partitioned into pairwise skew lines. Hence we have to give a set of $q^2 + 1$ pairwise skew lines in $\mathrm{PG}(3, q)$. The following example presents a geometric construction of such a set of lines.

Example 5.15. Let $f(X, Y) = X^2 + bXY + cY^2$ be an irreducible homogeneous quadratic polynomial over $\mathrm{GF}(q)$. For all $0 \neq t \in \mathrm{GF}(q)$ let $\mathcal{H}_t$ be the quadric with equation

$$\mathcal{H}_t: \quad f(X_0, X_1) + t \cdot f(X_2, X_3) = 0.$$

Let $g: X_0 = X_1 = 0$ and $h: X_2 = X_3 = 0$ be two lines. The irreducibility of f implies that $f(x, y) = 0$ if and only if $x = y = 0$. Thus if a point $S = (s_0 : s_1 : s_2 : s_3)$ of $\mathrm{PG}(3, q)$ is not in $g \cup h$, then S lies on exactly one quadric of type $\mathcal{H}_t$, namely on the quadric defined by $t = -f(s_0, s_1)/f(s_2, s_3)$. So $\mathcal{H}_{t_1} \cap \mathcal{H}_{t_2} = \emptyset$ if $t_1 \neq t_2$, and $\mathcal{H}_t \cap g = \mathcal{H}_t \cap h = \emptyset$ also hold for all $0 \neq t \in \mathrm{GF}(q)$. Thus

$$\left(\bigcup_{0 \neq t \in \mathrm{GF}(q)} \mathcal{H}_t \right) \cup \{g, h\}$$

is a partition of the set of points of $\mathrm{PG}(3, q)$.

As $\mathcal{H}_t \cap g = \emptyset$, the quadric $\mathcal{H}_t$ cannot contain a whole plane. According to the list of singular quadrics in $\mathrm{PG}(3, q)$ (see Exercise 4.4), this means that $\mathcal{H}_t$ is a line, a quadratic cone or a non-singular quadric, so it contains at most $(q+1)^2$ points for all $0 \neq t \in \mathrm{GF}(q)$. Thus the union of the $q-1$ quadrics has at most $(q-1)(q+1)^2$ points. As $g \cup h$ contains $2(q+1)$ points, the equation

$$\frac{(q^3 + q^2 + q + 1) - 2(q+1)}{q - 1} = (q+1)^2$$

implies that $\mathcal{H}_t$ must contain exactly $(q+1)^2$ points for all t. A quadratic cone contains $q^2 + q + 1$ points and an elliptic quadric contains $q^2 + 1$ points, so $\mathcal{H}_t$ is a hyperbolic quadric for all t. Hence, if $\mathcal{L}_t$ denotes one of the two sets of

$q + 1$ pairwise skew lines whose union contains all points of $\mathcal{H}_t$, then the set of lines

$$\left(\bigcup_{0 \neq t \in \mathrm{GF}(q)} \mathcal{L}_t \right) \cup \{g, h\}$$

is a line spread of $\mathrm{PG}(3, q)$.

Let us remark that if q is odd, then the determinant of the matrix belonging to $\mathcal{H}_t$ is equal to $t^2 \left(c - \frac{b^2}{4} \right)^2$. It is a square element in $\mathrm{GF}(q)$. By Exercise 4.6, this implies at once that $\mathcal{H}_t$ is a hyperbolic quadric.

Proposition 5.16. *Let Π_{r-1}^1, Π_{r-1}^2 and Π_{r-1}^3 be three mutually skew $(r-1)$-dimensional subspaces in $\mathrm{PG}(2r-1, q)$. Then there exist exactly $\Theta(r-1)$ lines having a non-empty intersection with each of Π_{r-1}^1, Π_{r-1}^2 and Π_{r-1}^3.*

Proof. Let $P_1 \in \Pi_{r-1}^1$ be a point. It follows from the dimension formula that any two of the three Π_{r-1}^i's generate $\mathrm{PG}(2r-1, q)$, and the r-dimensional subspace $\langle \Pi_{r-1}^2, P_1 \rangle$ meets Π_{r-1}^3 in a single point, say P_3. Then the line $P_1 P_3$ belongs to $\langle \Pi_{r-1}^2, P_1 \rangle$ and it is not contained in Π_{r-1}^2, hence it meets Π_{r-1}^2 in a single point, say P_2. So the line $P_1 P_2 P_3$ meets each of the three Π_{r-1}^i's in one point. If $P_1 \neq R_1 \in \Pi_{r-1}^1$ is another point, then we can construct the line $R_1 R_2 R_3$ in the same way. If the two points R_2 and P_2 were the same, then the r-dimensional subspace $\langle \Pi_{r-1}^1, P_2 \rangle$ would meet Π_{r-1}^3 in a line $P_3 R_3$ contradicting the fact $\langle \Pi_{r-1}^1, \Pi_{r-1}^3 \rangle = \mathrm{PG}(2r-1, q)$. In the same way, one can prove that $R_3 \neq P_3$. Thus distinct points of Π_{r-1}^1 give skew transversal lines which proves the proposition. □

The detailed description of regular spreads and the proof of the following generalization of Gallucci's theorem can be found in [93], Section 25.6.

Theorem 5.17. *Let Π_{r-1}^1, Π_{r-1}^2 and Π_{r-1}^3 be three mutually skew $(r-1)$-dimensional subspaces in $\mathrm{PG}(2r-1, q)$, and let $\mathcal{L} = \{\ell_1, \ell_2, \ldots, \ell_{\Theta(r-1)}\}$ be the set of all lines having a non-empty intersection with Π_{r-1}^1, Π_{r-1}^2 and Π_{r-1}^3. If an $(r-1)$-dimensional subspace $\Pi_{r-1} \subset \mathrm{PG}(2r-1, q)$ intersects at least three elements of $\mathcal{L}$ in exactly one point and does not contain any element of $\mathcal{L}$, then Π_{r-1} intersects all elements of $\mathcal{L}$.*

Thus the following definition is meaningful.

Definition 5.18. *Let Π_{r-1}^1, Π_{r-1}^2 and Π_{r-1}^3 be three mutually skew $(r-1)$-dimensional subspaces in $\mathrm{PG}(2r-1, q)$. The r-regulus $\mathcal{R}(\Pi_{r-1}^1, \Pi_{r-1}^2, \Pi_{r-1}^3)$, defined by the three subspaces, is the $(q+1)$-element set of those $(r-1)$-dimensional subspaces in $\mathrm{PG}(2r-1, q)$ whose intersection with each of the lines intersecting Π_{r-1}^1, Π_{r-1}^2 and Π_{r-1}^3 is a single point.*

In particular, in $\mathrm{PG}(3, q)$ a regulus $\mathcal{R}(\ell_1, \ell_2, \ell_3)$ is the set of $q+1$ mutually skew lines such that any line of $\mathrm{PG}(3, q)$ intersecting ℓ_1, ℓ_2 and ℓ_3 intersects all elements of $\mathcal{R}(\ell_1, \ell_2, \ell_3)$.

We have already seen reguli in the previous example. If e_1, e_2 and e_3 are three mutually skew lines lying on a hyperbolic quadric $\mathcal{H}_3$ in PG$(3,q)$ and a line f intersects each of them, then f is on $\mathcal{H}_3$, too. Hence the elements of the set $\{f_1, f_2, \ldots, f_{q+1}\}$ that consists of the lines meeting e_1, e_2 and e_3 are also on $\mathcal{H}_3$, and vice versa, the elements of the set $\{e_1, e_2, \ldots, e_{q+1}\}$ that consists of the lines meeting the elements of the set $\{f_1, f_2, \ldots, f_{q+1}\}$ are on $\mathcal{H}_3$, too. Each of these two sets of $q+1$ mutually skew lines lying on $\mathcal{H}_3$ is a regulus. One of them is $\mathcal{R}(e_1, e_2, e_3)$, the other one is $\mathcal{R}(f_1, f_2, f_3)$. They are called *opposite reguli*.

In Example 5.15 we constructed not only one, but 2^{q-1} different spreads, because we can choose any one of the two reguli lying on the hyperbolic quadrics $\mathcal{F}_t$ independently. These $2(q-1)$ reguli are divided into two classes in a natural way, such that each class contains one regulus from each quadric. The two spreads have an extra property: if three lines belong to the spread, then each line of the regulus defined by them belongs to the spread, too. The following definition generalizes this property.

Definition 5.19. *An $(r-1)$-spread $\mathcal{S}^{r-1}$ of PG$(2r-1,q)$ is called* regular, *if for any three distinct elements Π_{r-1}^1, Π_{r-1}^2 and Π_{r-1}^3 of $\mathcal{S}^{r-1}$ the whole regulus $\mathcal{R}(\Pi_{r-1}^1, \Pi_{r-1}^2, \Pi_{r-1}^3)$ is contained in $\mathcal{S}^{r-1}$.*

Under the Klein correspondence the q^2+1 lines of a spread of PG$(3,q)$ are mapped to a set of points of $\mathcal{H}_5$ having the property that the line joining any two elements of the set is a secant of $\mathcal{H}_5$. Such a set of points is an elliptic quadric $\mathcal{E}_3$; one can get it as the intersection of $\mathcal{H}_5$ and a 3-dimensional subspace whose polar line is skew to $\mathcal{H}_5$. In this case the $q+1$ points on any plane section of $\mathcal{E}_3$ correspond to $q+1$ lines of a regulus of PG$(3,q)$. In this way we can represent the regular spreads of PG$(3,q)$. There are other spreads of PG$(3,q)$, too. The simplest examples can be constructed by changing a regulus to its opposite one in some $\mathcal{H}_3$.

Spreads are closely connected to translation planes.

Example 5.20. Let $\mathcal{S}^{r-1}$ be an $(r-1)$-spread of PG$(2r-1,q)$. Embed PG$(2r-1,q)$ into PG$(2r,q)$ as its hyperplane at infinity and define a point-line incidence geometry in the following way.

- The *new points* are the points of the affine space AG$(2r,q) = $ PG$(2r,q) \setminus$ PG$(2r-1,q)$.

- The *new lines* are those r-dimensional subspaces, Π_r, of PG$(2r,q)$ for which $\Pi_r \cap PG(2r-1,q) \subset \mathcal{S}^{r-1}$ holds, so the points at infinity of Π_r form an element of the spread.

- The *incidence* is the set-theoretical inclusion.

This structure contains as many points as AG$(2r,q)$, that is q^{2r}. There are $q^r + 1$ distinct lines through each point, because $\mathcal{S}^{r-1}$ consists of $q^r + 1$

elements. There are q^r points on each line, because any r-dimensional subspace Π_r of $\mathrm{PG}(2r, q)$ contains q^r affine points. Thus the structure is an affine plane of order q^r.

The set $\mathcal{T}$ of collineations of $\mathrm{PG}(2r, q)$ fixing the hyperplane at infinity pointwise acts transitively on the points of $\mathrm{AG}(2r, q)$, hence $\mathcal{T}$ acts transitively on the points of the constructed incidence geometry, too. This means that the projective closure of the constructed structure is a translation plane. As the structure was built up over $\mathrm{GF}(q)$, the kernel of its coordinatizing quasifield obviously contains $\mathrm{GF}(q)$.

The process can be reversed. Starting from a translation plane one can construct a spread.

Example 5.21. Let Q be a quasifield of order q^r whose kernel is $\mathrm{GF}(q)$. Then, by Theorem 3.22, Q may be regarded as an r-dimensional vector space over $\mathrm{GF}(q)$. Let Π be a translation plane with translation line ℓ that is coordinatized by Q. Consider the affine plane $\mathcal{A} = \Pi \setminus \{\ell\}$. Its points can be identified with the elements of $Q \times Q$, hence with the elements of a $2r$-dimensional vector space V_{2r} over $\mathrm{GF}(q)$.

Take the lines of $\mathcal{A}$ through the point $O = (0,0)$. As the order of Π is q^r, the order of $\mathcal{A}$ is also q^r, hence there are $q^r + 1$ lines through O. Each of these lines contains $q^r - 1$ points other than O; the corresponding elements form an r-dimensional subspace in V_{2r}. Identify V_{2r} with $\mathrm{PG}(2r - 1, q)$. The r-dimensional vector subspaces of V_{2r} correspond to the $(r - 1)$-dimensional projective subspaces of $\mathrm{PG}(2r - 1, q)$. As any two vector subspaces intersect only in the origin, the corresponding projective subspaces are pairwise disjoint, hence they form an $(r - 1)$-spread of $\mathrm{PG}(2r - 1, q)$.

The more beautiful the geometric features of the spread, the stronger the algebraic properties of the quasifield. A detailed description of this correspondence can be found in the book [119].

Example 5.22. Suppose that $n + 1 = (s + 1)(r + 1)$. Let V_{n+1} be an $(n + 1)$-dimensional vector space over $\mathrm{GF}(q)$. Identify V_{n+1} and $\mathrm{PG}(n, q)$ in the usual way. Take the affine space $\mathrm{AG}(s + 1, q^{r+1})$. In this space the lines through the origin can be considered as $(r + 1)$-dimensional vector spaces over $\mathrm{GF}(q)$.

As any two of these vector subspaces intersect only in the origin, the corresponding r-dimensional projective subspaces $\mathrm{PG}(n, q)$ are pairwise disjoint. The number of these subspaces is $(q^{n+1} - 1)/(q^{r+1} - 1)$, hence they form an r-spread of $\mathrm{PG}(n, q)$.

We can construct spreads using linear transformations, too.

Example 5.23. Let V_{2r} be a $(2r)$-dimensional and W be an r-dimensional vector space over $\mathrm{GF}(q)$. Then V_{2r} can be considered as the direct sum $W \oplus W$. Suppose that for $i = 1, 2, \ldots, q^r - 1$ there exist linear transformations T_i of

W having the property that $T_j T_k^{-1}$ is fixed-point free for $j \neq k$. For $i = 1, 2, \ldots, q^r - 1$ let

$$W_i = \{(\mathbf{x}, \mathbf{x} T_i) : \mathbf{x} \in W\},$$

and define

$$W_0 = \{(\mathbf{0}, \mathbf{x}) : \mathbf{x} \in W\} \quad \text{and} \quad W_\infty = \{(\mathbf{x}, \mathbf{0}) : \mathbf{x} \in W\}.$$

These sets are r-dimensional subspaces of V_{2r}, and as $T_j T_k^{-1}$ is fixed-point free, the intersection of any two of them contains only the origin of V_{2r}. Thus the corresponding $(r-1)$-dimensional subspaces of $\mathrm{PG}(2r-1, q)$ form an $(r-1)$-spread.

Thus, as described in Example 5.20, from a family of linear transformations we are able to construct a translation plane. If a family of linear transformations $\{T_1, T_2, \ldots, T_{q^r-1}\}$ has the property that $T_j T_k^{-1}$ is fixed-point free for $j \neq k$, and S is an arbitrary non-singular linear transformation, then the family $\{T_1 S, T_2 S, \ldots, T_{q^r-1} S\}$ has the same property. Thus we may assume without loss of generality that T_1 is the identity. One can prove that the more regular the set of the T_i's, the more beautiful the coordinatizing quasifield of the translation plane.

Proposition 5.24. *Suppose that the set $\{T_1, T_2, \ldots, T_{q^r-1}\}$ of linear transformations satisfies the conditions of Example 5.23 and T_1 is the identity. Let Π be the translation plane constructed from the corresponding spread in the way described in Example 5.20 and let Q be the quasifield coordinatizing Π. Then*

- *Q is associative if and only if the linear transformations T_i form a group,*

- *Q is a field if and only if the linear transformations T_i form a cyclic group.*

We only sketch the proof. Its basic idea is the following. If a is a fixed non-zero element of $\mathrm{GF}(q)$, then the mapping on the elements of $\mathrm{GF}(q)$ defined as $x \mapsto ax$ can be considered as a linear transformation T_a.

According to the remark after Theorem 3.27, the multiplication of Q is associative if and only if Π is (Y, OX)-transitive. The action of the (Y, OX)-perspectivity of Π mapping the line $Y = aX$ to the line $Y = X$ can be considered on V_{2r}, too. The corresponding linear transformation sends the subspace $W_i = \{(\mathbf{x}, \mathbf{x} T_i) : \mathbf{x} \in W\}$ into the subspace $W_i' = \{(\mathbf{x}, \mathbf{x} T_i T_{a^{-1}}) : \mathbf{x} \in W\}$, thus this subspace must correspond to a line, too. Hence $T_i T_{a^{-1}} \in \{T_1, T_2, \ldots, T_{q^r-1}\}$ for all $a \in Q$. This proves the first part of the proposition, and it also implies that each T_i is of type T_a where $a \in Q$. The second part of the statement follows from the fact that T_g is a generator element of the group formed by the transformations if and only if g is a generator element in the multiplicative group of Q, and when the multiplicative group of a quasifield is cyclic, then the quasifield is a field.

Exercises

5.1. Prove that under the Klein correspondence

- the $q+1$ points on a fixed line of $\mathrm{PG}(3,q)$ go to the $q+1$ Latin planes through a fixed point of $\mathcal{H}_5$;

- the $q+1$ planes containing a fixed line of $\mathrm{PG}(3,q)$ go to the $q+1$ Greek planes through a fixed point of $\mathcal{H}_5$;

- the intersection of any two distinct planes in $\mathrm{PG}(3,q)$ is a line $\Leftrightarrow$ the intersection of any two Greek planes is a point on $\mathcal{H}_5$;

- there is a unique line through any two distinct points in $\mathrm{PG}(3,q)$ $\Leftrightarrow$ the intersection of any two Latin planes is a point on $\mathcal{H}_5$.

5.2. Show that the point $(1:0:0:0)$ of $\mathrm{PG}(3,q)$ corresponds to the Latin plane with equation $X_3 = X_4 = X_5 = 0$ on $\mathcal{H}_5$.

5.3. In $\mathrm{PG}(3,q)$ consider the conic $\mathcal{C}$ which contains the points $\{(0:1:t:t^2) : t \in \mathrm{GF}(q)\} \cup \{(0:0:0:1)\}$. Let $\mathcal{Q}$ be the quadratic cone whose vertex is $(1:0:0:0)$ and base is $\mathcal{C}$.

- Calculate the Plücker coordinates of the generators of $\mathcal{Q}$.

- Verify that the corresponding points of $\mathcal{H}_5$ form the conic with equation $X_0 X_2 = X_1^2$ in the Latin plane whose equation is $X_3 = X_4 = X_5 = 0$.

- Prove that the generators of any quadratic cone of $\mathrm{PG}(3,q)$ correspond to the points of a conic in a Latin plane on $\mathcal{H}_5$.

5.4. Let $(1:0:0:0:0:0)$ and $(0:0:0:1:0:0)$ be the Plücker coordinates of two skew lines in $\mathrm{PG}(3,q)$. Prove that the set of the $(q+1)^2$ transversals of the two lines corresponds to the points of the hyperbolic quadric with equation $X_1 X_4 + X_2 X_5 = 0$ in the 3-dimensional space $X_0 = X_3 = 0$.

5.5. Prove that the set of the $(q+1)$ transversals of three pairwise skew lines corresponds to the points of a conic which lies in a plane section of $\mathcal{H}_5$.

6

Arcs, ovals and blocking sets

In this chapter some combinatorially defined sets of points in projective and affine planes are considered. Arcs and blocking sets are interesting objects in their own right, and they also have several important applications to other areas of mathematics, for example, coding theory, cryptography and graph theory. Throughout this chapter the word "plane" without any attribute means any projective or affine plane.

Definition 6.1. *A set of points in a plane is called an* arc *if no three of its points are collinear. An arc is called a* k-arc, *if it contains exactly* k *points.*

A k-arc is complete *if it is not contained in any $(k + 1)$-arc.*

Definition 6.2. *Let $\mathcal{K}$ be a k-arc and ℓ be a line in a plane. Then ℓ is called*

- *a* secant *to $\mathcal{K}$ if $|\mathcal{K} \cap \ell| = 2$,*

- *a* tangent *to $\mathcal{K}$ if $|\mathcal{K} \cap \ell| = 1$,*

- *an* external *line to $\mathcal{K}$ if $|\mathcal{K} \cap \ell| = 0$.*

Obviously, each arc is a subset of some complete arcs, hence the really interesting objects are complete arcs. The first natural question regarding complete arcs is: what can we say about their size?

An arc cannot be too large, and a complete arc cannot be too small. First, we consider small complete arcs. The following simple observations are very useful in the study of arcs and blocking sets.

Proposition 6.3. *Suppose that a set $\mathcal{L}$ of m lines covers all points of a projective plane of order n. Then $m \geq n + 1$ and in case of equality the lines form a pencil.*

Proof. Let ℓ_0 be a fixed line of $\mathcal{L}$. It contains $n + 1$ points of the plane. Any other line $\ell \in \mathcal{L}$ meets ℓ_0 in a point, hence it covers only n points that are not covered by ℓ_0. So the union of m lines contains at most $n+1+(m-1)n$ points. Hence if the lines cover all points of the plane then $n+1+(m-1)n \geq n^2+n+1$. Rearranging this inequality we get $m \geq n + 1$.

In case of equality $\ell_0 \cap \ell$ must be the same point P for all $\ell_0 \neq \ell \in \mathcal{L}$. This means that $\mathcal{L}$ is the pencil with carrier P. $\qquad\square$

Proposition 6.4. *Let $\mathcal{K}$ be a complete k-arc in a finite plane. Then the secants to $\mathcal{K}$ cover all points of the plane.*

Proof. If there were a point P not covered by any secant to $\mathcal{K}$, then the set $\mathcal{K} \cup \{P\}$ would be a $(k+1)$-arc contradicting to the completeness of $\mathcal{K}$. □

Theorem 6.5 (Lunelli, Sce). *Let $\mathcal{K}$ be a complete k-arc in a finite projective plane of order n. Then $n < k(k-1)/2$.*

Proof. Consider the secants to $\mathcal{K}$. By Proposition 6.4, these lines cover all points of the plane. There are $k(k-1)/2$ secants to $\mathcal{K}$ thus, by Proposition 6.3, the inequality $k(k-1)/2 \geq n+1$ holds. This proves the therorem. □

This result could be improved by applying better estimates on the size of blocking sets, see Corollary 6.40.

It is worth mentioning that one can construct complete arcs in a very simple way by applying the following greedy algorithm. Take three non-collinear points; they form an arc $\mathcal{K}_1$. Suppose that we have an arc $\mathcal{K}_i$. Consider all the secants to $\mathcal{K}_i$. If they cover the whole plane, then $\mathcal{K}_i$ is complete. Otherwise, if a point P is not covered, then let $\mathcal{K}_{i+1} = \mathcal{K}_i \cup \{P\}$ and repeat this process step by step. At the end we always get a complete arc. The size of these complete arcs is not known in general. *Kim* and *Vu* [108] proved in a much more sophisticated way, using a probabilistic method, that any plane of order n contains a complete arc of size at most $\sqrt{n} \log^c n$, where c is an absolute constant.

Contrary to the lower bound, the following upper estimate on the size of an arc is tight in some cases.

Theorem 6.6 (Bose). *If there exists a k-arc in a finite plane of order n, then*

$$k \leq \begin{cases} n+1 & \text{if } n \text{ is odd,} \\ n+2 & \text{if } n \text{ is even.} \end{cases}$$

Proof. Suppose that the points $P_1, P_2, \ldots, P_k$ form a k-arc in a plane Π_n of order n. Then the lines $P_1 P_i$ are pairwise distinct lines through P_1. As in Π_n there are $n+1$ lines through P_1, and we get $k-1 \leq n+1$, so $k \leq n+2$.

Now assume that the points $P_1, P_2, \ldots, P_{n+2}$ form an $(n+2)$-arc, say $\mathcal{K}$. Then it follows from the previous argument that there are $n+1$ secants to $\mathcal{K}$ through P_i for $i = 1, 2, \ldots, n+2$. Hence $\mathcal{K}$ has no tangent line at all. Thus each line of Π_n meets $\mathcal{K}$ in either 0 or 2 points. Let R be a point not in $\mathcal{K}$ and consider the lines through R. These lines give a partition of the points of $\Pi_n \setminus \{R\}$ and each of them meets $\mathcal{K}$ in an even number of points. Hence $\mathcal{K}$ consists of an even number of points. So $n+2$ is even, but $n+2$ and n have the same parity, thus n must be even. □

Definition 6.7. *In a plane of order n an $(n+1)$-arc is called an* oval, *and an $(n+2)$-arc is called a* hyperoval.

Corollary 6.8. *There are no hyperovals in planes of odd order.*

Let $\mathcal{K}$ be a k-arc in a plane of order n. Then there are $k - 1$ secants through each point of $\mathcal{K}$, thus the number of tangents to $\mathcal{K}$ at any of its points is $t = n + 2 - k$. Hence, if $\mathcal{O}$ is an oval, there is a unique tangent to $\mathcal{O}$ at any of its points, while a hyperoval has no tangent at all.

The next theorem shows that the bound in Bose's theorem is tight for Galois planes.

Theorem 6.9. *There are ovals in* $\mathrm{PG}(2, q)$ *for all* q. *If* q *is even, then* $\mathrm{PG}(2, q)$ *contains hyperovals, too.*

Proof. We have already seen (Theorem 4.48 and Lemma 4.50) that each conic consists of $q + 1$ points and no three points of a conic are collinear. Hence, there are ovals in $\mathrm{PG}(2, q)$.

A typical example for conics is the normal parabola with affine equation $Y = X^2$. We can consider the corresponding projective curve in $\mathrm{PG}(2, q)$. Its equation is $X_1^2 = X_0 X_2$ and it contains the points

$$\mathcal{P} = \{(1 : t : t^2) : t \in \mathrm{GF}(q)\} \cup \{(0 : 0 : 1)\}.$$

The computation below explicitly shows that no three points of this set are collinear, thus $\mathcal{P}$ is an oval for all q. This obviously follows from the well-known collinearity condition (see Example 1.10), because if $t_i \neq t_j$, then

$$\begin{vmatrix} 1 & t_1 & t_1^2 \\ 1 & t_2 & t_2^2 \\ 1 & t_3 & t_3^2 \end{vmatrix} = (t_1 - t_2)(t_2 - t_3)(t_3 - t_1) \neq 0$$

and

$$\begin{vmatrix} 1 & t_1 & t_1^2 \\ 1 & t_2 & t_2^2 \\ 0 & 0 & 1 \end{vmatrix} = t_2 - t_1 \neq 0.$$

$\mathcal{P}$ is an oval, hence it has a unique tangent line at each of its points. Elementary calculation shows that the affine equation of the tangent line to $\mathcal{P}$ at the affine point (t_0, t_0^2) is

$$Y - t_0^2 = 2t_0(X - t_0), \tag{6.1}$$

while the tangent at $(0 : 0 : 1)$ is the line $X_0 = 0$. If q is even, then $2t_0 = 0$, hence Equation (6.1) becomes $Y = t_0^2$. Thus in this case each tangent to $\mathcal{P}$ contains the point at infinity $(0 : 1 : 0)$.

This means that no three points of the set $\mathcal{P} \cup \{(0 : 1 : 0)\}$ are collinear, so this set is a hyperoval. $\square$

An alternative proof of the extendability of $\mathcal{P}$ to a hyperoval follows from the fact that if q is even and $t_1 \neq t_2$, then

$$\begin{vmatrix} 1 & t_1 & t_1^2 \\ 1 & t_2 & t_2^2 \\ 0 & 1 & 0 \end{vmatrix} = (t_2^2 - t_1^2) = (t_2 - t_1)^2 \neq 0$$

and

$$\begin{vmatrix} 1 & t_1 & t_1^2 \\ 0 & 0 & 1 \\ 0 & 1 & 0 \end{vmatrix} = 1 \neq 0.$$

Not only the parabola can be extended to a hyperoval. The following theorem shows that the extendability of ovals is a combinatorial property in even order planes.

Theorem 6.10. *Let n be even and let $\mathcal{O}$ be an oval in a finite projective plane of order n. Then*

- *the tangents to $\mathcal{O}$ cover the plane,*

- *$\mathcal{O}$ is not complete; it can be extended to a hyperoval.*

Proof. The points of $\mathcal{O}$ are obviously covered by the tangents. Take an arbitrary point $A \notin \mathcal{O}$. As $n + 1$ is odd, $\mathcal{O}$ has an odd number of points. Each secant contains two points of $\mathcal{O}$. Thus among the lines joining A and the points of $\mathcal{O}$, there must be some tangents, too. This proves the first statement of the theorem.

There is a unique tangent to $\mathcal{O}$ at each of its points, hence there are $n + 1$ tangents to $\mathcal{O}$ altogether. So Proposition 6.3 implies that the tangents form a pencil. If N is the carrier of this pencil, then $\mathcal{O} \cup \{N\}$ is a hyperoval. $\square$

Definition 6.11. *Let n be even and let $\mathcal{O}$ be an oval in a projective plane of order n. The carrier of the pencil formed by the tangents to $\mathcal{O}$ is called the* nucleus *of $\mathcal{O}$.*

If n is odd, then the combinatorial properties of ovals are similar to the corresponding properties of conics.

Lemma 6.12. *Let n be odd and let $\mathcal{O}$ be an oval in the projective plane Π_n. Then any point of $\Pi_n \setminus \mathcal{O}$ lies on either 2 or 0 tangents to $\mathcal{O}$.*

Proof. Let ℓ be the tangent to $\mathcal{O}$ at a point P and let $P_1, P_2, \ldots, P_n$ be the other points of ℓ. Let t_i be the number of tangents to $\mathcal{O}$ through P_i. As $\mathcal{O}$ contains an even number, $n + 1$, of points, $t_i > 0$ must be even for all $i = 1, 2, \ldots, n$, too. There are $n+1$ tangents to $\mathcal{O}$; one of them is ℓ. Each of the n tangents distinct from ℓ, meets ℓ in a unique point, hence $\sum_{i=1}^{n}(t_i - 1) = n$. Thus, by the pigeonhole principle, $t_i - 1 = 1$ must hold for all i.

Hence if a point not on $\mathcal{O}$ lies on a tangent, then it lies on exactly two tangents to $\mathcal{O}$. $\square$

We will provide an interesting practical application of the proven combinatorial properties of ovals and hyperovals in Chapter 12 (see Examples 12.6 and 12.8).

Definition 6.13. *Let n be odd and let $\mathcal{O}$ be an oval in the projective plane Π_n. A point of $\Pi_n \setminus \mathcal{O}$ is called* external *or* internal *with respect to $\mathcal{O}$ according to whether it lies on 2 or 0 tangents to $\mathcal{O}$.*

Lemma 6.14. *Let n be odd and $\mathcal{O}$ be an oval in a projective plane Π_n. Then there are $\frac{(n+1)n}{2}$ external and $\frac{(n-1)n}{2}$ internal points with respect to $\mathcal{O}$.*

If the line ℓ is not a tangent to $\mathcal{O}$ then exactly half of the points of $\ell \setminus \mathcal{O}$ are external and half of them are internal with respect to $\mathcal{O}$.

Proof. There are $n+1$ tangents to $\mathcal{O}$; each of them contains n points from the set $\Pi_n \setminus \mathcal{O}$, and there are two tangents through any external point, so the number of external points is $(n+1)n/2$. Any point not on $\mathcal{O}$ is either external or internal, so the number of internal points is $n^2+n+1-(n+1)-(n+1)n/2 = (n-1)n/2$.

If ℓ is not a tangent to $\mathcal{O}$ and m denotes the number of points in $\mathcal{O} \cap \ell$, then $m = 0$ or 2. If ℓ contains e external points with respect to $\mathcal{O}$ then $n + 1 = m + 2e$, because each of the $n + 1$ tangents to $\mathcal{O}$ intersects ℓ in a unique point. Hence

$$e = \frac{n+1-m}{2} = \begin{cases} \frac{n-1}{2} & \text{if } m = 2, \\ \frac{n+1}{2} & \text{if } m = 0. \end{cases}$$

The number of internal points on ℓ is $n+1-m-e$. This number is $(n-1)/2$ if ℓ is a secant, and it is $(n + 1)/2$ if ℓ is an external line. Thus the numbers of external and of internal points on ℓ are equal in both cases. $\square$

The research on arcs was initiated in 1954–55 when *Segre* proved that if q is odd, then in $\mathrm{PG}(2,q)$ each oval is a conic. For proving this fundamental result we need a generalization of Ceva's theorem. In the Euclidean plane it gives a necessary and sufficient condition for three lines, each joining a vertex of a triangle and a point of the opposite side, to have a point in common. The condition is that the product of ratios of distances equals 1. The projective version of the theorem is the following.

Lemma 6.15. *In $\mathrm{PG}(2,q)$ the three lines $X_2 = \lambda_0 X_1$, $X_0 = \lambda_1 X_2$ and $X_1 = \lambda_2 X_0$ have a point in common if and only if $\lambda_0\lambda_1\lambda_2 = 1$.*

Proof. The homogeneous coordinates of the lines are $[0 : -\lambda_0 : 1]$, $[1 : 0 : -\lambda_1]$ and $[-\lambda_2 : 1 : 0]$. The three lines have a point in common if and only if

$$\begin{vmatrix} 0 & -\lambda_0 & 1 \\ 1 & 0 & -\lambda_1 \\ -\lambda_2 & 1 & 0 \end{vmatrix} = 0.$$

Expanding the determinant we get that its value is $1 - \lambda_0\lambda_1\lambda_2$. This proves the lemma. $\square$

The dual of this result is the generalization of Menelaus' theorem. For the sake of completeness we prove it.

Lemma 6.16. *In $\mathrm{PG}(2,q)$ the three points $(b_0 : 0 : 1)$, $(1 : b_1 : 0)$ and $(0 : 1 : b_2)$ are collinear if and only if $b_0 b_1 b_2 = -1$.*

Proof. The three points are collinear if and only if

$$\begin{vmatrix} b_0 & 0 & 1 \\ 1 & b_1 & 0 \\ 0 & 1 & b_2 \end{vmatrix} = 0.$$

Expanding the determinant we get that its value is $b_0 b_1 b_2 + 1$, so the points are collinear if and only if $b_0 b_1 b_2 = -1$. $\square$

The next lemma shows that a degenerate case of the classical Theorem of Brianchon about circumscribed hexagons of a conic is valid for ovals in PG$(2, q)$, too.

Lemma 6.17 (Segre). *Let q be odd and let $\mathcal{O}$ be an oval in PG$(2, q)$. If $P \in \mathcal{O}$ is a point, then let t_P denote the tangent to $\mathcal{O}$ at P. Take three points $A_1, A_2, A_3 \in \mathcal{O}$; consider the inscribed triangle formed by them and the corresponding circumscribed triangle with vertices $B_i = t_{A_j} \cap t_{A_k}$ where $\{i, j, k\} = \{1, 2, 3\}$. Then the three lines $A_1 B_1$, $A_2 B_2$ and $A_3 B_3$ have a point in common; in other words, the two triangles $A_1 A_2 A_3$ and $B_1 B_2 B_3$ are in perspective from a point.*

Proof. Choose the points A_1, A_2 and A_3 as base points of the coordinate system. Let $A_1 = (1 : 0 : 0)$, $A_2 = (0 : 1 : 0)$ and $A_3 = (0 : 0 : 1)$. The equation of lines through A_1, distinct from the sides of the base triangle can be written as $X_2 = \lambda_0 X_1$, where $\lambda_0 \neq 0$. In the same way, we can write the equation of lines through A_2 as $X_0 = \lambda_1 X_2$, and lines through A_3 into the form $X_1 = \lambda_2 X_0$. For the sake of brevity, from now on λ_i is called the coordinate of the corresponding line.

Create a $(q - 2) \times 3$ matrix $M = (m_{i,j})$ whose rows correspond to the points $\mathcal{O} \setminus \{A_1, A_2, A_3\} = \{P_1, P_2, \ldots, P_{q-2}\}$, and the columns correspond to the points $\{A_1, A_2, A_3\}$. For $i = 1, 2, \ldots, q - 2$ and $j = 1, 2, 3$ let the entry $m_{i,j}$ of M be the coordinate of the line $P_i A_j$.

It follows from Lemma 6.15 that the product of the three elements in each row of M is equal to 1, hence the product of all entries of M is equal to 1. On the other hand, in each column we see $q - 2$ pairwise distinct, non-zero elements of GF(q), because M has no zero entry and no three points of $\mathcal{O}$ are collinear. Hence if $i_1 \neq i_2$, then $m_{i_1,j} \neq m_{i_2,j}$ for $j = 1, 2, 3$.

The field GF(q) has $q - 1$ non-zero elements. In each column the missing element must be the coordinate β_j of the line t_{A_j}, because this line is the tangent to $\mathcal{O}$ at A_j, hence it does not contain any of the points $\{P_1, P_2, \ldots, P_{q-2}\}$. Add a $(q - 1)$-st row to M with entries β_1, β_2 and β_3. Then each column of the extended matrix contains all non-zero elements of GF(q), whose product is -1. (If q is a prime then this is the Theorem of Wilson. In general, the non-zero elements are the $q - 1$ roots of the polynomial $X^{q-1} - 1 = 0$, hence, by Vieta's formula, their product is $(-1)^{q-1}(-1) = -1$). So the product of all elements in the extended matrix is $(-1)^3 = -1$. As the product of all elements of M is 1, we get $\beta_1 \beta_2 \beta_3 = -1$.

The lines A_1B_1, A_2B_1 and A_3B_1 have a point in common, thus by Lemma 6.15, the product of their coordinates is 1. Hence the coordinate of A_1B_1 is $1/\beta_2\beta_3$. In the same way, we get that the coordinate of A_2B_2 is $1/\beta_1\beta_3$ and the coordinate of A_3B_3 is $1/\beta_1\beta_2$. Thus the product of the three coordinates is

$$\frac{1}{\beta_2\beta_3}\frac{1}{\beta_1\beta_3}\frac{1}{\beta_1\beta_2} = \frac{1}{(\beta_0\beta_1\beta_2)^2} = \frac{1}{(-1)^2} = 1.$$

So Lemma 6.15 implies that the three lines A_1B_1, A_2B_2 and A_3B_3 have a point in common. $\qquad\square$

From now on, we say that an oval $\mathcal{O}$ has *property* π, if any pair of an inscribed triangle and the corresponding circumscribed triangle of $\mathcal{O}$ are in perspective from a point.

The previous lemma strongly exploits that the base field is finite. We can extend the definition of ovals to any (not necessary finite) projective plane in the natural way: a set of points is called an oval, if no three of its points are collinear and there is a unique tangent to the set at any of its points. For example, in the classical projective plane the boundary points of a smooth convex body form an oval.

The next theorem is more general. It works for all planes over fields having the property that no three tangents to an oval have a point in common. This condition is satisfied for example by the real projective plane and all finite planes of odd order, but it is not satisfied by finite planes of even order.

Theorem 6.18. *Let $\mathbb{K}$ be a commutative field and $\mathcal{O}$ be an oval in $\mathrm{PG}(2,\mathbb{K})$. If $\mathcal{O}$ has property π and no three tangents to $\mathcal{O}$ have a point in common, then $\mathcal{O}$ is a conic.*

Proof. Let us choose the base points of the coordinate system so that the points $(0:0:1)$ and $(1:0:0)$ belong to $\mathcal{O}$, the tangents to $\mathcal{O}$ at these points are the lines $X_0 = 0$ and $X_2 = 0$, respectively, and the point $(1:1:1)$ also belongs to $\mathcal{O}$. Then the affine points of $\mathcal{O}$ can be considered as the graph of a function $f : \mathbb{K} \to \mathbb{K}$. These are the affine points $\{(x, f(x)) : x \in \mathbb{K}\}$, and we have $f(0) = 0$ and $f(1) = 1$.

Apply property π for the triangle $A_1 = (0:0:1)$, $A_2 = (1:0:0)$, $A_3 = (1:x:f(x))$ where $x \neq 0$. The tangents at these points are $X_0 = 0$, $X_2 = 0$ and the line t_x whose equation is not known at present. The vertices of the corresponding circumscribed triangle are $B_3 = (0:1:0)$, $B_2 = (0:1:m)$, where m is the slope of t_x and the point of intersection of the lines $X_2 = 0$ and t_x, that is denoted by B_1. As the affine equation of t_x is $Y = m(X-x)+f(x)$, the affine coordinates of B_1 are $(-f(x)/m + x, 0)$. We know that $m \neq 0$, because $m = 0$ would mean that three tangents to $\mathcal{O}$ would pass on the point $(0:1:0)$. Hence the affine equations of the lines A_1B_1, A_2B_2 and A_3B_3 are $X = (-f(x))/m + x$, $Y = mX$ and $Y = f(x)$, respectively. These three lines have a point in common if and only if $f(x) = -f(x) + mx$, thus we have $m = 2f(x)/x$. So we can calculate the slope of the tangent to $\mathcal{O}$ at any of its affine points. In particular, as the point $(1 : 1 : 1)$ belongs to $\mathcal{O}$, we have $f(1) = 1$, hence the affine equation of the tangent to $\mathcal{O}$ at the point $(1, 1)$ is $Y = 2X - 1$.

Now replace the point A_2 by the point $A_2' = (1 : 1 : 1)$ and apply property π for the triangle $A_1 A_2' A_3$. The vertices of the corresponding circumscribed triangle are B_1', that is the point of intersection of the lines t_x and $Y = 2X - 1$, B_2 and $B_3' = (0 : 1 : 2)$. The affine equation of the line $A_1 B_1'$ is $X = c$ where c is the solution of the equation $2X - 1 = m(X - x) + f(x)$. Thus $c = (mx - f(x) - 1)/(m - 2)$. We know that $m \neq 2$, because $m = 2$ would mean that three tangents to $\mathcal{O}$ would pass on the point $(0 : 1 : 2)$. The affine equations of the lines $A_2' B_2$ and $A_3 B_3'$ are $Y = mX - m + 1$ and $Y = 2X - 2x + f(x)$, respectively. Elementary calculation gives that the three lines have a point in common if and only if $m = (2f(x) - 2x)/(x - 1)$.

Comparing the two formulae for m we get that for all $x \in \mathbb{K}$, $0 \neq x \neq 1$:

$$\frac{2f(x) - 2x}{x - 1} = \frac{2f(x)}{x}.$$

Multiplying by $x(x - 1)$, rearranging and finally dividing by $2 \neq 0$ we get $f(x) = x^2$. The two excluded points, $(0, 0)$ and $(1, 1)$, also satisfy this equation, hence all affine points of $\mathcal{O}$ are on the curve with equation $Y = X^2$. The domain of the function f must contain all elements of $\mathbb{K}$, because if an element d were not in the domain, then the line with affine equation $X = d$ would be a second tangent to $\mathcal{O}$ at its point $(0 : 0 : 1)$.

Hence in the projective plane $\mathrm{PG}(2, \mathbb{K})$ the oval $\mathcal{O}$ coincides with the conic $X_0 X_2 = X_1^2$. $\qquad\square$

It is worth mentioning that in the case of the classical projective plane if the function f is differentiable, then we do not need the second part of the calculation. In this case the slope of the tangent is $f'(x)$, hence from the first part of the proof we get the differential equation $f'(x) = 2f(x)/x$. As we also know $f(0) = 0$, $f(1) = 1$ and $f'(0) = 0$, its unique solution is $f(x) = x^2$.

Now we are ready to prove the Theorem of Segre.

Theorem 6.19 (Segre). *If q is odd, then each oval of the finite plane $\mathrm{PG}(2, q)$ is a conic.*

Proof. By Lemma 6.17 we know that the oval has property π. As q is odd, no three tangents to the oval have a point in common, hence Theorem 6.18 gives that the oval is a conic. $\qquad\square$

The Theorem of Segre is not valid in finite planes of even order. Let $\mathcal{C}$ be a conic (for example the parabola constructed in the proof of Theorem 6.9) and let N be the nucleus of $\mathcal{C}$. Take a point $P \in \mathcal{C}$ and consider the set $(\mathcal{C} \setminus \{P\}) \cup \{N\}$. This set of points is clearly an oval. But if $q > 4$, then it could not be a conic, because, by Pascal's theorem, two distinct conics have at most 4 points in common. A natural question is whether each hyperoval is the union of a conic and its nucleus. The answer to this question is no; a more detailed explanation of this problem and further properties of hyperovals can be found in Chapter 11.

The core of the proof of Segre's theorem is Lemma 6.17. This lemma is usually called the *Lemma of Tangents* and it has several applications in finite geometry. Let us look one of them.

Proposition 6.20 (Bichara, Korchmáros). *Let $\mathcal{K}$ be a set of $q+2$ points in the plane $\mathrm{PG}(2,q)$. Suppose that the points $A_1, A_2, A_3 \in \mathcal{K}$ have the property that each line through at least one of them meets $\mathcal{K}$ in exactly two points. Then q is even.*

Proof. We proceed in the same way as in the proof of Lemma 6.17.

Choose the points A_1, A_2 and A_3 as base points of the coordinate system. Let $A_1 = (1:0:0)$, $A_2 = (0:1:0)$ and $A_3 = (0:0:1)$. Create a $(q-1) \times 3$ matrix $M = (m_{i,j})$ whose rows correspond to the points $\mathcal{K} \setminus \{A_1, A_2, A_3\} = \{P_1, P_2, \ldots, P_{q-1}\}$, and the columns correspond to the points $\{A_1, A_2, A_3\}$. For $i = 1, 2, \ldots, q-1$ and $j = 1, 2, 3$, let the entry $m_{i,j}$ of M be the coordinate of the line $P_i A_j$.

It follows from Lemma 6.15 that the product of the three elements in a row is equal to 1, hence the product of all entries of M is equal to 1. On the other hand, in each of the three columns we see the $q-1$ non-zero elements of $\mathrm{GF}(q)$, because for $i_1 \neq i_2$ the points P_{i_1}, P_{i_2} and A_j are not collinear. Thus the product of the elements in each column is equal to -1, hence the product of all entries of M is also equal to -1. This means that $1 = -1$, so q must be even. $\square$

The Theorem of Segre suggests the natural question whether in Galois planes of odd order there are complete arcs other than conics. The answer is affirmative.

Theorem 6.21 (Lombardo–Radice). *Let $q \equiv 3 \pmod 4$. Then the following set of points $\mathcal{K}$ is a complete $(q+5)/2$-arc in $\mathrm{PG}(2,q)$:*

$$\left\{ \left(1 : z : \frac{1}{z}\right) \; : \; z = w^2, 0 \neq w \in \mathrm{GF}(q)\right\} \cup \{(0:0:1), (0:1:0), (1:0:0)\}.$$

Proof. All but one point of $\mathcal{K}$ belong to the conic $X_0^2 = X_1 X_2$. We know that conics are arcs, hence if three points of $\mathcal{K}$ are collinear, then one of them must be $(1:0:0)$. Both of the lines $X_1 = 0$ and $X_2 = 0$ meet $\mathcal{K}$ in two points. From now on, we use affine coordinates. The affine part of $\mathcal{K}$ contains the origin and the $(q-1)/2$ points from the hyperbola $XY = 1$ whose coordinates are square elements in $\mathrm{GF}(q)$. Suppose that the points $(0,0)$, $(u, 1/u)$ and $(v, 1/v)$ are collinear. Elementary calculation gives that $u^2 = v^2$, hence $u = \pm v$. As $q \equiv 3 \pmod 4$, the element -1 is not a square in $\mathrm{GF}(q)$, so $u = -v$ is impossible. Thus $u = v$, hence $\mathcal{K}$ is an arc.

Now we show that $\mathcal{K}$ is complete. By Proposition 6.4, we have to prove that the secants of $\mathcal{K}$ cover the whole projective plane. If at least one homogeneous coordinate of a point is 0, then the point is obviously covered. Consider the affine point $A = (a, b)$ where $ab \neq 0$. If either a or b is a square element in $\mathrm{GF}(q)$, then the secant line having affine equation $X = a$ or $Y = b$ covers the point. Finally, suppose that both a and b are non-squares. Then the element a/b is a square, say $a/b = u^2$. Then $a/b = (-u)^2$ also holds. Exactly one of u and $-u$ is a square, because -1 is a non-square. We may assume without loss of generality that u is a square. Then the point $(u, 1/u)$ is in $\mathcal{K}$ and the secant joining $(0, 0)$ and $(u, 1/u)$ has equation $X = u^2 Y$, thus it contains the point A. □

This construction can be extended to planes of order $q \equiv 1 \pmod{4}$, $q \geq 9$, with a slight modification. Take a conic $\mathcal{C}$ and one of its internal points, say P. Consider the secants to $\mathcal{C}$ through P and choose one of the two points of intersection on each secant. These points and P clearly form a $(q + 3)/2$-arc. This arc maybe not complete, but one can extend it to a complete arc $\mathcal{K}$. The cardinality of $\mathcal{K}$ is strictly less than $q + 1$, because if it were $q + 1$, then, by the Theorem of Segre, $\mathcal{K}$ would be a conic. This is impossibile, because two distinct conics have at most 4 points in common and the condition $q \geq 9$ implies $|\mathcal{C} \cap \mathcal{K}| \geq 5$.

This construction does not work if q is even. In this case there are other, quite similar methods; we sketch one of them in the following theorem.

Theorem 6.22 (Tallini Scafati). *Let q be even and $m \in \mathrm{GF}(q)$ be an arbitrary element that cannot be written as $u^2 + u$ with $u \in \mathrm{GF}(q)$. Then the set of points*

$$\mathcal{K} = \{(1 : z^2 + z : (z^2 + z)^2) : z \in \mathrm{GF}(q)\} \cup \{(0 : 0 : 1), (0 : 1 : m)\}$$

is a complete $(q + 4)/2$-arc in $\mathrm{PG}(2, q)$.

Proof. Consider the additive group of $\mathrm{GF}(q)$. Its subset $\{x^2 + x : x \in \mathrm{GF}(q)\}$ is clearly a subgroup, because q is even. The mapping $x \mapsto x^2 + x$ is a linear transformation over $\mathrm{GF}(2)$ whose kernel is $\{0, 1\}$. The kernel is 1-dimensional, thus the image is a 1-codimensional subspace, hence it contains $q/2$ elements.

All but one point of $\mathcal{K}$ belong to the parabola $X_1^2 = X_0 X_2$, hence these points form an arc. The line joining the affine points (u, u^2) and (v, v^2) has slope $u + v$, thus it can be written as $x^2 + x$ with $x \in \mathrm{GF}(q)$. As m cannot be written in this form, the point $(0 : 1 : m)$ does not lie on any secant of $\mathcal{K} \setminus \{(0 : 1 : m)\}$, hence $\mathcal{K}$ is a $(q + 4)/2$-arc.

The proof of the completeness of $\mathcal{K}$ is more complicated (see [164]), so we skip it. Let us remark that the arc $\mathcal{K} \setminus \{(0 : 1 : m)\}$ can be extended to a complete arc in $q/2$ ways, because we can add any point $(0 : 1 : m)$ if m cannot be written as $u^2 + u$ with $u \in \mathrm{GF}(q)$. □

Our next construction gives complete arcs in planes of square order. It was discovered independently by *Ebert* [59]; *Boros* and *Szőnyi* [32]; and *Fisher*,

Hirschfeld and *Thas* [67] in 1985-86. We will prove in Chapter 8 that these arcs are the second largest complete arcs in some cases. Actually, these arcs were found earlier by *Kestenband* [106] without proving their completeness. For the construction we use the algebraic representation of PG(2, q) in GF(q^3) as desribed in Theorem 2.15. Recall that in this model two non-zero elements g and f of GF(q^3) correspond to the same point of PG(2, q) if and only if $f/g \in$ GF(q), that is $f^{q-1} = g^{q-1}$. First, we prove a lemma about the existence of cyclic arcs in this model.

Lemma 6.23. *Let N be a divisor of $q^2 + q + 1$ and consider the $N(q-1)$st roots of unity in GF(q^3). These elements correspond to a set $\mathcal{N}$ of N points in PG(2, q). This set of points is an arc if and only if the following holds. Let $u^N = v^N = 1$ and $u \neq 1 \neq v$, then the equality*

$$u\frac{u^q - 1}{u - 1} = v\frac{v^q - 1}{v - 1} \tag{6.2}$$

implies $u = v$.

Proof. Suppose that three distinct points from the set $\mathcal{N}$ are collinear in PG(2, q). In this model, multiplication by any $N(q-1)$st root of unity corresponds to a collineation that fixes the $N(q-1)$st roots of unity setwise. Thus we may assume without loss of generality that the element 1 corresponds to one of the three chosen points. Let x and y be the two other $N(q-1)$st roots of unity corresponding to the elements of the collinear triple. Then there exist $0 \neq a, b \in$ GF(q) for which $a + bx = y$ holds. Hence $a + bx^q = y^q$ also holds because $a^q = a$ and $b^q = b$. Subtracting these two equations and rearranging it we get $b(x^q - x) = y^q - y$. Taking the $(q-1)$st power (in the form $z^{q-1} = \frac{z^q}{z}$) and applying $b^{q-1} = 1$ we have

$$\frac{x^{q^2} - x^q}{x^q - x} = \frac{y^{q^2} - y^q}{y^q - y},$$

so

$$\frac{x^{q^2-1} - x^{q-1}}{x^{q-1} - 1} = \frac{y^{q^2-1} - y^{q-1}}{y^{q-1} - 1}.$$

Let $x^{q-1} = u$ and $y^{q-1} = v$; then this is the same as Equation (6.2).

The points represented by 1, x and y are pairwise distinct if and only if 1, u and v are pairwise distinct. As each N-th root of unity can be written in the form z^{q-1} with a suitable z, the non-existence of a collinear triple in the corresponding set of points is equivalent to the condition that if u and v satisfy Equation (6.2), then $u = v$. □

Theorem 6.24. *Let $q > 4$ be a square. Then there exist complete $(q - \sqrt{q} + 1)$-arcs in PG(2, q).*

Proof. Let $\mathcal{A}$ be the set of points in PG(2, q) corresponding to the $(q - \sqrt{q} + 1)(q - 1)$st roots of unity in GF($q^3$). We claim that $\mathcal{A}$ is a complete arc. Take the involutory field automorphism $\sigma : x \mapsto x^{q\sqrt{q}}$. As $q - \sqrt{q} + 1$ divides $q\sqrt{q} + 1$, the equality $z^{q-\sqrt{q}+1} = 1$ implies $z^{q\sqrt{q}} = \frac{1}{z}$. Thus σ maps any $(q - \sqrt{q} + 1)$st root of unity to its multiplicative inverse. Hence if $z^{q-\sqrt{q}+1} = 1$, then

$$\left(z\frac{z^q - 1}{z - 1}\right)^{\sigma} = \frac{1}{z}\frac{\left(\frac{1}{z}\right)^q - 1}{\left(\frac{1}{z}\right) - 1} = \frac{1}{z}\frac{1 - z^q}{z^{q-1} - z^q} = \frac{1}{z^q}\frac{z^q - 1}{z - 1}.$$

This means that applying σ to both sides of Equation (6.2) we get

$$\frac{1}{u^q}\frac{u^q - 1}{u - 1} = \frac{1}{v^q}\frac{v^q - 1}{v - 1}.$$

Dividing it by Equation (6.2) we have $u^{q+1} = v^{q+1}$. This implies $u = v$ because $q + 1$ and $(q - \sqrt{q} + 1)$ are coprimes. Hence, by Lemma 6.23, $\mathcal{A}$ is an arc.

Suppose that $\mathcal{A}$ is not complete, that is, it can be extended by the point corresponding to an element w. As the arc is cyclic, this implies that $\mathcal{A}$ can also be extended by any point corresponding to wx where x is an arbitrary $(q - \sqrt{q} + 1)(q - 1)$st root of unity. Of course, we do not know whether $\mathcal{A}$ can be extended by w and wx simultaneously. But we do know that the points corresponding to the elements of type wx also form an arc, because this set is nothing else than the set of elements corresponding to $\mathcal{A}$ multiplied by w, and multiplication defines a collineation.

If $\mathcal{A}$ can be extended by a point P, then there are $q - \sqrt{q} + 1$ tangents to $\mathcal{A}$ through P. We know that $\mathcal{A}$ has exactly $(q - \sqrt{q} + 1)(\sqrt{q} + 1)$ tangents, $\sqrt{q} + 1$ through any of its points. On the other hand, there are at least $q - \sqrt{q} + 1$ points of type P and there is a $(q - \sqrt{q} + 1)$-arc among them. The number of tangents to $\mathcal{A}$ through a point of type P is $q - \sqrt{q} + 1$. No three points of an arc are collinear, hence any tangent to $\mathcal{A}$ is counted at most twice, so the total number of tangents to $\mathcal{A}$ is at least $(q - \sqrt{q} + 1)^2/2$. But if $q > 9$, then

$$\frac{(q - \sqrt{q} + 1)^2}{2} > (q - \sqrt{q} + 1)(\sqrt{q} + 1).$$

This contradiction proves the theorem if $q > 9$. For $q = 9$ elementary calculation gives that the secants of the corresponding 7-arc covers the plane $PG(2, 9)$.   □

In the second part of this chapter blocking sets are investigated.

Definition 6.25. *A set of points $\mathcal{B}$ of a plane is called a* blocking set *if $\mathcal{B}$ meets every line of the plane.*

Lemma 6.26. *Let $\mathcal{B}$ be a blocking set in a projective plane of order n. Then $|\mathcal{B}| \geq n + 1$. If equality holds, then $\mathcal{B}$ is a line.*

Proof. Let $P \notin \mathcal{B}$ an arbitrary point. There are $n + 1$ lines through P; each of them intersects $\mathcal{B}$, hence $|\mathcal{B}| \geq n + 1$.

We have already proved the second part of the statement, because that is the dual of Proposition 6.3.   □

In projective planes any line is a blocking set; these are trivial examples. If $\mathcal{B}$ is a blocking set and $\mathcal{P}$ is an arbitrary set of points of the plane, then $\mathcal{B} \cup \mathcal{P}$ is a blocking set, too. The really interesting blocking sets do not contain lines and they are minimal with respect to containment.

Definition 6.27. *A blocking set is called* non-trivial *if it does not contain any line completely.*

A blocking set is called minimal *if none of its proper subsets is a blocking set.*

The definition implies that the complement of a non-trivial blocking set is also a non-trivial blocking set. Note that if $\mathcal{B}$ is a minimal blocking set and $P \in \mathcal{B}$ is a point, then there exists at least one line ℓ_P such that ℓ_P is a tangent to $\mathcal{B}$ at P.

It is easy to see that there is no non-trivial blocking set in the Fano plane. For $n > 2$ the following set, called the *vertexless triangle*, is a minimal non-trivial blocking set of size $3(n-1)$ in any projective plane of order n. Let P_1, P_2 and P_3 be three non-collinear points and let $\mathcal{B} = (P_1 P_2 \cup P_2 P_3 \cup P_3 P_1) \setminus \{P_1, P_2, P_3\}$. A slight modification of this construction gives a non-trivial minimal blocking set of size $2n$ for all $n > 2$: let P_1, P_2 and P_3 be three non-collinear points, R be a point on the line $P_2 P_3$ distinct from both P_2 and P_3, and let $\mathcal{B} = (P_1 P_2 \cup P_1 P_3 \cup \{R\}) \setminus \{P_2, P_3\}$.

Our goal is to construct smaller non-trivial blocking sets. Let us remark that we have already seen large non-trivial minimal blocking sets. The unitals meet every line, hence they are blocking sets, and there is a unique tangent line at each of its points, thus they are minimal ones.

Lemma 6.28. *Let $\mathcal{B}$ be a non-trivial blocking set in a projective plane Π_n of order n. Then each line of Π_n meets $\mathcal{B}$ in at most $|\mathcal{B}| - n$ points.*

Proof. Let ℓ be a line and $P \in \ell \setminus \mathcal{B}$ be a point. There are n other lines through P, each of them contains at least one point of $\mathcal{B}$, hence $|\mathcal{B} \setminus \ell| \geq n$ and this implies

$$|\mathcal{B} \cap \ell| \leq |\mathcal{B}| - n.$$

$\square$

The starting point of the investigation of blocking sets was Bruck's theorem (Theorem 1.12). Its proof in Chapter 1 contains that if a projective plane Π_n of order n has a subplane $\Pi_{\sqrt{n}}$ of order $\sqrt{n}$ then the sum of the number of tangents to $\Pi_{\sqrt{n}}$ and the number of secants of $\Pi_{\sqrt{n}}$ is exactly the total number of lines in Π_n. Hence the subplane is a blocking set. This blocking set is obviously non-trivial and it is minimal, because we have also proved that there are $n - \sqrt{n}$ tangents to $\Pi_{\sqrt{n}}$ at each of its points. The extended version of the theorem can be formulated in the following way.

Theorem 6.29 (Bruck). *If a projective plane Π_n of order n contains a subplane Π_s of order $s < n$, then either $n = s^2$ or $n \geq s^2 + s$.*

If $n = s^2$, then Π_s is a minimal non-trivial blocking set in Π_n.

Let us recall that a subplane of order $\sqrt{n}$ of a projective plane of order n is called a Baer subplane.

Theorem 6.30 (Bruen, Pelikán). *Let $\mathcal{B}$ be a non-trivial blocking set in the projective plane Π_n of order n. Then $\mathcal{B}$ contains at least $n + \sqrt{n} + 1$ points. Equality holds if and only if $\mathcal{B}$ is a Baer subplane.*

Proof. Let the lines of Π_n be $\ell_1, \ell_2, \ldots, \ell_{n^2+n+1}$ and for $1 \leq i \leq n^2 + n + 1$ let $m_i = |\ell_i \cap \mathcal{B}|$. The standard double countings of the pairs (point in $\mathcal{B}$, line through it) and the triples (two points in $\mathcal{B}$, line through them) give

$$\sum_{i=1}^{n^2+n+1} m_i = |\mathcal{B}|(n+1) \quad \text{and} \quad \sum_{i=1}^{n^2+n+1} m_i(m_i - 1) = |\mathcal{B}|(|\mathcal{B}| - 1).$$

As $\mathcal{B}$ is a non-trivial blocking set, by Lemma 6.28, we have $1 \leq m_i \leq |\mathcal{B}| - n$. Applying these inequalities we get

$$|\mathcal{B}|(|\mathcal{B}| - 1) = \sum_{i=1}^{n^2+n+1} m_i(m_i - 1) \leq (|\mathcal{B}| - n) \sum_{i=1}^{n^2+n+1} (m_i - 1)$$

$$= (|\mathcal{B}| - n)(|\mathcal{B}|(n+1) - (n^2 + n + 1)).$$

Rearranging it gives

$$0 \leq |\mathcal{B}|^2 - (2n + 2)|\mathcal{B}| + n^2 + n + 1.$$

The solution of this quadratic inequality is

$$|\mathcal{B}| \leq n - \sqrt{n} + 1 \quad \text{or} \quad n + \sqrt{n} + 1 \leq |\mathcal{B}|.$$

By Lemma 6.26, $n + 1 \leq |\mathcal{B}|$, so we get the first part of the statement.

If equality holds, then for all i either $m_i = 1$ or $m_i = |\mathcal{B}| - n$. Thus the lines intersecting $\mathcal{B}$ in at least two points meet it in exactly $|\mathcal{B}| - n = \sqrt{n} + 1$ points. There is a unique secant line joining any pair of points of $\mathcal{B}$. As $|\mathcal{B}| = n + \sqrt{n} + 1$ and each secant line contains $\sqrt{n} + 1$ points, there are exactly $\sqrt{n} + 1$ secant lines through each point, thus any two secants meet in a unique point of $\mathcal{B}$. Hence the points of $\mathcal{B}$ and its secant lines form a subplane of order $\sqrt{n}$. □

Let us look at another proof of the first statement of the theorem.

Proof. The notation is the same as in the first proof. If there exists a line ℓ_i for which $|\mathcal{B} \cap \ell_i| > \sqrt{n} + 1$, then, by Lemma 6.28, $|\mathcal{B}| > n + \sqrt{n} + 1$. If for all i the inequalities $1 \leq m_i \leq \sqrt{n} + 1$ hold, then

$$\sum_{i=1}^{n^2+n+1} (m_i - 1)(m_i - \sqrt{n} - 1) \leq 0.$$

From this we get

$$|\mathcal{B}|(|\mathcal{B}| - 1) - |\mathcal{B}|(n+1)(\sqrt{n}+1) + (n^2+n+1)(\sqrt{n}+1) \leq 0.$$

The solution of this quadratic inequality is $n + \sqrt{n} + 1 \leq |\mathcal{B}| \leq n\sqrt{n} + 1$. If $|\mathcal{B}| = n + \sqrt{n} + 1$, then for all i either $m_i = 1$ or $m_i = \sqrt{n} + 1$. $\qquad\square$

In the rest of this chapter we consider only the Galois planes $\mathrm{PG}(2, q)$. In this case there are stronger results than the Theorem of Bruen and Pelikán.

In the late 1960's *Di Paola* [58] investigated blocking sets of planes having small order. For $q = 4$ and 9 she proved that the smallest blocking set has size 7 and 13, respectively. These results are consistent with the previous theorem. For $q = 3, 5$ and 7 she found the sizes $6, 9$ and 12, respectively. She conjectured that if p is an odd prime, then a blocking set in $\mathrm{PG}(2, p)$ has size at least $3(p+1)/2$. This conjecture was proved by *Blokhuis* [24] 25 years later (see Theorem 6.38). The bound is tight as Example 6.45 shows. If $p = 7$ or 13, then the plane $\mathrm{PG}(2, p)$ contains two projectively non-equivalent non-trivial blocking sets of size 12 and 21, respectively. *Blokhuis, Brouwer* and *Wilbrink* [26] proved in 2003 by an exhaustive computer search that for $p \leq 31$ these are the only exceptional planes. For all other values of p the non-trivial blocking set of size $3(p+1)/2$ is projectively equivalent to the blocking set constructed in Example 6.45, Part 1.

For proving Blokhuis' result we need some definitions and theorems about polynomials over finite fields.

Definition 6.31. *A commutative ring with an identity is a set R together with two binary operations $+$ (addition) and $\cdot$ (multiplication) satisfying the following axioms:*

- *$(R, +)$ is an abelian group;*

- *multiplication is associative and commutative;*

- *the operations satisfy the distributive law:*

$$(a + b) \cdot c = a \cdot c + b \cdot c$$

  for all a, b, $c \in R$;

- *there exists a multiplicative identity $1 \in R$, for which $1 \cdot a = a$ for all $a \in R$.*

For an arbitrary field $\mathbb{K}$ the polynomials with coefficients from $\mathbb{K}$ in an indeterminate X form the *polynomial ring* $\mathbb{K}[X]$. The degree of a polynomial $p(X)$, written as $\deg p$, is the largest k such that the coefficient of X^k is not zero. The zero polynomial has no degree. A polynomial $p(X) \in \mathbb{K}[X]$ with $\deg p = d > 0$ is *reducible* if it can be written as a product of polynomials of smaller degree, that is there exist polynomials $r(X)$, $s(X) \in \mathbb{K}[X]$ with

$0 < \deg r$, $\deg s < d$ such that $p(X) = r(X)s(X)$. Otherwise $p(X)$ is irreducible. Constant polynomials are neither reducible nor irreducible. A polynomial is called *monic*, if its leading coefficient is 1. The polynomial ring $\mathbb{K}[X]$ is not only a commutative ring with an identity, but it is also a *unique factorization domain*. This means that any non-zero polynomial $p(X)$ can be written as

$$p(X) = c \prod_{i=1}^{m} f_i(X)$$

where c is a non-zero constant and each $f_i(X)$ is an irreducible monic polynomial in $\mathbb{K}[X]$. This factorization is unique up to the order of the polynomials $f_i(X)$.

Let $f(X), g(X) \in \mathbb{K}[X]$. The Euclidean algorithm, just as in the case of the integers, works in $\mathbb{K}[X]$. If $g(X)$ is not the zero polynomial, then there are unique polynomials $q(X)$, $r(X) \in \mathbb{K}[X]$ such that

$$f(X) = g(X)q(X) + r(X), \quad \text{and} \quad \deg r < \deg g.$$

We say that g *divides* f, and write $g|f$, if there exists $h(X) \in \mathbb{K}[X]$ such that $f(X) = g(X)h(X)$. If at least one of $f(X)$ and $g(X)$ is not the zero polynomial, then

- $f(X)$ and $g(X)$ have a unique monic greatest common divisor, denoted by $(f(X), g(X))$;

- there exist polynomials $u(X)$, $v(X) \in \mathbb{K}[X]$ such that

$$(f(X), g(X)) = u(X)f(X) + v(X)g(X).$$

A useful tool for studying factorization properties of polynomials with real coefficients is the derivative. This notion can be extended in the natural way.

Definition 6.32. *Let* $f(X) = \sum_{i=0}^{n} a_i X^i$ *be a polynomial in the ring* $\mathrm{GF}(q)[X]$. *The* formal derivative *of* $f(X)$ *is the polynomial*

$$f'(X) = \sum_{i=0}^{n-1} (i+1)a_{i+1}X^i,$$

where $(i+1)a_{i+1}$ *is the element of* $\mathrm{GF}(q)$, *defined as* $a_{i+1} + \ldots + a_{i+1}$ *where the number of summands in the sum is* $i + 1$.

It is easy to check that many of the properties of classical derivatives are true for formal derivatives. We leave the proof of the next proposition as an exercise.

Proposition 6.33. *For any two polynomials* $f(X), g(X) \in \mathrm{GF}(q)[X]$ *and* $a, b \in \mathrm{GF}(q)$ *the following hold:*

- *The formal derivative is linear:*

$$(af + bg)'(X) = af'(X) + bg'(X).$$

- *The formal derivative satisfies Leibniz's rule:*

$$(f \cdot g)'(X) = f'(X)g(X) + f(X)g'(X).$$

- *The formal derivative satisfies the chain rule:*

$$(f(g(X)))' = f'(g(X)) \cdot g'(X).$$

Let p be the characteristic of $\mathrm{GF}(q)$. A multiple root of a polynomial with multiplicity m is a root of its formal derivative with multiplicity $m - 1$ if m is not divisible by p, and with multiplicity at least m if m is divisible by p. It follows from the definition that $(X^p)' = 0$, thus it could happen that the formal derivative of a non-constant polynomial is 0. But we can control this situation because it happens if and only if each exponent in $f(X)$ is divisible by p. In this case $f(X)$ can be considered as a polynomial of X^p.

We would like to describe the intersection properties of sets of affine points and lines by polynomials. To achieve this goal we introduce the Rédei polynomial. Most of the following theorems arose from the book written by *Rédei* [145] about lacunary polynomials.

Definition 6.34. *Let $U \subset \mathrm{AG}(2, q)$ be a set of points, say $U = \{(a_i, b_i) : i = 1, 2, \ldots, |U|\}$. The* Rédei polynomial *of U is defined as*

$$H(X, Y) = \prod_{i=1}^{|U|}(X + a_i Y - b_i) = \sum_{j=0}^{|U|} h_j(Y) X^{|U|-j}.$$

Note that $h_j(Y)$ is a polynomial of degree at most j in Y and $h_0(Y) \equiv 1$.

Lemma 6.35. *Let $H(X, Y)$ be the Rédei polynomial of the set of points U and let $m \in \mathrm{GF}(q)$. Then $X = k$ is a root of the monic polynomial $H(X, m)$ with multiplicity r if and only if the affine line with equation $Y = mX + k$ meets U in exactly r points.*

Proof. Exactly one linear factor of the Rédei polynomial belongs to each point $(a_i, b_i) \in U$. As $k + a_i m - b_i = 0$ if and only if the line with equation $Y = mX + k$ contains the point (a_i, b_i), the value $X = k$ is a root with multiplicity r if and only if r linear factors vanish at (a_i, b_i). It happens if and only if U has exactly r points on the line $Y = mX + k$. $\qquad\square$

The next result about lacunary polynomials is due to *Blokhuis*. We prove it only for prime fields. The general theorem and the complete proof can be found in [88, Theorem 13.17]. First we need a definition.

Definition 6.36. *The polynomial* $f \in \mathrm{GF}(q)[X]$ *is* fully reducible *over* $\mathrm{GF}(q)$ *if it factorizes completely into linear factors over* $\mathrm{GF}(q)$.

Theorem 6.37 (Blokhuis). *Let p be a prime and let $f \in \mathrm{GF}(p)[X]$ be fully reducible over $\mathrm{GF}(p)$. Suppose that $f(X) = X^p g(X) + h(X)$ where g and h have no common factor. If $f(X) \neq a(X^p - X)$ and $f(X) \neq aX^p + b$ then at least one of g and h has degree greater than or equal to $(p+1)/2$.*

Proof. We may assume that both $\deg g$ and $\deg h$ are less than p; otherwise the theorem is proven. If $f'(X) = 0$, then either $f(X) = aX^p + b$, or at least one of g and h has degree greater than or equal to p. So we may also assume that $f'(X) \neq 0$.

Let $s(X)$ be the product of all linear factors of $f(X)$ exactly once and let $t(X)$ be the rest. Then the roots of $s(X)$ are the simple roots of $f(X)$, hence $s(X)$ is the greatest common divisor of $f(X)$ and $X^p - X$. The roots of $t(X)$ are the multiple roots of $f(X)$, hence $t(X)$ is the greatest common divisor of $f(X)$ and $f'(X)$. Since $s(X)$ divides both $X^p - X$ and $f(X) = X^p g(X) + h(X)$, we get that $s(X)$ divides $Xg(X) + h(X)$. If $Xg(X) + h(X) = 0$, then $g(X) = a$, since $g(X)$ and $h(X)$ are coprime. Using $g(X) = a$ and $Xg(X) + h(X) = 0$ we get $h(X) = -aX$ and $f(X) = a(X^p - X)$.

Since $t(X)$ divides $f(X) = X^p g(X) + h(X)$ and $f'(X) = X^p g'(X) + h'(X)$, we have

$$t(X) \mid f(X)g'(X) - f'(X)g(X) = h(X)g'(X) - h'(X)g(X).$$

We claim that the polynomial on the right-hand side is the zero polynomial if and only if both $g(X)$ and $h(X)$ are constant polynomials. The equality $h(X)g'(X) = h'(X)g(X)$ and $(g, h) = 1$ imply that $h(X)$ divides $h'(X)$ and $g(X)$ divides $g'(X)$, so $h'(X) = g'(X) = 0$. Hence $f'(X) = 0$, but we have already excluded this possibility.

Thus

$$f(X) = s(X)t(X) \mid (Xg(X) + h(X))(h(X)g'(X) - h'(X)g(X)),$$

and the polynomial on the right-hand side is not the zero polynomial, so its degree is greater than or equal to $\deg f$. We distinguish three cases.

First, consider the case $\deg g = \deg h = k$. Then $\deg f = p + k$, $\deg(Xg + h) = k + 1$ and $\deg(hg' - h'g) \leq 2k - 2$, because the highest order terms cancel in their difference. So $p + k \leq k + 1 + 2k - 2$, hence $k \geq (p+1)/2$. If $k = \deg g > \deg h$, then $\deg(Xg+h) = k+1$ and $\deg(hg' - h'g) \leq 2k - 2$, hence $k \geq (p+1)/2$. Finally, if $k = \deg h > \deg g = s \leq k - 1$, then $\deg(Xg+h) \leq k$ and $\deg(hg' - h'g) \leq k + s - 1$, while $\deg f = p + s$. Hence $p + s \leq k + k + s - 1$, so $k \geq (p+1)/2$ holds in this case, too. $\square$

In the book of *Rédei* only the case $f(X) = X^q + g(X)$ was considered, and he proved the previous theorem in this particular case. The next theorem proves the conjecture of di Paola for the planes $\mathrm{PG}(2,p)$, p prime. Let us remark that the corresponding result for blocking sets of Rédei type (see Definition 6.42 later) was already proved in Chapter 36 of the book of *Rédei* [145].

Theorem 6.38 (Blokhuis). *Let $\mathcal{B}$ be a non-trivial blocking set in $\mathrm{PG}(2,p)$, $p > 2$ prime. Then $\mathcal{B}$ contains at least $3(p+1)/2$ points.*

Proof. Consider $\mathrm{PG}(2,p)$ as $\mathrm{AG}(2,p) \cup \ell_\infty$. We may assume without loss of generality that $\mathcal{B}$ is minimal, it contains $p+k$ points, one of them is (∞) and $|\mathcal{B} \cap \ell_\infty| > 1$. Let

$$\mathcal{B} \cap \ell_\infty = \{(\infty), (m_1), \ldots, (m_{k-1-s})\}$$

and let

$$U = \{(a_i, b_i) \ : \ i = 0, \ldots, p+s-1\}$$

be the affine part of $\mathcal{B}$.

Consider the Rédei polynomial $H(X,Y)$ of U. If $(m) \notin \mathcal{B}$, then each affine line of slope m must contain at least one point from U. Thus Lemma 6.35 implies that if $(m) \notin \mathcal{B}$ then $H(X,m)$ is divisible by $(X^p - X)$. This means that for such an m the polynomial $H(X,m)$ is lacunary; there is a long sequence of zeros in its coefficients. Namely, each coefficient between X^p and X^{s+1} equals to 0. Hence the polynomials $h_{s+1}(Y), \ldots, h_{p-1}(Y)$ vanish when $m \neq m_r$ for $r = 1, 2, \ldots, k-1-s$. There are $p+s-k+1$ choices for m, the degree of $h_j(Y)$ is at most j, so the polynomial $h_j(Y)$ is the zero polynomial if $j \leq p+s-k$.

Now, consider $H(X,m)$ when $m = m_r$ for $r = 1, \ldots, k-1-s$. The coefficients of $X^{p-1}, \ldots, X^k$ are zeros because $h_j(Y) \equiv 0$. Hence for $m = m_r$ we get

$$H(X, m_r) = X^p g(X) + h(X), \quad \deg g = s, \quad \deg h \leq k-1.$$

This is almost the situation considered in Theorem 6.37, because $H(X, m_r)$ is fully reducible by definition. It could happen that $\gcd(g,h) \neq 1$, but in this case we can divide by the greatest common divisor of g and h and get the polynomial $f_1(X) = X^p g_1(X) + h_1(X)$ where $\gcd(g_1, h_1) = 1$, $\deg g_1 \leq s$ and $\deg h_1 \leq k-1$. If $f_1(X) = aX^p + b = (aX + b)^p$, then the line with equation $Y = m_r X - b/a$ would be completely contained in $\mathcal{B}$, contradicting the non-triviality condition, while if $H(X, m_r)$ were divisible by $(X^p - X)$, then $\mathcal{B} \setminus \{(m_r)\}$ would also be a blocking set contradicting the minimality of $\mathcal{B}$.

So all conditions of Theorem 6.37 are satisfied. As $k-1 \geq s$, from Theorem 6.37 we get $k-1 \geq (p+1)/2$, hence $|\mathcal{B}| = p+k \geq 3(p+1)/2$. $\square$

The Rédei polynomial can also be applied to the investigation of blocking sets in the affine planes. In this there are no trivial examples, since a line is not a blocking set. The clumsiest way of constructing blocking sets is to take a line and choose one point on each of the $q-1$ lines parallel to it. This yields a blocking set of size $2q-1$ in an affine plane of order q. The surprising result of *Jamison* [97], and *Brouwer* and *Schrijver* [34] shows that we cannot do better if the plane is AG$(2, q)$.

Theorem 6.39 (Jamison; Brouwer, Schrijver). *If $\mathcal{B}$ is a blocking set in* AG$(2, q)$, *then* $|\mathcal{B}| \geq 2q - 1$.

Proof. Let us assume $(0, 0) \in \mathcal{B}$, and let

$$\mathcal{B} \setminus \{(0, 0)\} = \{(a_i, b_i) : i = 1, 2, \ldots, |\mathcal{B}| - 1\}.$$

The equation of a line not through the origin can be written as $CX + DY = 1$, where C, D are not both zero. Write the polynomial $H(X, Y) = \prod(a_i X + b_i Y - 1)$. As $\mathcal{B}$ is an affine blocking set, the polynomial $H(X, Y)$ is zero for every $(x, y) \neq (0, 0)$. For $x = 0$ and $y = 0$, we have $H(0, 0) \neq 0$. Now reduce the polynomial by $X^q - X$ and $Y^q - Y$. Then we get a polynomial $K(X, Y)$, in which the X-degree and the Y-degree is at most $q - 1$. The polynomial $XK(X, Y)$ is zero everywhere, and its Y-degree is at most $q - 1$, hence it is divisible by $X^q - X$. This implies $X^{q-1} - 1 | K(X, Y)$. Similarly, by considering $YK(X, Y)$, we get $Y^{q-1} - 1 | K(X, Y)$, so $(X^{q-1} - 1)(Y^{q-1} - 1) | K(X, Y)$. This shows $\deg K \geq 2(q - 1)$, which implies $|\mathcal{B}| \geq 2(q - 1) + 1$.   □

The above proof works only for the affine Galois plane AG$(2, q)$. There are non-desarguesian planes where Theorem 6.39 is not true, see *Bruen* and *de Resmini* [35], and *De Beule, Héger, Szőnyi* and *Van de Voorde* [53].

Corollary 6.40. *Let $\mathcal{K}$ be a complete arc in* PG$(2, p)$ $p > 2$ *prime. Then $\mathcal{K}$ contains at least $\sqrt{3p}$ points.*

Proof. It follows from Proposition 6.5 that the secants to $\mathcal{K}$ form a blocking set $\mathcal{B}$ in the dual plane.

First, suppose that $\mathcal{B}$ contains a complete line ℓ. Then, by duality, there is a point L in the plane such that all lines of the pencil with carrier L are secants to $\mathcal{K}$. Hence $|\mathcal{K}| \geq p + 2 > \sqrt{3p}$ in this case. If $\mathcal{B}$ is non-trivial, then, by Blokhuis' theorem, $\binom{k}{2} \geq 3(p + 1)/2$ which proves a bit stronger inequality than the statement of the theorem.   □

With more sophisticated calculations Theorem 6.37 can be extended to the case $q = p^h$. Applying these results one can prove the following generalization of Theorem 6.38 (the proof can be found in [88, Theorem 13.18]).

Theorem 6.41 (Blokhuis). *Let $\mathcal{B}$ be a non-trivial blocking set in* PG$(2, q)$, *where $q = p^h$, $h \geq 2$. Then*

- $|\mathcal{B}| \geq q + \sqrt{q} + 1$, *if h is even;*

- $|\mathcal{B}| \geq q + \sqrt{pq} + 1$, *if h is odd.*

If q is a square, then this bound is the same as the (combinatorial) bound in Theorem 6.30, while for h odd, this bound is much better than the combinatorial bound. It is known that in the case $q = p^3$ the bound is also tight; the corresponding blocking set of size $p^3 + p^2 + 1$ will be constructed in Example 6.46.

The most studied type of blocking sets is the class of Rédei type blocking sets. These objects are closely connected with the so-called *direction problem*, that is finding the number of directions determined by a function $f : \mathrm{GF}(q) \to \mathrm{GF}(q)$.

Definition 6.42. *A blocking set $\mathcal{B} \subset \mathrm{PG}(2, q)$ is of* Rédei type *with respect to the line ℓ if $|\mathcal{B} \cap \ell| = |\mathcal{B}| - q$.*

Let us remark that, according to Lemma 6.28, the maximum size of a collinear set of points in $\mathcal{B}$ is $|\mathcal{B}| - q$.

Definition 6.43. *A function $f : \mathrm{GF}(q) \to \mathrm{GF}(q)$ determines the direction m if there exist $x \neq y \in \mathrm{GF}(q)$ such that*

$$m = \frac{f(x) - f(y)}{x - y}.$$

A set of points $U \subset \mathrm{AG}(2, q)$ determines the direction $m \in \mathrm{GF}(q) \cup \{\infty\}$ if there exist points $X, Y \in U$ such that in $\mathrm{PG}(2, q) = \mathrm{AG}(2, q) \cup \ell_\infty$ the line joining X and Y contains $(m) \mathrm{I} \ell_\infty$.

Let us remark that if U contains more than q points, then, by the pigeonhole principle, U determines all directions.

Proposition 6.44. 1. *Let $U \subset \mathrm{AG}(2, q)$ be a set of q points and let $D \subset \ell_\infty$ be the set of directions determined by U. Then $\mathcal{B} = U \cup D$ is a blocking set in $\mathrm{PG}(2, q)$. This blocking set is non-trivial if the points of U are not collinear and $|D| \leq q$. The blocking set $\mathcal{B}$ is of Rédei type with respect to ℓ_∞.*

2. *If a blocking set $\mathcal{B} \subset \mathrm{PG}(2, q)$ contains $q + k$ points, $(1 < k \leq q)$, and the line ℓ_∞ contains k points of $\mathcal{B}$, then in the affine plane $\mathrm{PG}(2, q) \setminus \ell_\infty$ the set of points $U = \mathcal{B} \setminus \ell_\infty$ contains q points. If D is the set of directions determined by U, then $D \subset \mathcal{B} \cap \ell_\infty$. Thus if $\mathcal{B}$ is a minimal blocking set, then $\mathcal{B} = U \cup D$. Let us remark that U can be considered as a graph of a function $\mathrm{GF}(q) \to \mathrm{GF}(q)$, because $k \leq q$.*

Proof. If $(m) \notin D$, then each affine line with point at infinity (m) contains at most one point of U. As U contains q points, by the pigeonhole principle, each of these lines must meet U in a single point. This proves that $\mathcal{B} = U \cup D$ is a blocking set. The non-triviality and the Rédei type property obviously follow from the conditions.

For proving Part 2, take a line e that meets U in at least two points. This implies that the point $M = e \cap \ell_\infty$ belongs to $\mathcal{B}$, because the $q - 1$ other affine lines through M cannot be blocked by the remaining $q - 2$ points of U. In another words, each direction determined by U must be in $\mathcal{B} \cap \ell_\infty$. Thus $\mathcal{B}$ can be constructed from U in the way described in Part 1, and maybe we have to delete some points of the set $\mathcal{B} \cap \ell_\infty$ because of the minimality of $\mathcal{B}$. $\qquad\square$

The previous proposition and Theorem 6.38 give that any set $U \subset AG(2,p)$ of p non-collinear points determines at least $(p+3)/2$ directions if $p > 2$ prime. For prime powers, the Theorem of Bruen–Pelikán implies that any set of q non-collinear points determines at least $\sqrt{q} + 1$ directions, and if a set of q points determines exactly $\sqrt{q}+1$ directions then the points and the determined directions must belong to a Baer subplane. But any two Baer subplanes of $PG(2,q)$ are projectively equivalent, hence we proved that there is only one function (up to linear transformations) $f : GF(q) \to GF(q)$ which determines $\sqrt{q} + 1$ directions. The function $f(x) = x^{\sqrt{q}}$ has this property, so this is the unique function (up to linear transformations) determining so few directions.

In the next examples we consider some sets of affine points and functions with few determined directions and the corresponding small blocking sets.

Example 6.45 (Megyesi). The following are minimal non-trivial blocking sets $\mathcal{B} \subset PG(2,q)$ where $\mathcal{B}$ is constructed from the affine set of points U as in Proposition 6.44.

1. Let A be a subgroup of the multiplicative group of $GF(q)$ with index d. Let
$$U = \{(a,0) : a \in A\} \cup \{(0,b) : b \notin A\} \cup \{(0,0)\}.$$
Then U determines the directions
$$D = \{(m) : \ -m \notin A\} \cup \{(0), (\infty)\}.$$
Hence the cardinality of D is $q + 1 - |A|$. This example is called a *projective triangle*.

2. Let A be a subgroup of the additive group of $GF(q)$, and let
$$U = \{(0,a) : a \in A\} \cup \{(1,b) : b \notin A\}.$$
It is easy to see that U determines $q + 1 - |A|$ directions, too.

Example 6.46. Let $q = p^h$ and $q_1 = p^e$, where e is a divisor of h. Then the following examples are minimal non-trivial blocking sets $\mathcal{B} \subset PG(2,q)$, where $\mathcal{B}$ is constructed from the function f as in Proposition 6.44.

1. Let $f : GF(q) \to GF(q_1)$ be the trace function defined as
$$f(x) = \text{Tr}(x) = x + x^{q_1} + x^{q_1^2} + \ldots + x^{q/q_1}.$$
This function determines $1 + q/q_1$ directions, because $(\text{Tr}(x + y) - \text{Tr}(x))/y = \text{Tr}(y)/y$, hence $\text{Tr}(y) = \text{Tr}(\lambda y)$ for $\lambda \in GF(q_1)$. Thus $\mathcal{B}$ is a blocking set of size $q + 1 + q/q_1$.

2. Let $f(x) = x^{q_1}$. Then f is an automorphism of $\mathrm{GF}(q)$ and hence it determines $\dfrac{q-1}{q_1-1}$ directions. The corresponding blocking set contains $q + \dfrac{q-1}{q_1-1}$ points.

Let us remark that in Example 6.45, Part 1, the size of $\mathcal{B}$ is $3(q+1)/2$ if $|A| = (q-1)/2$, hence in the case $q = p$ we get that the bound in Theorem 6.38 is sharp. Similarly, in Example 6.46, Part 1, if $q = p^3$ then $\mathcal{B}$ has size $p^3 + p^2 + 1$, thus the bound in Theorem 6.41 is also sharp if $q = p^3$.

Another interesting problem about blocking sets is to determine the size of the largest minimal blocking sets. The proof of the following theorem is purely combinatorial.

Theorem 6.47 (Bruen, J. Thas). *Let $\mathcal{B}$ be a minimal blocking set in a projective plane Π_n of order n. Then $|\mathcal{B}| \le n\sqrt{n} + 1$.*
Equality holds if and only if $\mathcal{B}$ is a unital.

Proof. Let t denote the number of tangent lines to $\mathcal{B}$. The minimality of $\mathcal{B}$ implies $|\mathcal{B}| \le t$. Let $N = n^2 + n + 1 - t$ and let $\{\ell_1, \ell_2, \ldots, \ell_N\}$ be the set of secant lines to $\mathcal{B}$. Again, for $j = 1, \ldots, N$, let $m_j = |\ell_j \cap B|$. The same double countings as in the proof of the Theorem of Bruen–Pelikán give

$$\sum_{j=1}^{N} m_j = |\mathcal{B}|(n+1) - t \quad \text{and} \quad \sum_{j=1}^{N} m_j(m_j - 1) = |\mathcal{B}|(|\mathcal{B}| - 1), \qquad (6.3)$$

hence

$$\sum_{j=1}^{N} m_j^2 = \sum_{j=1}^{N} m_j + \sum_{j=1}^{N} m_j(m_j - 1) = |\mathcal{B}|^2 + |\mathcal{B}|n - t. \qquad (6.4)$$

Applying the inequality between the arithmetic and quadratic means we get

$$\left(\sum_{j=1}^{N} m_j\right)^2 \le (n^2 + n + 1 - t) \cdot \sum_{j=1}^{N} m_j^2.$$

Substituting (6.3) and (6.4) gives

$$(|\mathcal{B}|n + |\mathcal{B}| - t)^2 \le (n^2 + n + 1 - t)(|\mathcal{B}|^2 + |\mathcal{B}|n - t),$$

which becomes

$$(|\mathcal{B}|^2 + n^2 + n + 1 - |\mathcal{B}|n - 2|\mathcal{B}|)t - |\mathcal{B}|(n^3 + n^2 + n - |\mathcal{B}|n) \le 0.$$

We can substitute t for $|\mathcal{B}|$, because the coefficient of t is positive and $t \ge |\mathcal{B}|$. This gives

$$|\mathcal{B}|^3 - 2|\mathcal{B}|^2 + |\mathcal{B}| \le |\mathcal{B}|n^3,$$

so finally we get

$$|\mathcal{B}| \leq n\sqrt{n} + 1.$$

If equality holds, then $t = |\mathcal{B}|$ and m_j is a constant. Elementary calculation gives that $m_j = \sqrt{n} + 1$. Hence there is a unique tangent to $\mathcal{B}$ at each of its points and each secant contains $\sqrt{n} + 1$ points of $\mathcal{B}$, so $\mathcal{B}$ is a unital. □

With a slight modification of the previous proof one can give an estimate of the number of lines not containing any point of a given set, and also of the cardinality of a set of points having at least as many tangent lines as points.

Baer subplanes and unitals have a joint combinatorial characterization.

Lemma 6.48. *Let n be a square and $\mathcal{B}$ be a set of points in a projective plane Π_n of order n. If every line intersects $\mathcal{B}$ either in 1 or in $\sqrt{n} + 1$ points, then $\mathcal{B}$ is either a Baer subplane or a unital.*

Proof. Copy the proof of Theorem 6.30 again. Let the lines of Π_n be $\ell_1, \ell_2, \ldots, \ell_{n^2+n+1}$ and for $1 \leq i \leq n^2 + n + 1$ let $m_i = |\ell_i \cap \mathcal{B}|$. Now, the double countings give

$$\sum_{i=1}^{n^2+n+1} m_i = |\mathcal{B}|(n+1) \quad \text{and} \quad \sum_{i=1}^{n^2+n+1} m_i(m_i - 1) = |\mathcal{B}|(|\mathcal{B}| - 1),$$

hence

$$\sum_{i=1}^{n^2+n+1} m_i^2 = |\mathcal{B}|^2 + n|\mathcal{B}|.$$

We know that $m_i = 1$ or $m_i = \sqrt{n} + 1$ for all $i = 1, 2, \ldots, n^2 + n + 1$, thus

$$0 = \sum_{i=1}^{n^2+n+1} (m_i - 1)(m_i - \sqrt{n} - 1)$$

$$= \sum_{i=1}^{n^2+n+1} m_i^2 - (\sqrt{n} + 2) \sum_{i=1}^{n^2+n+1} m_i + (n^2 + n + 1)(\sqrt{n} + 1)$$

$$= |\mathcal{B}|^2 - \left(n\sqrt{n} + n + \sqrt{n} + 2\right)|\mathcal{B}| + (n^2 + n + 1)(\sqrt{n} + 1).$$

This is a quadratic equation on $|\mathcal{B}|$, so it has at most two roots. By Theorems 6.30 and 6.47, $\sqrt{n} + 1$ and $n\sqrt{n} + 1$ satisfy the equation, hence it has no other root. Theorems 6.30 and 6.47 also give the characterizations in the cases $|\mathcal{B}| = \sqrt{n} + 1$ and $|\mathcal{B}| = n\sqrt{n} + 1$, respectively. This proves the statement. □

Blocking sets appear in several other problems, too. As we have already seen, we can give bounds on the size of small complete arcs using estimates on the size of blocking sets. In the next theorem we show that line spreads of $PG(3, q)$ are also connected to blocking sets. The spreads were defined when we considered translation planes (see Definition 5.13). A natural question regarding spreads of $PG(3, q)$ is the following: When can a given set of pairwise skew lines in $PG(3, q)$ be extended to a line spread of the space? Before answering this question, we define partial spreads.

Definition 6.49. *In* $PG(3, q)$ *a set of pairwise skew lines is called a* partial spread.

A partial spread is maximal *if it is not contained properly in a larger partial spread.*

It follows from the definition that a maximal partial spread meets all lines of $PG(3, q)$.

Theorem 6.50. *Let b be the size of the smallest non-trivial blocking set in* $PG(2, q)$. *If a partial spread in* $PG(3, q)$ *contains more than $q^2 + q + 1 - b$ lines, then it can be extended to a spread.*

Proof. Obviously, it is enough to prove the statement for maximal partial spreads. Let $\mathcal{P}$ be a maximal partial spread in $PG(3, q)$ and suppose that it is not a spread. Then $\mathcal{P}$ contains at most q^2 lines, hence there are at most $|\mathcal{P}|(q + 1) \leq q^3 + q^2$ planes which contain a line of $\mathcal{P}$. This is less than the total number of planes. Let Π be a plane which does not contain any line of $\mathcal{P}$. Consider the set of points $\mathcal{B}' = \{r \cap \Pi : r \in \mathcal{P}\}$. It is a blocking set in Π, because $\mathcal{P}$ meets all lines of the space; in particular, $\mathcal{B}'$ meets all lines of Π. We claim that $\mathcal{B}'$ is non-trivial. Suppose to the contrary that $\mathcal{B}'$ contains the line ℓ. Then there exist $q+1$ lines $\ell_1, \ell_2, \ldots, \ell_{q+1}$ such that each of them belongs to $\mathcal{P}$ and meets ℓ. For $i = 1, 2, \ldots, q + 1$ the planes $\langle \ell, \ell_i \rangle$ are pairwise different; each of them is different from Π but contains ℓ. This is a contradiction because there are only $q+1$ planes containing ℓ. Hence $\mathcal{B}'$ is a non-trivial blocking set, so its complement, $\Pi \setminus \mathcal{B}'$, is also a non-trivial blocking set. Thus $|\Pi \setminus \mathcal{B}'| \geq b$, so $|\mathcal{B}'| \leq q^2 + q + 1 - b$. $\square$

If q is a prime, then there exist maximal partial spreads whose size is close to this bound. Take a non-regular line spread $\mathcal{R}$ of $PG(3, p)$ and let $\ell \notin \mathcal{P}$ be a line such that the lines of $\mathcal{R}$ meeting ℓ do not form a regulus. Delete these elements from $\mathcal{R}$ and add ℓ to the remaining set of lines. In this way we constructed a partial spread $\mathcal{P}$. We show that $\mathcal{P}$ can be extended to a maximal spread by adding at most one line to it. Suppose to the contrary that $\mathcal{P} \cup \{\ell_1, \ell_2\}$ is a partial spread. Then each line of $\mathcal{R}$ meeting ℓ also meets both ℓ_1 and ℓ_2, hence these lines form a regulus, a contradiction.

In this way we can construct maximal partial spreads containing $q^2 - q + 1$ or $q^2 - q + 2$ lines. The difference of these numbers and the size of a spread is q or $q - 1$, while the upper bound in the case $q = p$ is $q^2 - \frac{q+1}{2}$, so the order of magnitudes of the differences are the same. In particular, *Heden* [81] proved that for $q = 7$ there are maximal partial spreads containing $45 \ (= q^2 - \frac{q+1}{2})$ lines, hence in this case the bound of Theorem 6.50 is tight.

For further reading about arcs and blocking sets we recommend the book of *Hirschfeld* [88], the paper of *Blokhuis* [25] and the survey paper of *Hirschfeld* and *Storme* [91].

Exercises

6.1. A *semioval* in a projective plane Π is a non-empty subset $\mathcal{S}$ of points with the property that for every point $P \in \mathcal{S}$ there exists a unique line ℓ such that $\mathcal{S} \cap \ell = \{P\}$. Show that if the order of Π is n, then $n + 1 \leq |\mathcal{S}| \leq n\sqrt{n} + 1$.

6.2. Show that in $\mathrm{PG}(2,3)$ the number of 3-arcs is equal to the number of 4-arcs.

6.3. Let $\mathcal{K}$ be a 5-arc in $\mathrm{PG}(2,5)$ and i_P denote the number of secants to $\mathcal{K}$ through the point P. Prove in a combinatorial way that

- if ℓ is a secant to $\mathcal{K}$, then there is a unique point $P \in \ell$ with $i_p = 1$;
- the secants to $\mathcal{K}$ cover all but one point of the plane;
- $\mathcal{K}$ can be uniquely extended to a 6-arc.

6.4. Calculate the number of 5-arcs in a projective plane of order n.

6.5. Show that in $\mathrm{PG}(2,q)$ there is a unique conic through the points of a 5-arc.

6.6. Calculate the number of conics in $\mathrm{PG}(2,q)$.

6.7. Show that if $\mathcal{H}_1$ and $\mathcal{H}_2$ are hyperovals in a projective plane of order n and $|\mathcal{H}_1 \cap \mathcal{H}_2| > \frac{n+2}{2}$ then $\mathcal{H}_1 = \mathcal{H}_2$.

6.8. (Bagchi and Sastry [4].) Let n be even and $\mathcal{O}_1, \mathcal{O}_2, \ldots, \mathcal{O}_n$ be n pairwise disjoint ovals in a projective plane of order n. Prove that these ovals have a common tangent line.

6.9. Let n be odd and $\mathcal{O}$ be an oval in a projective plane Π_n. Show that if the point P is not on $\mathcal{O}$, then exactly half of the non-tangent lines through P are secants to $\mathcal{O}$ and half of them are external lines with respect to $\mathcal{O}$.

6.10. Prove Theorem 6.24 for $q = 9$.

6.11. Prove the three properties of formal derivatives listed in Proposition 6.33.

6.12. Show that in $\mathrm{PG}(2,p^2)$, p prime, the size of a complete arc is at least $\sqrt{3p^2}$.

6.13. Prove that the vector space V of functions (polynomials in X, Y) on $AG(2, q)$ (with values in $GF(q)$) has dimension q^2. The polynomials of the form $X^i Y^j$, $0 \le i, j \le q - 1$ form a basis.

6.14. Let $f(X, Y)$ be a polynomial, for which $f(0, 0) \ne 0$ and $f(x, y) = 0$ for every $(x, y) \ne (0, 0)$. Then $f(X - a, Y - b)$ vanishes everywhere except (a, b), hence the q^2 polynomials $f(X - a, Y - b)$, $(a, b) \in GF(q)^2$ span the vector space V.

6.15. Use Exercises 6.13 and 6.14 to prove Theorem 6.39.

7

(k, n)-arcs and multiple blocking sets

In the previous chapters planes of order n and planes of order q were distinguished; the former one referred to arbitrary planes of the given order, while the latter one to desarguesian planes only. In this chapter, in accordance with the customs of the literature, the letter n is reserved for another purpose, so Π_q stands for an arbitrary projective plane of order q and $\mathrm{PG}(2, q)$ denotes the Galois plane over the field $\mathrm{GF}(q)$. Similarly to Chapter 6, the word "plane" without any attribute means any projective or affine plane throughout this chapter, too.

We are going to examine the two objects in the title, as well as to assess their relationship with each other. Besides counting arguments, our main tool will be polynomials over $\mathrm{GF}(q)$. We will explain how to get estimates on the sizes of these sets of points and how to describe their combinatorial structure by using algebraic properties of polynomials.

Let us start with the definitions.

Definition 7.1. *A set of points $\mathcal{K}$ in a plane of order q is a (k, n)-arc if $|\mathcal{K}| = k$, each line of the plane contains at most n points of $\mathcal{K}$ and there exists a line which intersects $\mathcal{K}$ in exactly n points.*

Definition 7.2. *A set of points $\mathcal{B}$ in a plane of order q is a t-fold blocking set if each line of the plane contains at least t points of $\mathcal{B}$ and there exists a line which intersects $\mathcal{B}$ in exactly t points.*

It obviously follows from the definitions that these two objects are the complements of each other in Π_q if $n + t = q + 1$. If a line ℓ contains no point of a (k, n)-arc $\mathcal{K}$, then we can take the complement of $\mathcal{K}$ in the affine plane $\Pi_q \setminus \{\ell\}$. The resulting set is a t-fold blocking set with $t = q - n$ in the affine plane. Usually (k, n)-arcs are considered when n is small, and t-fold blocking sets are investigated when t is small.

First, we look at some combinatorial properties.

Proposition 7.3 (Barlotti). *Let $\mathcal{K}$ be a (k, n)-arc in a plane of order q. Then*

$$k \le nq - q + n,$$

and if equality holds, then n is a divisor of q.

Proof. Let $P \in \mathcal{K}$ be a point and consider the lines through P. Each of these lines contains at most $(n-1)$ points of $\mathcal{K} \setminus \{P\}$, thus $k \le 1 + (q+1)(n-1) = nq - q + n$.

If equality holds, then each line that intersects $\mathcal{K}$ must contain exactly n points of $\mathcal{K}$. In another words, each line meets $\mathcal{K}$ either in 0 or in n points. Take now a point $P' \notin \mathcal{K}$ and consider the lines through it. These lines give a partition of the points of the plane distinct from P'. The number of points of $\mathcal{K}$ on each line is divisible by n, hence the total number of points of $\mathcal{K}$ is also divisible by n. As $|\mathcal{K}| = nq + n - q$, we get that n divides q. $\quad\square$

The corresponding statement for t-fold blocking sets is the following.

Proposition 7.4. *Each t-fold blocking set $\mathcal{B}$ in a plane of order q contains at least $t(q+1)$ points.*

Proof. Let $P \notin \mathcal{B}$ be a point and consider the $(q+1)$ lines through P. Each of them meets $\mathcal{B}$ in at least t points, hence $\mathcal{B}$ contains at least $(q+1)t$ points. $\quad\square$

Changing to the complement of a t-fold blocking set we get a (k,n)-arc $\mathcal{K}$ where $n = q + 1 - t$ or $n = q - t$ according to whether the plane is projective or affine. If $\mathcal{K}$ has $t(q+1)$ points, then Barlotti's result says that n divides q.

If a (k,n)-arc meets all lines, then we can prove another estimate on its size.

Lemma 7.5 (Ball, Hirschfeld [12]). *Let $\mathcal{K}$ be a (k,n)-arc in a projective plane Π_q of order q. Suppose that $\mathcal{K}$ is a blocking set, too. Then*

$$k < (n-1)q + \left\lceil \frac{n^2}{q} \right\rceil.$$

Proof. Write k as $k = (n-1)q + \varepsilon$. First, we prove that each line of Π_q meets $\mathcal{K}$ in at least ε points. Let $P \in \mathcal{K}$ be a point and consider the lines through P. If one of them contains s points of $\mathcal{K}$, then

$$|\mathcal{K}| = k \le 1 + (s-1) + q(n-1) = (n-1)q + s,$$

hence $s \ge \varepsilon$.

Let r_i denote the number of lines intersecting $\mathcal{K}$ in exactly i points. Then the standard double countings of the pairs (point in $\mathcal{K}$, line through it) and the triples (two points in $\mathcal{K}$, line through them) give

$$\sum_{i=\varepsilon}^{n} r_i = q^2 + q + 1,$$

$$\sum_{i=\varepsilon}^{n} i r_i = k(q+1), \tag{7.1}$$

$$\sum_{i=\varepsilon}^{n} i(i-1) r_i = k(k-1).$$

As each line meets $\mathcal{K}$ in at least ε and in at most n points, the inequality

$$0 \le \sum_{i=\varepsilon}^{n}(n - i)(i - \varepsilon)r_i = \sum_{i=\varepsilon}^{n}(-i^2 r_i + (n + \varepsilon)i r_i - \varepsilon n r_i)$$

holds, because each summand in the first sum is non-negative. Substituting from Equations (7.1) we get

$$0 \le -k(k + 1) + (\varepsilon + n - 1)k(q + 1) - \varepsilon n(q^2 + q + 1).$$

Writing $n(q - 1) + \varepsilon$ instead of k, rearranging and dividing by q we conclude that

$$0 \le n^2 - \varepsilon q + \varepsilon(\varepsilon - n) - n. \tag{7.2}$$

If $\varepsilon \ge n^2/q$, then $n^2 - \varepsilon q \le 0$ and $\varepsilon(\varepsilon - n) - n < 0$, so Equation (7.2) gives a contradiction, hence the lemma is proved. $\square$

Corollary 7.6 (Chao). *Let $\mathcal{K}$ be a (k, n)-arc in a projective plane Π_q of order $q > 3$ and suppose that $\mathcal{K}$ is a blocking set, too. If $n \le \sqrt{2q}$, then the inequality*

$$k \le (n - 1)q + 1$$

also holds.

Earlier it was conjectured by *Lunelli* and *Sce* [121] that the inequality $k \le n(q-1)+1$ holds in general provided that n and q are coprimes. This conjecture is false, as was shown by *Hill* and *Mason* [85]; a typical counterexample is the complement of the union of some disjoint Baer subplanes (see Exercises 7.5 and 7.6).

Definition 7.7. *A (k, n)-arc in a plane of order q is called* maximal *if it contains $nq - q + n$ points.*

The next result shows an important connection between a maximal arc and its dual.

Proposition 7.8 (Cossu [47]). *Let Π_q be a projective plane of order q and suppose that there exists a $(qn - q + n, n)$-arc $\mathcal{K}$ in Π_y. Then the set of external lines to $\mathcal{K}$ forms a $(q^2/n - q + q/n, q/n)$-arc in the dual plane Π_q^*.*

Proof. We have to show that there are 0 or q/n external lines to $\mathcal{K}$ through any point in Π_q. As $\mathcal{K}$ is a maximal arc, each line meets it in 0 or in n points. So if $P \notin \mathcal{K}$ is an arbitrary point, then the number of n-secants to $\mathcal{K}$ through P is $(qn - q + n)/n$, hence there are $q + 1 - (qn - q + n)/n = q/n$ external lines to $\mathcal{K}$ through P. As each external line contains $q + 1$ points, the number of the pairs (external line, a point on it) is $e \cdot (q + 1)$ where e denotes the number of external lines. On the other hand, there are $q^2 + q + 1 - (qn - q + n)$ points not in $\mathcal{K}$, so the number of the pairs is $(q^2 + q + 1 - (qn - q + n)) \cdot q/n$. Thus

$$e(q + 1) = \left(q^2 + q + 1 - (qn - q + n)\right)\frac{q}{n},$$

so

$$e = \frac{(q^2 + q + 1 - (qn - q + n))q}{(q+1)n}$$

$$= \frac{(q^2 + q)q}{(q+1)n} - \frac{(qn + n)q}{(q+1)n} + \frac{(q+1)q}{(q+1)n} = \frac{q^2}{n} - q + \frac{q}{n},$$

which proves the statement. $\qquad\square$

If $\mathcal{K}$ is a hyperoval in a plane of even order, then $n = 2$ and, by the above proposition, we get a maximal arc with $k = \frac{q^2}{2} - \frac{q}{2}$ and $n = q/2$. We will show later (see Theorem 7.18) that if q is even and n is a non-trivial divisor of q, then the planes $AG(2, q)$ and $PG(2, q)$ contain maximal $(qn - q + n, n)$-arcs.

In the rest of this chapter we only consider the desarguesian planes $PG(2, q)$ and $AG(2, q)$.

Definition 7.9. *Let $\mathcal{S}$ be an arbitrary set of points. A point $P \notin \mathcal{S}$ is said to be a t-fold nucleus of $\mathcal{S}$ if each line through P contains at least t points of $\mathcal{S}$. A 1-fold nucleus is called a* nucleus.

The name is motivated by the notion of the nucleus of an oval in a plane of even order, because that is a nucleus in this sense as well. Obviously, if a set of points $\mathcal{S}$ has a t-fold nucleus, then $|\mathcal{S}| \geq t(q+1)$. Our goal is to give upper bounds on the number of t-fold nuclei of a given set of points. The following method works if there exists a line that is disjoint from the set of points $\mathcal{S}$. If it happens, then we can choose this line as the line at infinity and $\mathcal{S}$ becomes part of an affine plane. It follows from the definition that each t-fold nucleus of $\mathcal{S}$ is in the affine plane, too.

Before the proofs of the estimates we recall a classical theorem about the modulo p value of binomial coefficients. Let us remark that for $c < d$, by definition, $\binom{c}{d} = 0$.

Theorem 7.10 (Lucas). *Let p be a prime number. Write the integers a and b in base p numeral system, let $a = a_0 + a_1 p + \ldots + a_n p^n$ and $b = b_0 + b_1 p + \ldots + b_n p^n$, with $0 \leq a_i, b_i \leq p - 1$. Then*

$$\binom{a}{b} \equiv \prod_{i=0}^{n} \binom{a_i}{b_i} \pmod{p}.$$

Proof. It is enough to prove that for $0 \leq c, e \leq p - 1$,

$$\binom{c + dp}{e + fp} \equiv \binom{c}{e}\binom{d}{f} \pmod{p},$$

because we get the statement of the theorem by repeating this process. Consider the polynomial $f(x) = (1 + x)^{c+dp}$ over the field $GF(p)$. As taking the p-th power is an automorphism in $GF(p)$, we get

$$(1 + x)^{c+dp} = (1 + x)^c (1 + x^p)^d,$$

hence

$$\sum_{i=0}^{c+dp}\binom{c+dp}{i}x^i = \left(\sum_{i=0}^{c}\binom{c}{i}x^i\right)\left(\sum_{i=0}^{d}\binom{d}{i}x^{ip}\right).$$

Consider the coefficient of x^{e+fp}. It is $\binom{c+dp}{e+fp}$ on the left-hand side. The inequalities $c, e \le p-1$ imply that x^{e+fp} appears on the right-hand side if and only if we multiply the power x^e from the first factor by the power x^{fp} from the second factor. Thus the coefficient of x^{e+fp} is $\binom{c}{e}\binom{d}{f}$. As the coefficients on the two sides must be equal, the statement is proved. $\qquad\square$

Theorem 7.11 (Blokhuis). *Let $q = p^h$, p prime, and let $\mathcal{S} \subset \mathrm{AG}(2,q)$ be a set of points. If $|\mathcal{S}| = t(q+1) + m - 1$ with $m < q$, and $\binom{|\mathcal{S}|}{m} \not\equiv 0 \pmod{p}$ then $\mathcal{S}$ has at most $m(q-1)$ t-fold nuclei.*

Proof. Identify the points of $\mathrm{AG}(2,q)$ with the elements of $\mathrm{GF}(q^2)$ in the usual way (see Example 1.24). Consider $\mathcal{S}$ as a subset of $\mathrm{GF}(q^2)$ and define the polynomial

$$F(X,T) = \prod_{s \in \mathcal{S}}(T - (X - s)^{q-1}).$$

If x is a t-fold nucleus of $\mathcal{S}$, then the multiset $\{(x - s)^{q-1} : s \in \mathcal{S}\}$ contains each of the $(q+1)$-st roots of unity at least t times. This means that the polynomial $F(x,T)$ divisible by $(T^{q+1} - 1)^t$ for all x corresponding to a t-fold nucleus. Let $\sigma_j(X)$ denote the j-th elementary symmetric polynomial of the elements $(X-s)^{q-1}$. As a polynomial of X, the degree of σ_j is at most $j(q-1)$, and the bound is tight if $\binom{|\mathcal{S}|}{j} \neq 0$ in $\mathrm{GF}(q)$. In particular, in the case $j = m$, by Lucas' theorem, $\deg \sigma_m = m(q-1)$ if and only if the binomial coefficient $\binom{|\mathcal{S}|}{m}$ is not divisible by p.

If we expand $F(X,T)$ as a polynomial in T, then the coefficients are the σ_j's. More precisely

$$F(X,T) = \sum_{j=0}^{|\mathcal{S}|}(-1)^j\sigma_j(X)T^{|\mathcal{S}|-j}.$$

If we substitute a t-fold nucleus $X = x$, then we get

$$F(x,T) = (T^{q+1} - 1)^t(T^k + \ldots).$$

The coefficient of $T^{t(q+1)-1}$ on the right-hand side is 0 because $m < q$, while the same coefficient on the left-hand side is $(-1)^m\sigma_m(x)$. Thus each t-fold nucleus corresponds to a root of $\sigma_m(X)$. According to our assumption, the main coefficient of $\sigma_m(X)$ is not 0 in $\mathrm{GF}(q)$. So the number of its roots cannot exceed its degree, hence the number of t-fold nuclei is not greater than the degree of $\sigma_m(X)$, that is $m(q-1)$. $\qquad\square$

Corollary 7.12 (Blokhuis). *Let $\mathcal{B}$ be a t-fold blocking set in $\mathrm{AG}(2, q)$, $q = p^h$, and assume that $(t, q) = 1$. Then $|\mathcal{B}| \geq (t+1)q - 1$.*

In particular, for $t = 1$ each blocking set in $\mathrm{AG}(2, q)$ contains at least $2q-1$ points.

Proof. Suppose to the contrary that $\mathcal{B}$ is a t-fold blocking set and it contains only $(t+1)q - 2 = t(q+1) + q - t - 2$ points. If $P \notin \mathcal{B}$ is an arbitrary point, then P is a t-fold nucleus of $\mathcal{B}$. Thus the number of t-fold nuclei is $q^2 - (t+1)q + 2$. According to Lucas' theorem, the binomial coefficient $\binom{(t+1)q-2}{q-t-1}$ is 0 in $\mathrm{GF}(q)$ if and only if t is divisible by p, which does not happen in our case. Hence, by Theorem 7.11, $\mathcal{B}$ has at most $(q-t-1)(q-1)$ t-fold nuclei. But $(q-t-1)(q-1) < q^2 - (t+1)q + 2$; this contradiction proves the statement. $\qquad\square$

Let us remark that the result for $t = 1$ was proved in the 1970's independently by *Jamison* [97] and by *Brouwer* and *Schrijver* [34], see Theorem 6.39 in Chapter 6.

Theorem 7.13 (Ball). *Let $\mathcal{B}$ be a t-fold blocking set in $\mathrm{AG}(2, q)$, $q = p^h$, and let $e(t)$ be the maximal exponent for which $p^{e(t)}$ divides t. Then $|\mathcal{B}| \geq (t+1)q - p^{e(t)}$.*

Proof. Write $|\mathcal{B}|$ into the form $t(q+1) + m - 1$ as in Theorem 7.11. We show that if $m = q - t - p^{e(t)}$ (so it is one less than the bound), then $\mathcal{B}$ cannot be a t-fold blocking set. Let $t = cp^{e(t)}$ where c is not divisible by p. Consider the binomial coefficient

$$\binom{t(q+1) + m - 1}{m} = \binom{t(q+1) + m - 1}{t(q+1) - 1} = \binom{tq + q - p^{e(t)} - 1}{tq + t - 1}.$$

Applying Lucas' theorem (Theorem 7.10), we get

$$\binom{tq + q - p^{e(t)} - 1}{tq + t - 1} \equiv \binom{q - p^{e(t)} - 1}{cp^{e(t)} - 1} \pmod{p}.$$

Here

$$\binom{q - p^{e(t)} - 1}{cp^{e(t)} - 1} = \binom{q - 2p^{e(t)} + p^{e(t)} - 1}{(c-1)p^{e(t)} + p^{e(t)} - 1}$$

$$\equiv \binom{q/p^{e(t)} - 2}{c - 1}\binom{p^{e(t)} - 1}{p^{e(t)} - 1} \equiv \binom{q/p^{e(t)} - 2}{c - 1} \pmod{p}.$$

If we write $q/p^{e(t)} - 2$ in base p numeral system, then each digit is $(p-1)$, except the last one which is $(p-2)$. While if we write $c - 1$ in the base p numeral system, then the condition $c \not\equiv 0 \pmod{p}$ implies that the last digit is not $(p-1)$. Thus $\binom{q/p^{e(t)}-2}{c-1}$ is not 0 modulo p.

Hence Theorem 7.11 gives that the number of t-fold nuclei of $\mathcal{B}$ is at most $k(q-1)$. According to our assumption, the number of points in $\mathcal{B}$ is at most $(t+1)q - p^{e(t)} - 1$, the sum of these two numbers is less than q^2, thus $\mathcal{B}$ cannot be a t-fold blocking set. This proves the theorem. $\qquad\square$

Corollary 7.14. *Let $\mathcal{K}$ be a (k, n)-arc in $\mathrm{AG}(2, q)$, $q = p^h$, and let $e(n)$ be the maximal exponent for which $p^{e(n)}$ divides n. Then $|\mathcal{K}| \leq (n-1)q + p^{e(n)}$.*

Proof. Let $\mathcal{B}$ be the complement of $\mathcal{K}$ in $\mathrm{AG}(2, q)$. This is a t-fold blocking set with $t = (q - n)$. The statement follows from Theorem 7.13 at once. □

We have equality in the above corollary for maximal arcs, hence we have equality in Ball's theorem for complements of maximal arcs.
Corollaries 7.6 and 7.14 give the following result.

Corollary 7.15. *Let $\mathcal{K}$ be a (k, n)-arc in $\mathrm{PG}(2, q)$, $q = p^h$, and let $e(n)$ be the maximal exponent for which $p^{e(n)}$ divides n. If $n \leq \sqrt{2q}$, then*

$$k \leq (n-1)q + p^{e(n)}.$$

Proof. If $q = 2$ or 3, then the statement follows from Bose's theorem. If $\mathcal{K}$ is not a blocking set, then Corollary 7.14 gives the result. Finally, if $\mathcal{K}$ is a blocking set and $q > 3$, then the condition $\sqrt{2q} \leq q - 1$ is satisfied, hence we can apply Corollary 7.6 which gives $k \leq (n-1)q + 1$. □

In particular, if $n = \sqrt{q} + 1$, then the bound on the size is $k \leq q\sqrt{q} + 1$. This is tight because unitals are $(q\sqrt{q} + 1, \sqrt{q} + 1)$-arcs. *Lunelli* and *Sce* conjectured that in $\mathrm{PG}(2, q)$, if $2 < n < q$, then the size of any (k, n)-arc satisfies the inequality $k \leq (n-1)q + 1$. It is not true in general; *Hill* and *Mason* constructed counterexamples which are unions of disjoint Baer subplanes. If n and q are coprime and either there exists a line having empty intersection with the (k, n)-arc, or n is small compared to q, then Corollaries 7.14 and 7.15 prove the Lunelli–Sce Conjecture.

Let us return to nuclei. In the particular case $t = 1$ and $k = 0$ we get the following simple result.

Theorem 7.16 (Blokhuis, Wilbrink [30]). *In $\mathrm{PG}(2, q)$, if a set of points $\mathcal{S}$ of size $q + 1$ is not a line, then it has at most $q - 1$ nuclei.*

Proof. As $\mathcal{S}$ is not a line, it cannot be a blocking set. Thus there exists a line ℓ with empty intersection with $\mathcal{S}$. Let us choose ℓ as the line at infinity. Then $\mathcal{S}$ is in $\mathrm{AG}(2, q)$ and all of its nuclei are in $\mathrm{AG}(2, q)$, too. In this case a very simple polynomial gives the desired result. Let $f(X) = \sum_{s \in \mathcal{S}} (X - s)^{q-1}$. If x is a nucleus of $\mathcal{S}$, then each $(q + 1)$-st root of unity appears as $(x - s)^{q-1}$ exactly once. The sum of the $(q + 1)$-st roots of unity is 0, hence each nucleus is a root of $f(X)$. The degree of f is $q - 1$, so $\mathcal{S}$ has at most $q - 1$ nuclei. □

It is very instructive to see another proof of the previous theorem. We need the following lemma which applies the trick of Segre.

Lemma 7.17 (Segre, Korchmáros). *In $\mathrm{PG}(2, q)$ let $\mathcal{S}$ be a set of points of size $q + 1$ and A_0, A_1 and A_2 be three non-collinear nuclei of $\mathcal{S}$. Let $B_i = A_j A_k \cap \mathcal{S}$ where $\{i, j, k\} = \{0, 1, 2\}$. Then the points B_0, B_1 and B_2 are collinear.*

Proof. Let us choose the points A_0, A_1 and A_2 as points of reference and let $B_0 = (0 : 1 : b_0)$, $B_1 = (b_1 : 0 : 1)$ and $B_2 = (1 : b_2 : 0)$. Then $b_0 b_1 b_2 \neq 0$ obviously holds. Apply the trick of Segre for the lines joining the base points and the points of the set $\mathcal{S} \setminus \{B_0, B_1, B_2\} = \{P_1, P_2, \ldots, P_{q-2}\}$. Create a $(q - 2) \times 3$ matrix $M = (m_{i,j})$ whose rows correspond to the points in $\mathcal{S} \setminus \{B_0, B_1, B_2\}$ and columns correspond to the points A_0, A_1 and A_2. For $i = 1, 2, \ldots, q - 2$ and $j = 1, 2, 3$ let $m_{i,j}$ be the coordinate of the line $P_i A_j$. More precisely, for $P_i \in \mathcal{S} \setminus \{B_0, B_1, B_2\}$ write the equation of the line $P_i A_j$ into the form $X_2 = \lambda_0 X_1$, or $X_0 = \lambda_1 X_2$ or $X_1 = \lambda_2 X_0$ and call the corresponding λ the coordinate of the line (in the same way as we did in the proof of Lemma 6.17). Theorem 6.15 gives that the product of the three entries in each row is 1. On the other hand, in each of the three columns we see $q - 2$ pairwise distinct, non-zero elements of $\mathrm{GF}(q)$, because if $i_1 \neq i_2$, then the points P_{i_1}, P_{i_2} and A_j are not collinear, hence $m_{i_1} \neq m_{i_2}$. In each column exactly one non-zero element is missing, the coordinate of the line $A_i B_i$, that is b_i. As the product of all elements of M is 1 and the product of all non-zero elements of $\mathrm{GF}(q)$ is -1, we get $b_0 b_1 b_2 = -1$. By Lemma 6.16, this implies that the points B_1, B_2 and B_3 are collinear. $\square$

The following beautiful proof of Theorem 7.16 is based on the Segre-Korchmáros lemma. It is due to *Blokhuis* and *Mazzocca*.

Proof. Let $\mathcal{S}$ be the set of $q + 1$ points, suppose that $\mathcal{S}$ is not a line, and let $\mathcal{A}$ be the set of nuclei of $\mathcal{S}$. In the same way as in the original proof, let ℓ be a line having empty intersection with $\mathcal{S}$. Let us choose ℓ as line at infinity, then both $\mathcal{S}$ and $\mathcal{A}$ are in $\mathrm{AG}(2, q)$.

Define a function $f : \mathcal{A} \to \mathrm{GF}(q)^*$ in the following way: choose a base point $A \in \mathcal{A}$ and let $f(A) = 1$. If $A \neq A_i \in \mathcal{A}$ then let $B_i \in \mathcal{S}$ be the unique point of intersection of the line AA_i and the set $\mathcal{S}$. Let the cartesian coordinates of the points be $A = (a^1, a^2)$, $A_i = (a_i^1, a_i^2)$ and $B_i = (b_i^1, b_i^2)$. As the points A, A_i and B_i are collinear, there uniquely exists $\lambda \in \mathrm{GF}(q) \setminus \{0, 1\}$ for which

$$(b_i^1, b_i^2) = \lambda(a^1, a^2) + (1 - \lambda)(a_i^1, a_i^2)$$

holds. Then let $f(A_i) = \lambda/(\lambda - 1)$. It follows from the definition of f that in the 3-dimensional affine space, $\mathrm{AG}(3, q)$, the points $(a^1, a^2, f(A))$, $(a_i^1, a_i^2, f(A_i))$ and $(b_i^1, b_i^2, 0)$ are collinear. From the Segre–Korchmáros lemma we get that if the point $B' \in \mathcal{S}$ is on the line $A_i A_j$, then the points B_i, B_j and B' are collinear in $\mathrm{AG}(2, q)$, hence the points $(a_i^1, a_i^2, f(A_i))$, $(a_j^1, a_j^2, f(A_j))$ and $(b'^1, b'^2, 0)$ are also collinear in $\mathrm{AG}(3, q)$. In particular, if $A_i \neq A_j$, then $f(A_i) \neq f(A_j)$. This means that the function f is injective, hence the cardinality of its domain is less than or equal to the cardinality of its range, so $|\mathcal{A}| \leq q - 1$. $\square$

In the next part of the chapter we investigate the existence problem of maximal (k, n)-arcs. As we have already seen, hyperovals are maximal arcs in $\mathrm{PG}(2, q)$ if q is even. The following construction of *Denniston* [57] can be considered as a generalization of hyperovals.

Theorem 7.18 (Denniston). *Let $q = 2^r$ and let $1 < n < q$ be an arbitrary divisor of q. Then there exists a $(qn - q + n, n)$-arc in $\mathrm{AG}(2, q)$ (hence in $\mathrm{PG}(2, q)$, too).*

Proof. We construct the arc in the affine plane. Choose $\beta \in \mathrm{GF}(q)$ such that the polynomial $X^2 + \beta X + 1$ is irreducible over $\mathrm{GF}(q)$. For each $\lambda \in \mathrm{GF}(q)$ consider the quadratic curve $\mathcal{C}_\lambda$ defined by the affine equation $X^2 + \beta XY + Y^2 = \lambda$. It is easy to check that if $\lambda \neq 0$, then $\mathcal{C}_\lambda$ is a conic; it contains $q + 1$ affine points and its nucleus is the point $(0, 0)$, while if $\lambda = 0$ then $\mathcal{C}_\lambda$ consists of the single point $(0, 0)$.

If n is a divisor of q, then $n = 2^h$ with $1 \leq h < r$. Let H be a subgroup of order n of the additive group of $\mathrm{GF}(q)$. We show that the set of points

$$\mathcal{M} = \bigcup_{\lambda \in H} \mathcal{C}_\lambda$$

is a $(qn - q + n, n)$-arc. Obviously, $|\mathcal{M}| = 1 + (n-1)(q+1) = nq - q + n$, because $\mathcal{C}_{\lambda_i} \cap \mathcal{C}_{\lambda_j} = \emptyset$ for any pair $\lambda_i \neq \lambda_j$.

Any line through the origin is a tangent to each $\mathcal{C}_\lambda$ if $\lambda \neq 0$, hence it meets $\mathcal{M}$ in exactly n points. Now, consider a line ℓ with equation $Y = mX + b$ where $b \neq 0$. Substitute Y to the equation of $\mathcal{C}_\lambda$. Then we get $uX^2 + vX + w = 0$ with $w = b^2 \neq 0$. The polynomial $uX^2 + vX + w$ has a non-zero constant term, and it is irreducible, because its reducibility would mean that the line with equation $Y = mX + b$ is a secant to $\mathcal{C}_0$, but this is impossible. In other words, if we consider the additive subgroup $U = \{uX^2 + vX : X \in \mathrm{GF}(q)\}$ then $w \notin U$. As the mapping $x \mapsto ux^2 + vx$ is linear over $\mathrm{GF}(2)$ and its kernel has dimension 1, its image has codimension 1, hence $|U| = q/2$. Thus the elements $\{uX^2 + vX + w : X \in \mathrm{GF}(q)\}$ form a coset of U. This coset intersects the subgroup H either in 0 or in $n/2$ elements, which means that the line ℓ meets $\mathcal{M}$ either in 0 or in n points. $\square$

In the late 1990's *Mathon* [124] proved that if q is even and $2 \neq n \neq q/2$ is a divisor of q, then there are other constructions for $(qn - q + n, n)$-arcs. Ho also used sets of conics having the same nucleus.

If q is odd, then the situation is completely different. As *Ball, Blokhuis* and *Mazzocca* showed, in this case there are no maximal arcs except the trivial ones. The proof of the following theorem is beyond the scope of this book; it can be found in [10]. For $n = 3$ it was proved by J. Thas [166].

Theorem 7.19 (Ball, Blokhuis, Mazzocca). *If q is odd and $1 < n < q$, then there are no $(qn - q + n, n)$-arcs in $\mathrm{PG}(2, q)$.*

If there are no maximal arcs, then the natural question arises: if n divides q, then what can we say about the size of the largest, or the second largest (k, n)-arc if q is odd or even, respectively? Our next theorem partially answers this question.

Theorem 7.20. *Let K be a (k,n)-arc in $\mathrm{PG}(2,q)$. If n divides q and $n \leq q/4$, then $k \leq nq - q + 2n/3$.*

Proof. Let the size of K be $k = nq - q + n - \varepsilon$ and suppose that $0 < \varepsilon \leq n/3$. We divide the proof into two main parts.

Part 1. For each point P let $l_i(P)$ be the number of $(n-i)$-secants to K through P. If a line ℓ is an $(n-i)$-secant with $0 < i < n$, then ℓ is called a *short* line. Let the *index* of the point P be defined as $\mathrm{ind}(P) = \sum_{0<i<n} i l_i(P)$. If $P \in K$, then obviously $\mathrm{ind}(P) = \varepsilon$. If $P \notin K$ and there are $(q/n) - t$ external lines of K through P then comparing the number of points of K on the secants through P and $qn - q + n$, we get $\mathrm{ind}(P) = tn + \varepsilon$. We claim that in this equality $t = 0$ is the only possible value. First, we give a lower bound on the number of points having index ε. Let T_i be the number of $(n-i)$-secants to K. Then the standard double counting gives

$$\sum_{i=1}^{n-1} iT_i = \begin{cases} \sum_{P \in \ell} \mathrm{ind}(P), & \text{if } \ell \text{ is 0-secant or } n\text{-secant,} \\ \sum_{P \in \ell} \mathrm{ind}(P) - jq, & \text{if } \ell \text{ is } (n-j)\text{-secant for some } 0 < j < n. \end{cases}$$

(7.3)

As ε is small, we can estimate $\sum iT_i$ by counting the weighted (point in K, short line through it) pairs, where the weight means that if a short line is an $(n-j)$-secant, then we count it j times. On the one hand, we get that the number of these pairs is $|K|\varepsilon$, because there are ε weighted short lines through each point of K. On the other hand, this number is $\sum_{j=1}^{n-1}(n-j)jT_j$, because in each point of an $(n-j)$-secant we count j. Applying the inequality $j \leq \varepsilon$, we get that this number is at least $(n-\varepsilon)\sum jT_j$. If the line ℓ is a 0-secant or an n-secant, then from $\varepsilon \leq n/3$ we get

$$\sum_{P \in \ell} \mathrm{ind}(P) \leq \frac{(qn - q + n - \varepsilon)\varepsilon}{n - \varepsilon} < \frac{3}{2}(q+1)\varepsilon.$$

(7.4)

As each point on ℓ has index either ε or at least $n + \varepsilon \geq 4\varepsilon$, we get that at least $5/6$ of the points on ℓ satisfy the equality $\mathrm{ind}(P) = \varepsilon$. In particular, the number of points having index greater than ε is at least $n + 1$, which means that there are 0-secants to K, too. In the next step one of the 0-secants will be chosen as the line at infinity.

Part 2. First, applying the Rédei polynomial we determine the short lines through a point having index ε. We may assume that K is in the affine plane. Let $K = \{(a_i, b_i) : i = 1, 2, \ldots, nq - q + n - \varepsilon\}$, then its Rédei polynomial is

$$H(X, Y) = \prod_{i=1}^{nq-q+n-\varepsilon} (X + a_i Y - b_i) = \sum_{j=0}^{nq-q+n-\varepsilon} h_j(Y) X^{nq-q+n-\varepsilon-j}.$$

We are looking for a polynomial $a(X, Y)$ for which the product $H(X, Y)a(X, Y)$ has the property that if $Y = m$ corresponds to a point with index ε then each root of $H(X, m)a(X, m)$ has multiplicity n. As n

divides q, this means that for such an m, the coefficient of a power of X must be 0 if the exponent is not divisible by n in $H(X, m)a(X, m)$. Because of $\varepsilon < n$, we get a set of ε linear equations for the coefficients of $a(X, Y)$. If $a(X, Y) = \sum_{i=0}^{\varepsilon} a_i(Y) X^{\varepsilon-i}$, then these equations are

$$a_1(Y) = - h_1(Y),$$
$$a_2(Y) = - h_2(Y) - a_1(Y) h_1(Y),$$
$$\vdots$$
$$a_i(Y) = - h_i(Y) - \sum_{k=1}^{i-1} a_k(Y) h_k(Y),$$
$$\vdots$$

(7.5)

If the point (m) has index ε, then we show that $H(X, m)a(X, m)$ is an n-th power. Consider the short lines through (m). If the line $Y = mX + b$ is an $(n - j)$-secant with $0 < j < n$, then take the corresponding factor $(X - b)^j$. For a fixed m the product of these factors is a polynomial $a^*(X, m)$ with degree ε, and the product $H(X, m)a^*(X, m)$ is an n-th power, because, by definition, each of its roots has multiplicity n. This implies that the coefficients of $a^*(X, m)$ give a solution of the set of equations (7.5). But the solution is obviously unique, hence $a^*(X, m) = a(X, m)$. We know that the degree of $h_i(Y)$ is at most i, so the total degree of $a(X, Y)$ is ε, too. This means that $H(X, m)a(X, m)$ is an n-th power if the point (m) has index ε.

We know that at least 5/6 of the points have index ε. Applying this we show that if the index of a point is not ε, then the index must be close to q. We use the Rédei polynomial again. Let s denote the number of points with index ε on the line at infinity. For such points the polynomial $H(X, Y)a(X, Y)$, as a polynomial of X, is an n-th power. If the index of a point (m) is not ε, then let $p^e < n$ be the greatest exponent for which $H(X, m)a(X, m)$ is a p^e-th power as a polynomial of X.

Let $W(U, V)$ denote the polynomial for which $W(X^{p^e}, m) = H(X, m)a(X, m)$. If c is a root with multiplicity n of the polynomial $H(X, m)a(X, m)$, then $\bar{c} = c^{p^e}$ is a root with multiplicity n/p^e of the polynomial $W(U, m)$. The condition $p^e < n$ implies that this multiplicity is divisible by p, hence $\bar{c}$ is a root with the same multiplicity of the polynomial $W'(U, m)$. We show that the degree of W' is much smaller than the degree of W. Let

$$H(X, Y)a(X, Y) = X^{nq-q+n} + u_1(Y) X^{qn-q+n-1} + \ldots + u_{qn-q+n}(Y)$$

where $\deg u_i \le i$ for $i = 1, 2, \ldots, qn - q + n$. If the index of the point (y) is ε, then $u_i(y) = 0$, so among the coefficient polynomials $u_1, \ldots, u_{s-1}$, only those are non-zeros whose exponents are divisible by n. But in these cases $\deg W' \le \deg W - s$. Let us consider the factors of W and W' and count the difference of the degrees. If the multiplicity of the factor $U - \bar{b}$ of W is not n/p^e, then the corresponding factor $(X - b)^{p^e}$ with $b = \bar{b}^{n/p^e}$ of $H(X, m)a(X, m)$ has multiplicity either at most $n - p^e$, or at least $n + p^e$. In the first case, the line having equation $Y = mX + b$ is short, while in the second case, the factor comes from the polynomial $a(X, m)$. Thus the number of lines belonging to the first case is at most $\mathrm{ind}((m))/p^e$, and the number of lines belonging to the second case is at most $\deg(a)/p^e$, hence altogether it is at most ε/p^e. So we have $\deg(W') \geq (qn - q + n - \mathrm{ind}((m)) - \varepsilon)/p^e$. Comparing it with our previous observation $\deg W' \leq \deg W - s$, we conclude that the number of points having index ε is at least $s - \varepsilon$.

Thus, by Inequality (7.4), we get

$$\frac{3}{2}(q+1)\varepsilon > \sum \mathrm{ind}(P) \geq s\varepsilon + (q+1-s)(s-\varepsilon) \geq (q+1)\varepsilon + (q+1-s)(s-2\varepsilon).$$

Because of $\varepsilon \leq n/3 \leq q/6$ and $s \geq 5(q+1)/6$ this implies that $(q+1-s) < \varepsilon$, hence $s > q + 1 - \varepsilon$. Applying again that the number of points whose index is not ε is at least $s - \varepsilon$, we get that the number of these points is at least $q + 1 - 2\varepsilon > q - n + \varepsilon$. We know from Part 1 that if there is a 0-secant through a point, then the index of the point is at most $q - n + \varepsilon$. This gives a contradiction if there were points having index different from ε. Thus if there is at least one 0-secant through a point, then the number of 0-secants through that point is exactly q/n. This means that the 0-secants of $\mathcal{K}$ form a $(q \cdot q/n - q + q/n, q/n)$-arc. Then Cossu's result (Theorem 7.8) gives that the dual of this arc is a $(qn - q + n, n)$-arc which contains $\mathcal{K}$. $\square$

The best result in this direction comes from *Ball* and *Blokhuis* [9]. It is quite complicated; it roughly states that the embeddability result of the previous theorem is still valid if q/n is large enough and $\varepsilon < cn$ with $c < 2/3$. More precisely, if $n = q/2$ and $\varepsilon < 0.381n$, or $q/n = 3$ and $\varepsilon < 0.476n$, or $q/n \geq 4$ and $\varepsilon < n/2$, then each (k, n)-arc with $k = qn - q + n - \varepsilon$ points can be embedded into a maximal arc having $qn - q + n$ points. Because of the non-existence result of Ball–Blokhuis–Mazzocca (Theorem 7.19), these results imply upper bounds on the sizes of the largest (k, n)-arcs. For more details we refer the reader to the survey paper [91].

Let us return to t-fold blocking sets.

Proposition 7.21. *Let $\mathcal{B}$ be a t-fold blocking set in $\mathrm{PG}(2, q)$. If $\mathcal{B}$ contains a line ℓ, then $|\mathcal{B}| \geq (t + 1)q - t + 2$.*

Proof. The set of points $\mathcal{B} \setminus \ell$ is a $(t - 1)$-fold blocking set in the affine plane $\mathrm{PG}(2, q) \setminus \ell$. Thus the statement follows from Theorem 7.13. $\square$

Theorem 7.22. *Let $\mathcal{B}$ be a t-fold blocking set in $\mathrm{PG}(2, q)$ and suppose that $q = p$ prime. Then*

$$|\mathcal{B}| \geq \max \left\{ tq + \frac{q+3}{2}, tq + t \right\}.$$

Proof. We may assume that no proper subset of $\mathcal{B}$ is a t-fold blocking set. We know from Proposition 7.4 that $|\mathcal{B}| \geq t(q+1)$, so we may also assume that $t \leq (q+1)/2$, otherwise we are done. Let $|\mathcal{B}| = tq + k$. If $k \geq q$ then there is nothing to prove, so we may also assume that $k < q$. Choose a line ℓ_∞ for which $|\mathcal{B} \cap \ell_\infty| = s > t$ holds. As usual, take the affine plane $\mathrm{AG}(2,q) = \mathrm{PG}(2,q) \setminus \ell_\infty$, let $\mathcal{U} = \mathcal{B} \setminus \ell_\infty$ and choose the system of reference such that $(\infty) \in \mathcal{B}$. Let $\mathcal{U} = \{(a_i, b_i) : i = 1, 2, \ldots, |\mathcal{B}| - s\}$ and consider the Rédei polynomial of $\mathcal{U}$

$$H(X,Y) = \prod_{i=1}^{|\mathcal{B}|-s} (X + a_i Y - b_i) = \sum_{j=0}^{|\mathcal{B}|-s} h_j(Y) X^{|\mathcal{B}|-s-j}$$

where $\deg h_j \leq j$.

If a point at infinity $(m) \notin \mathcal{B}$, then the polynomial $H(X, m)$ is divisible by $(X^q - X)^t$. This implies that the equalities $h_{s+1}(m) = \ldots = h_{q-1}(m) = 0$ hold for such m, so $h_{s+1}(Y) \equiv \ldots \equiv h_{q-s}(Y) \equiv 0$. Now, consider the polynomial $H(X, m)$ with $(m) \in \mathcal{B}$. Then $H(X, m)$ is also divisible by $(X^q - X)^{t-1}$. Let $H(X, m) = (X^q - X)^{t-1} f(X)$. Then $f(X)$ can be written as $X^q g(X) + h(X)$ with $\deg g = |\mathcal{B}| - tq - s$ and $\deg h \leq |\mathcal{B}| - tq - 1$. After dividing by the greatest common factor of g and h we can apply Lemma 6.37, and we get either $|\mathcal{B}| - tq - 1 \geq (q+1)/2$ which proves the theorem, or that $f(X)$ is divisible by $(X^q - X)$. In the latter case we have two possibilities again. Either $\mathcal{B} \setminus \{(m)\}$ is a t-fold blocking set, or $f'(x) = 0$. The first case contradicts our assumption given at the beginning of the proof, while in the second case $\mathcal{B}$ contains a whole line. If it happens, then from Proposition 7.21 we get that $|\mathcal{B}| \geq (t+1)q - t + 2$, which is better than the bound of this theorem. □

With a more sophisticated method the result can be improved significantly for planes of prime order. We present the best known estimate; its proof can be found in [6].

Theorem 7.23 (Ball). *Let $\mathcal{B}$ be a t-fold blocking set in $\mathrm{PG}(2,q)$, $q > 3$ prime. If $t \leq (q+1)/2$, then*

$$|\mathcal{B}| \geq tq + t + \frac{q+1}{2}.$$

It is worth mentioning that this bound is tight if q is any odd prime power and $t = (q+1)/2$. Let $\mathcal{C}$ be a conic and $\mathcal{E}$ be the set of external points of $\mathcal{C}$. Then Lemma 6.14 implies at once that the set $\mathcal{C} \cup \mathcal{E}$ is a t-fold blocking set in $\mathrm{PG}(2,q)$.

A "trivial" 2-fold blocking set is a triangle (union of three non-concurrent lines). This is a set of size $3q$. The bound in Theorem 7.23 gives $2q + \frac{q+5}{2}$, that is roughly $2.5q$, when q is a prime. It is surprisingly difficult to construct 2-fold blocking sets of size less than $3q$. The first 2-fold blocking set of size $3q-1$ for $q = 13$ was found by *Braun, Kohnert* and *Wassermann* [33]. Recently *Csajbók* and *Héger* constructed blocking sets of size $3q - 1$ for $q = 13, 19, 31, 37, 43$ (and also for $q = 16, 25, 27$), see [50].

There are several results in the general case $q = p^h$, too. We refer the reader to the survey paper [91]. Let us remark that for $t > 1$ one cannot prove sharp bounds in a purely combinatorial way.

Exercises

7.1. Prove that a 2-fold blocking set in Π_q has at least $2q + \sqrt{2q} + 1$ points. Generalize this for 2-fold blocking sets in Π_q.

7.2. A (k, n)-arc is *complete* if it is maximal with respect to containment. Find a lower bound on the size of a complete (k, n)-arc in a projective plane of order q.

7.3. Take two disjoint Baer subplanes in $\mathrm{PG}(2, q^2)$. Show that there is no line meeting their union in $2(q + 1)$ points.

7.4. Show that $\mathrm{PG}(2, q^2)$ can be partitioned in disjoint Baer subplanes.

7.5. Prove that the intersection of the union of t disjoint Baer subplanes and a line consists of either t or $q + t$ points.

7.6. Show that the complement of the union of 2 disjoint Baer subplanes is a counterexample for the Lunelli-Sce conjecture.

For which values of f is the complement of the union of t disjoint Baer subplanes a counterexample for the Lunelli-Sce conjecture?

7.7. A *set of type* (m, n) is a set of points with the property that each line meets the set in either m or n points. Let $\mathcal{S}$ be a set of type (m, n) in a projective plane of order q, and let $m < n$ and suppose that $\mathcal{S}$ consists of k points. Show that

- $k^2 - k(q(n + m - 1) + n + m) + mn(q^2 + q + 1) = 0$;
- q is divisible by $n - m$;
- the complement of $\mathcal{S}$ is a set of type $(q + 1 - n, q + 1 - m)$;
- $mq + n \leq k \leq (n - 1)q + m$;
- we have equality in the previous inequality if $\mathcal{S}$ is a Hermitian curve or a subgeometry (or their complement).

7.8. Let $\mathcal{C}$ be a conic and $\mathcal{I}$ be the set of internal points of $\mathcal{C}$ in $\mathrm{PG}(2, q)$, q odd. Prove that $\mathcal{I}$ is a set of type $(0, m, n)$ with $m = \frac{q-1}{2}$ and $n = \frac{q+1}{2}$.

7.9. A non-empty set of points $\mathcal{S}$ is called a *regular semioval* with character a if it is a set of type $(0, 1, a)$ and for each point $P \in \mathcal{S}$ there is a unique line t_p (called the tangent to $\mathcal{S}$ at P) such that $\mathcal{S} \cap t_p = \{P\}$. Let $\mathcal{S}$ be a regular semioval in a projective plane of order q, Π_q. Prove that

- (*de Finis* [54]) if a divides $q - 1$, then the tangents to $\mathcal{S}$ form a semioval in the dual plane;

- $\mathcal{S}$ is either an oval ($a = 2$) or a divides $q - 1$, and in the latter case any point not in $\mathcal{S}$ is incident with 0 or a tangents;

- if $\Pi_q = \mathrm{PG}(2, q)$ and $a \neq q - 1$, then a and $q - 1$ are coprimes, and the tangents at collinear points of $\mathcal{S}$ are concurrent (hint: apply the Rédei polynomial).

8

Algebraic curves and finite geometry

In this chapter two applications of algebraic curves to the theory of finite geometries are presented.

In the first part we concentrate on the size of the second largest complete arc in $\mathrm{PG}(2, q)$. Most of the methods are based on the work of Segre, but we also give a brief overview of the latest results. Let us start with some basic definitions and theorems from the theory of algebraic curves and surfaces.

Definition 8.1. *Let $\mathbb{K}$ be an arbitrary field and consider the ring of polynomials $\mathbb{K}[\mathbf{X}] = \mathbb{K}[X_0, X_1, \ldots, X_m]$. Two homogeneous polynomials f, $g \in \mathbb{K}[\mathbf{X}]$ are called equivalent if there exists $0 \neq \lambda \in \mathbb{K}$ for which $\lambda f = g$ holds. In the m-dimensional projective space, $\mathrm{PG}(m, \mathbb{K})$, an equivalence class of degree n homogeneous polynomials in $\mathbb{K}[\mathbf{X}]$ is called a* hypersurface of degree n. *In particular, if $m = 2$ or 3, then a hypersurface is called a* curve *or a* surface, *respectively. The irreducible factors of the corresponding polynomials are the* components *of the hypersurface.*

Let us remark that according to this definition, multiple components are allowed, which is sometimes forbidden in the literature.

We apply the usual geometric terminology: "a point P is on the hypersurface f", or "the hypersurface f passes through the point" means that if P has homogeneous coordinates $(x_0 : x_1 : \cdots : x_m)$, then $f(x_0, x_1, \ldots, x_m) = 0$. We also say that "$f = 0$ is the equation of the hypersurface", or "the hypersurface is defined by the equation $f = 0$".

If f is a homogeneous polynomial of degree n with $m + 1$ variables, then f can be written as

$$f(X_0, X_1, \ldots, X_m) = \sum_{i_j \geq 0,\ i_0 + i_1 + \ldots + i_m = n} a_{i_0, i_1, \ldots, i_m} X_0^{i_0} X_1^{i_1} \ldots X_m^{i_m}. \quad (8.1)$$

So f can be identified with the vector $(a_{n,0,\ldots,0}, a_{n-1,1,0,\ldots,0}, \ldots, a_{0,0,\ldots,0,n})$. Hence the equivalence classes of homogeneous polynomials can be considered as points of the projective space over $\mathbb{K}$ having dimension

$$N = \binom{n + m}{m} - 1, \quad (8.2)$$

where the dimension comes from the well-known fact that the equation $i_0 + i_1 + \ldots + i_m = n$ has $\binom{n+m}{m}$ non-negative integer solutions. In particular, if $n = m = 2$, then $N = 5$, as we know from the theory of conics (planar quadratic curves). Intuitively, the dimension gives us the number of linear conditions that we can prescribe on the coefficients of the hypersurface. In the rest of this chapter, when we consider hypersurfaces of degree n in $\mathrm{PG}(m, \mathbb{K})$, the letter N always denotes the number defined in Equation (8.2).

Definition 8.2. *A set of hypersurfaces of degree n in $\mathrm{PG}(m, \mathbb{K})$ whose coefficient vectors form a d-dimensional subspace of $\mathrm{PG}(N, \mathbb{K})$ is called a* linear system *of dimension d.*

In other words, the elements of a d-dimensional linear system are the polynomials

$$\{\lambda_0 F_0 + \ldots + \lambda_d F_d : \lambda_i \in \mathbb{K}, i = 0, \ldots, d\}$$

where $F_0, \ldots, F_d$ are fixed, linearly independent polynomials of degree n.

A linear system of dimension one is a pencil *and that of dimension two is a* net.

Proposition 8.3. *Let $\mathcal{S} = \{P_1, P_2, \ldots, P_r\}$ be a set of points in $\mathrm{PG}(m, \mathbb{K})$. Then the set of hypersurfaces passing through each element of $\mathcal{S}$ is a linear system whose dimension is at least $N - r$.*

Proof. Being a point on a hypersurface imposes a linear condition on the coordinates of the corresponding point of $\mathrm{PG}(N, \mathbb{K})$. If we substitute the homogeneous coordinates of P_i to Equation (8.1) for $i = 1, 2, \ldots, r$, then we get a set of r linear equations on the coefficients of f. These equations are not necessarily linearly independent, hence the dimension of the subspace defined by them could be greater than $N - r$. $\qquad\square$

Let us recall some basic properties of polynomials. The element a is a root of $f(X)$ with multiplicity r if and only if $(X - a)^r$ divides $f(X)$, but $(X - a)^{r+1}$ does not divide $f(X)$. The weak form of the Fundamental Theorem of Algebra states that any one-variable polynomial $f \in \mathbb{K}[X]$ of degree n has at most n roots, and if $\mathbb{K}$ is algebraically closed and the roots are counted with multiplicities, then it has exactly n roots. If we apply a linear transformation $x \mapsto ux + v$, then the polynomial $f(X)$ and its roots change, but the multiplicities of the roots are invariants. A degree n homogeneous polynomial in two variables $f(X_1, X_2) = \sum_{i=0}^{n} a_i X_1^{n-i} X_2^i$ can be mapped to a polynomial in one variable by dividing it with X_2^n. The degree of the resulting polynomial $\hat{f}(U) = \sum_{i=0}^{n} a_i U^{n-i}$ is at most n.

The intersection multiplicity of two plane curves at a point will be defined in general in the second part of this chapter. Now we only need a particular case of this general notion. This definition is much simpler than the general one and it gives a simple rule for calculating it, too.

Definition 8.4. *Let F be a curve of degree n and L be a line (a curve of degree one). Suppose that both F and L pass through the point P. Substitute* *the equation of L into the equation of F, then dehomogenize the resulting homogeneous polynomial in two variables to get a polynomial in one variable, and dehomogenize the coordinates of P in the same way. This coordinate of P is a root of the dehomogenized polynomial in one variable. The multiplicity of this root is the* intersection multiplicity *of F and L at the point P.*

This may depend on the way the equation of the line is given, and also on the way of dehomogenization. It can be proved that this is not the case. One can also prove (see for example [90]) that this value does not depend on the choice of the system of reference; it is invariant under projective transformations.

As an example, let the equation of the curve F be

$$X_0^4 + X_0(X_1 + X_2)^3 + X_1X_2(X_1 + X_2)^2 = 0,$$

let the line be $L : X_0 + X_1 + X_2 = 0$ and consider the point $P = (0 : -1 : 1)$. Writing $-X_1 - X_2$ instead of X_0 we get the homogeneous polynomial F^* : $(X_1 + X_2)^2 X_1 X_2$ in two variables, hence the corresponding polynomial in one variable is $\hat{F} : (U + 1)^2 U$, while the coordinates of the point P correspond to the field element $-1/1 = -1$. The element $u = -1$ is a root with multiplicity 2 of the polynomial, hence the intersection multiplicity of F and L at P is 2.

It follows from the Fundamental Theorem of Algebra that the sum of the intersection multiplicities of a curve of degree n and a line at their points of intersection cannot exceed n. In other words, if the sum of multiplicities is greater than n, then the line must be a component of the curve.

If we apply a linear transformation of the variables, then the coefficients of the resulting polynomials are linear combinations of the coefficients of the original polynomials. It means that we can choose the system of reference in an appropriate way when studying intersection multiplicities. Let us look at an example again.

Consider the action of the collineation defined by $(x_0 : x_1 : x_2) \mapsto (x_0 : x_2 : x_1 + x_2)$ on the 5-dimensional space of plane quadratic curves. It acts on the terms in the following way: $x_0^2 \mapsto x_0^2$, $x_0x_1 \mapsto x_0x_2$, $x_0x_2 \mapsto x_0x_1 + x_0x_2$, $x_1^2 \mapsto x_2^2$, $x_1x_2 \mapsto x_1x_2 + x_2^2$ and $x_2^2 \mapsto x_1^2 + 2x_1x_2 + x_2^2$. Thus its action on the coefficient vector is

$(a_{2,0,0} : a_{1,1,0} : a_{1,0,1} : a_{0,2,0} : a_{0,1,1} : a_{0,0,2}) \mapsto$
$(a_{2,0,0} : a_{1,0,1} : a_{1,1,0} + a_{1,0,1} : a_{0,0,2} : a_{0,1,1} + a_{0,0,2} : a_{0,2,0} + a_{0,0,2} + 2a_{0,1,1}).$

Proposition 8.5. *Let $r > 0$ be a given integer. Consider a point P and a line ℓ through P. Then the set of curves of degree n whose intersection multiplicity with ℓ at P is at least r, is a linear system of dimension $N - r$.*

Proof. As the intersection multiplicity is invariant under linear transformations, we may assume without loss of generality that $P = (0 : 0 : 1)$ and the equation of ℓ is $X_0 = 0$. Substituting it into the polynomial defining a curve we get the homogeneous polynomial $F^*(X_1, X_2) = \sum_{i=0}^{n} a_{0,n-i,i}X_1^{n-i}X_2^i$, and the corresponding polynomial in one variable is $\hat{F}(U) = \sum_{i=0}^{n} a_{0,n-i,i}U^{n-i}$.

If the intersection multiplicity at P is at least r, then 0 is a root of $\hat{F}$ with multiplicity at least r. This means that $a_{0,n-i,i} = 0$ if $i = n, n-1, \ldots, n-r+1$. Thus we get r independent linear conditions on the coefficients of F, hence the corresponding curves form a linear system of dimension $N - r$. □

Now we are ready to prove the generalization of Menelaus' theorem for curves of degree n.

Theorem 8.6 (Segre). *Let $A_0 = (1 : 0 : 0)$, $A_1 = (0 : 1 : 0)$ and $A_2 = (0 : 0 : 1)$ be the vertices of the fundamental triangle. Let $P_0^i = (0 : b_i : 1) \in A_1 A_2$, $P_1^i = (1 : 0 : c_i) \in A_0 A_2$ and $P_2^i = (a_i : 1 : 0) \in A_0 A_1$ be points distinct from the vertices of the fundamental triangle. Suppose that there are multiplicities m_j^i assigned to the points P_j^i such that $\sum_i m_j^i = n$ holds for $j = 0, 1, 2$. Then there exists a curve F_n of degree n which meets the line $A_r A_s$ at P_t^i with multiplicity m_t^i for $\{r, s, t\} = \{0, 1, 2\}$, if and only if*

$$\prod_i a_i^{m^i_2} \prod_j b_j^{m^j_0} \prod_k c_k^{m^k_1} = (-1)^n. \tag{8.3}$$

This complicated condition can be explained in the following way. Consider the given points on the sides of the fundamental triangle. If we multiply their coordinates with the prescribed multiplicities then the result is $(-1)^n$. According to this interpretation, there is a product of $3n$ terms on the left-hand side of Equation (8.3).

Proof. For the sake of brevity we say that a curve is well-intersecting if it meets the three sides of the fundamental triangle exactly at the given points with the required multiplicities. First, suppose that there exists a well-intersecting curve F_n of degree n. Let

$$F_n(X_0, X_1, X_2) = \sum_{i+j+k=n} e_{i,j,k} X_0^i X_1^j X_2^k$$

$$= e_{n,0,0} X_0^n + e_{n-1,1,0} X_0^{n-1} X_1 + \ldots + e_{0,n,0} X_1^n + e_{0,0,n} X_2^n + \ldots,$$

as in Equation (8.1). The vertices of the fundamental triangle are not on the curve, hence $e_{n,0,0} e_{0,n,0} e_{0,0,n} \neq 0$. Consider the points P_0^i. These points are on the line $X_0 = 0$, hence their coordinates are the roots of the polynomial

$$e_{0,n,0} X_1^n + e_{0,n-1,1} X_1^{n-1} X_2 + \ldots + e_{0,0,n} X_2^n = 0.$$

In the same way, the coordinates of the points P_1^i and P_2^i are the roots of the polynomials

$$e_{n,0,0} X_0^n + e_{n-1,0,1} X_0^{n-1} X_2 + \ldots + e_{0,0,n} X_2^n = 0$$

and

$$e_{n,0,0} X_0^n + e_{n-1,1,0} X_0^{n-1} X_1 + \ldots + e_{0,n,0} X_1^n = 0,$$

respectively. When we dehomogenize and introduce the new variables $X_1/X_2 = U$, $X_2/X_0 = V$ and $X_0/X_1 = W$, the corresponding equations become

$$\sum_{i=0}^{n} e_{0,i,n-i} U^i = e_{0,n,0} U^n + \ldots + e_{0,0,n} = 0,$$

$$\sum_{i=0}^{n} e_{n-i,0,i} V^i = e_{0,0,n} V^n + \ldots + e_{n,0,0} = 0,$$

$$\sum_{i=0}^{n} e_{i,n-i,0} W^i = e_{n,0,0} w^n + \ldots + e_{0,n,0} = 0.$$

Since F_n is well-intersecting, the solutions of these equations are the co-ordinates b_j, c_k and a_i with the prescribed multiplicities. Thus, by Vieta's formula, we get

$$\prod_j b_j^{m^j_0} = (-1)^n \frac{e_{0,0,n}}{e_{0,n,0}}, \quad \prod_k c_k^{m^k_1} = (-1)^n \frac{e_{n,0,0}}{e_{0,0,n}}, \quad \prod_i a_i^{m^i_2} = (-1)^n \frac{e_{0,n,0}}{e_{n,0,0}}.$$

The product of these three equations is exactly Equation (8.3), so we proved the first part of the theorem.

Now, suppose that the coordinates of the points satisfy Equation (8.3). Delete the point $P_2^1 = (a_1 : 1 : 0)$ (or decrease its multiplicity by one). Consider the linear system $\mathcal{L}_n$ of curves of degree n passing through the remaining $3n - 1$ points. According to Proposition 8.3, the dimension of $\mathcal{L}_n$ is at least

$$\binom{n+2}{2} - 1 - (3n - 1) = \binom{n-1}{2}.$$

Of course, this system contains not only well-intersecting curves, but also the curves containing the components X_0, X_1 and X_2. These polynomials are divisible by $X_0 X_1 X_2$, hence they can be written as $X_0 X_1 X_2 G_{n-3}$, where G_{n-3} is an arbitrary curve of degree $(n-3)$. Thus the dimension of this linear subsystem is $\binom{n-3+2}{2} - 1$, that is one less than the dimension of $\mathcal{L}_n$. Hence $\mathcal{L}_n$ must contain a curve F_n which is not divisible by $X_0 X_1 X_2$.

We prove that F_n is well-intersecting. No vertex of the fundamental triangle is on F_n, because if a vertex were on F_n, then at least one of the two sides of the fundamental triangle would meet F_n in at least $n+1$ points. Thus this side would be a component of F_n, hence another vertex of the triangle would lie on F_n. Repeating this reasoning we conclude that all sides of the fundamental triangle are components of F_n, so F_n would be divisible by $X_0 X_1 X_2$, giving a contradiction. Thus F_n meets each side of the fundamental triangle in n points. Among these (altogether $3n$) intersections $3n - 1$ coincide with the prescribed ones. The remaining one, say P_2', is on the line $X_2 = 0$, and its homogeneous coordinates are $(a_1' : 1 : 0)$. We claim that $a_1' = a_1$. Apply the first part of the theorem to the curve F_n. This gives

$$\left(a'_1 a_1^{m_2^1 - 1} \prod_{i>1} a_i^{m_2^i}\right) \prod_j b_j^{m^j_0} \prod_k c_k^{m^k_1} = (-1)^n.$$

But Equation (8.3) also holds, hence $a'_1 = a_1$, so F_n is well-intersecting, and the theorem is proved. $\qquad\square$

Now we extend this theorem to $k > 3$ lines (like the Theorem of Helly in classical geometry). The original version of the following theorem was proved by Segre. He considered only the case when $\mathcal{K}$ is a dual arc. We prove a more general result.

Theorem 8.7 (Segre; J. Thas, Cameron, Blokhuis). *Let $\mathcal{K} = \{L_1, L_2, \ldots, L_k\}$ be a line-set in $\mathrm{PG}(2, \mathbb{K})$. Let P^i_j be points on L_j distinct from the points $L_j \cap L_t$. Suppose that there are given multiplicities m^i_j assigned to the points P^i_j such that $\sum_i m^i_j = n$ holds for $j = 1, 2, \ldots, k$. We say that a curve of degree n is well-intersecting for a subset of $\mathcal{K}$ if it meets each line of the subset of $\mathcal{K}$ exactly at the given points with the required multiplicities. Suppose that for any pencil $\mathcal{P}$ of lines, for each subset of type $\mathcal{P} \cap \mathcal{K}$, and for each triangle whose sides belong to $\mathcal{K}$ there exists a well-intersecting curve of degree n. Then*

- *there exists a curve F_n of degree n which is well-intersecting for the whole line-set $\mathcal{K}$,*

- *if $k > n$, then F_n is unique.*

Proof. We follow the original paper by *J. Thas, Cameron* and *Blokhuis* [171], and use the same notation. We prove the theorem by induction on k. If all the lines are concurrent or $k = 3$, then there is nothing to prove. Hence let $k \geq 4$ and assume that $\mathcal{K}$ is not a subset of a pencil of lines.

Let $Y_i = L_i \cap L_k$ for $i = 1, \ldots, k-1$. We may assume without loss of generality that $Y_1 \neq Y_2$. Choose the system of reference so that the equation of the line L_k is $X_0 = 0$, and let $Y_i = (0 : c_i : d_i)$ for $i = 1, \ldots, k-1$. By induction, there exists a curve F_i of degree n which is well-intersecting $\mathcal{K} \setminus \{L_i\}$. Of course, F_i and F_j could be different curves if $i \neq j$.

Take the curves $F_1(0, X_1, X_2) = 0$ and $F_2(0, X_1, X_2) = 0$. Both are well-intersecting for the line L_k, thus if we dehomogenize the polynomials $F_1(0, X_1, X_2)$ and $F_2(0, X_1, X_2)$, then the resulting polynomials have the same roots with the same multiplicities, so they only differ by a constant factor. Thus we may assume that $F_1(0, X_1, X_2) = F_2(0, X_1, X_2)$.

As the curves F_1 and F_k are well-intersecting for the line L_2, the point Y_2 does not lie on F_1 or on F_k, hence $F_1(0, c_2, d_2) \neq 0 \neq F_k(0, c_2, d_2)$. Let $\beta' = -F_k(0, c_2, d_2)/F_1(0, c_2, d_2)$ and define the polynomial

$$F'(X_0, X_1, X_2) = F_k(X_0, X_1, X_2) + \beta' F_1(X_0, X_1, X_2).$$

If $F' = 0$, then we are done because in this case both F_1 and F_k are well-intersecting for $\mathcal{K}$.

Consider the points of intersection of the curve F' and the line L_2. As both F_k and F_1 meet L_2 at P_2^i with multiplicity m_2^i, their linear combination F' also meets L_2 at P_2^i and the intersection multiplicity is at least m_2^i. But F' also contains the point Y_2, hence L_2 is a component of F'. In the same way, for each $j = 3, 4, \ldots, k-1$, we get that L_j is a component of F', too. Thus if $R_i = d_i X_1 - c_i X_2$, then $R_i(Y_i) = 0$ and

$$F'(0, X_1, X_2) = F_k(0, X_1, X_2) + \beta' F_1(0, X_1, X_2)$$

is divisible by $\prod_{i=2}^{k-1} R_i$.

Now interchange the roles of L_1 and L_2 and repeat the whole process. Let $\beta'' = -F_k(0, c_1, d_1)/F_2(0, c_1, d_1)$ and define the polynomial $F'' = F_k + \beta'' F_2$. As before, if $F'' = 0$, then the theorem is proved. Consider the polynomial $F' - F''$. We get

$$\begin{aligned} F'(0, X_1, X_2) - F''(0, X_1, X_2) &= \beta' F_1(0, X_1, X_2) - \beta'' F_2(0, X_1, X_2) \\ &= (\beta' - \beta'') F_1(0, X_1, X_2), \end{aligned} \tag{8.4}$$

because $F_1(0, X_1, X_2) = F_2(0, X_1, X_2)$. Hence $F'(0, X_1, X_2) - F''(0, X_1, X_2)$ vanishes at the point P_i^k with multiplicity m_i^k, and also vanishes at the point Y_3. Thus it has at least $n + 1$ roots, which is more than its degree, so it is the zero polynomial. As $F_1(0, X_1, X_2)$ is not the zero polynomial we get $\beta' = \beta''$, so

$$\begin{aligned} F''(0, X_1, X_2) &= F_k(0, X_1, X_2) + \beta'' F_2(0, X_1, X_2) \\ &= F_k(0, X_1, X_2) + \beta' F_1(0, X_1, X_2) \end{aligned}$$

is divisible by $R_1 R_3 R_4 \ldots R_{k-1}$. As R_1 is not a constant multiple of R_2, this means that $F'(0, X_1, X_2) = F''(0, X_1, X_2)$ is divisible by $\prod_{i=1}^{k-1} R_i$. Hence

$$F_k(0, X_1, X_2) + \beta' F_1(0, X_1, X_2) = R_1 R_2 \ldots R_{k-1} R.$$

Let $L_i : e_i X_0 + R_i = 0$ for $i = 1, \ldots, k-1$, and let F be the curve with equation

$$F(X_0, X_1, X_2) = F_k(X_0, X_1, X_2) - L_1 L_2 \ldots L_{k-1} R = 0.$$

Then F is well-intersecting for each of the lines $L_1, \ldots, L_{k-1}$, because F_k is well-intersecting for these lines. We get the points of intersections of F and L_k by substituting $X_0 = 0$. This results in $\beta' F_1(0, X_1, X_2)$, hence F is well-intersecting for L_k, too.

The proof of the uniqueness of the curve is quite simple. Let $k > n$ and suppose to the contrary that both of the curves F and G are well-intersecting for $\mathcal{K}$. This means that none of them contains Y_1. Let $\alpha = -F(0, c_1, d_1)/G(0, c_1, d_1)$ and let H be the curve defined by the polynomial

$$H(X_0, X_1, X_2) = F(X_0, X_1, X_2) + \alpha G(X_0, X_1, X_2).$$

According to the assumption $H(X_0, X_1, X_2)$ is not the zero polynomial and its degree is at most n. But H contains the point Y_1, hence the sum of the intersection multiplicities of H and the line L_k is at least $n + 1$. Thus L_k is a component of H. This means that the point Y_i is also on H for $i = 2, 3, \ldots, k - 1$, hence the line L_i is also a component of H. So H contains at least $k > n$ linear components, which is a contradiction. $\qquad\square$

Let us remark that one can easily construct a set of $2 \times (q + 1)$ points, two of them on each line of a pencil, such that the 6 points on any triple of lines are on a conic, but there is no conic containing all points (see Exercise 8.1). This example shows that the condition "there exists a well-intersecting curve for each subset of type $\mathcal{P} \cap \mathcal{K}$" cannot be weakened if $n = 2$. This counterexample can be generalized for $n > 2$, too.

Notation 8.8. *Let $m_2'(2, q)$ denote the size of the second largest complete arc in* $\mathrm{PG}(2, q)$.

We will study dual arcs instead of arcs, because the description of the corresponding algebraic curves is more natural in this setting. For the sake of completeness we recall the definition of a dual arc.

Definition 8.9. *The set of lines $\{L_1, \ldots, L_k\}$ is a* dual k-arc *if no three of its elements have a point in common.*

A dual tangent *of a dual k-arc $\mathcal{K}$ is a point which is incident with exactly one line of $\mathcal{K}$.*

Theorem 8.10 (Segre). *Let $\mathcal{K} = \{L_1, L_2, \ldots, L_k\}$ be a dual k-arc in $\mathrm{PG}(2, q)$ and let $\mathcal{T}$ be the set of points formed by the tk dual tangents of $\mathcal{K}$, where $t = q + 2 - k$. Then the following hold:*

(a) *If q is even, then the points of $\mathcal{T}$ are on a curve C_t of degree t. If $T \in \mathcal{T}$ is an arbitrary point and $L_i \in \mathcal{K}$ is an arbitrary line then the intersection multiplicity of C_t and L_i at T is one. If $k > (q + 2)/2$, then the curve C_t is unique.*

(b) *If q is odd, then the points of $\mathcal{T}$ are on a curve C_{2t} of degree $2t$. If $T \in \mathcal{T}$ is an arbitrary point and $L_i \in \mathcal{K}$ is an arbitrary line then the intersection multiplicity of C_{2t} and L_i at T is two. In this case C_{2t} may contain components of multiplicity at most two, but does not consist entirely of double components. If $k > 2(q + 2)/3$, then C_{2t} is unique.*

The curves C_t and C_{2t} do not contain any point $L_i \cap L_j$ for $1 \le i < j \le k$.

Proof. By the previous theorem, it is enough to show that for each triangle of $\mathcal{K}$, the coordinates of the dual tangents on its sides satisfy the condition given in Equation (8.3).

We apply the method of Lemma 6.17 again. Let us choose the system of reference so that the three elements of $\mathcal{K}$ are $L_1 : X_0 = 0$, $L_2 : X_1 = 0$ and

$L_3 : X_2 = 0$, and take the homogeneous coordinates of the other points on these lines as $(0 : 1 : a)$, $(b : 0 : 1)$ and $(1 : c : 0)$, respectively. By Menelaus' theorem (Theorem 8.6), three points from distinct lines are collinear if and only if $abc = -1$. Create a $(k - 3) \times 3$ matrix M whose rows correspond to the lines $\mathcal{K} \setminus \{L_1, L_2, L_3\}$, and the columns correspond to the lines $\{L_1, L_2, L_3\}$. For $i = 1, 2, \ldots, k - 3$ and $j = 1, 2, 3$ let the (i, j) entry $m_{i,j}$ of M be the coordinate of the point $L_{i+3} \cap L_j$. Then the product of the three elements in any row of M is (-1). In each column of M there are $k - 3$ pairwise distinct, non-zero elements of GF(q). Extend M with a $(q + 2 - k) \times 3$ matrix M' whose entries are the coordinates of the dual tangents to $\mathcal{K}$. Then all non-zero elements of GF(q) appear exactly once in each column of the extended matrix. This means that in the j-th column of M' we see the coordinates of the dual tangents lying on the line L_j. So the product of all elements in the extended matrix is $(-1)^3 = -1$. As the product of all elements of M is $(-1)^{k-3}$, we get that the product of all elements of M' is $(-1)^{4-k}$.

If q is even, then $(-1)^{4-k} = (-1)^{q+2-k} = 1$, so the coordinates of the dual tangents satisfy Equation (8.3). Thus, by Theorem 8.7, there exists a curve C_t of degree t which is well-intersecting for $\mathcal{K}$. If q is odd, then $4 - k$ and $q + 2 - k$ have opposite parity, so $(-1)^{4-k} \neq (-1)^{q+2-k}$. But if we consider each dual tangent with multiplicity 2, then Equation (8.3) is satisfied again, because $(-1)^{2(4-k)} = (-1)^{2(q+2-k)} = 1$. So, by Theorem 8.7, there exists a curve C_{2t} of degree $2t$ for which if $T \in \mathcal{T}$ is an arbitrary point and $L_i \in \mathcal{K}$ is an arbitrary line then the intersection multiplicity of C_{2t} and L_i at T is two.

The uniqueness of the curves also follows from Theorem 8.7, because in the case q even $k > t = q + 2 - k$ if $k > (q + 2)/2$, while in the case q odd $k > 2t = 2(q + 2 - k)$ if $k > 2(q + 2)/3$.

Finally, in the latter case, there is no curve of degree t which contains each point of $\mathcal{T}$, because the product of the coordinates of the points of $\mathcal{T}$ does not satisfy Equation (8.3). Hence the curve C_{2t} cannot consist entirely of double components. $\square$

Corollary 8.11. *Let $\mathcal{K}$ be a k-arc in* PG$(2, q)$, *q even. Suppose that the point P lies on a secant of $\mathcal{K}$. Then there are at most $t = q + 2 - k$ tangents to $\mathcal{K}$ through P.*

Proof. There are exactly t tangents to $\mathcal{K}$ at any of its points. Consider the dual plane. Let $\mathcal{K}'$ be a dual k-arc and ℓ be a line through a point of intersection of two elements of $\mathcal{K}'$. Let C_t be the curve associated to $\mathcal{K}'$ according to Theorem 8.10. Then C_t contains each of the t dual tangents lying on the line ℓ, and we also know that ℓ is not a component of C_t (because ℓ contains a point of intersection of two elements of $\mathcal{K}'$). Thus, by the Fundamental Theorem of Algebra, $|C_t \cap \ell| \leq t$. $\square$

Proposition 8.12. *Let $\mathcal{K}$ be a dual k-arc in* PG$(2, q)$ *and $t = q + 2 - k$. Suppose that the associated curve C_t or C_{2t}, according to q being even or odd, has a linear component L. Then $\mathcal{K} \cup \{L\}$ is a dual $(k + 1)$-arc.*

Conversely, if $k > t$, or $k > 2t$, according to q being even or odd, and $\mathcal{K}$ can be extended to a dual $(k+1)$-arc by adding the line L, then L must be a component of the curve. If q is odd, then each linear component of C_{2t} is a double component of the curve.

Proof. Let $\mathcal{K} = \{L_1, L_2, \ldots, L_k\}$ and C denote the curve C_t or C_{2t}, according to whether q is even or odd. Then C contains no point $L_i \cap L_j$ for $1 \le i < j \le k$. Hence any linear component L^* of C meets the lines of $\mathcal{K}$ in pairwise distinct points, so $\mathcal{K} \cup \{L^*\}$ is a dual $(k+1)$-arc.

In the opposite direction, if $\mathcal{K} \cup \{L^*\}$ is a dual $(k+1)$-arc, then L^* meets the associated curve C in at least k distinct points, because each point of intersection lies on at most one line of $\mathcal{K}$. This means that $|L^* \cap C| \ge k$, hence, because of $k > t$ (or $k > 2t$), the Fundamental Theorem of Algebra gives that C is divisible by L^*.

If q is odd, then if we multiply the coordinates of the points of intersections of L^* and the sides of any triangle chosen from $\mathcal{K}$, by Menelaus' theorem, we get -1. So if we delete these points and consider only the remaining ones, then for any triangle chosen from $\mathcal{K}$, the product of their coordinates is $(-1)^{2t-1}$. Thus there exists a curve C^* of degree $(2t-1)$ containing these points. The curve C^* contains the points $L_i \cap L^*$ for $i = 1, 2, \ldots, k$. But $k > 2t - 1$ implies that L^* is a component of C^*, thus L^* is a double component of C. □

The next result was proved in higher dimensions by *Blokhuis, Bruen* and *J. Thas* [27]. The planar case can be found in *Szőnyi* [161]. In the odd case his bound is 1 larger than the planar case of the general bound. The improvement was also pointed out by *Chao* and *Kaneta* [41].

Theorem 8.13. *If $\mathcal{K}$ is a k-arc in $\mathrm{PG}(2, q)$ with*

$$k \ge \begin{cases} (q+4)/2, & \text{if } q \text{ is even,} \\ (2q+5)/3, & \text{if } q \text{ is odd,} \end{cases}$$

then there is a unique complete arc containing $\mathcal{K}$.

Proof. We apply the dual point of view again and distinguish two cases.

Case 1, q even. Let $\mathcal{K}$ be a dual k-arc and consider the set of lines that could extend $\mathcal{K}$ to a $(k+1)$-arc. Let

$$\mathcal{L} = \{L \in \mathrm{PG}(2, q) : \mathcal{K} \cup \{L\} \text{ is a dual arc}\}.$$

We claim that $\mathcal{K} \cup \mathcal{L}$ is also an arc. First, we show that for all $L_1, L_2 \in \mathcal{L}$ the set $\mathcal{K}^* = \mathcal{K} \cup \{L_1, L_2\}$ is a dual $(k+2)$-arc. Suppose to the contrary that $\mathcal{K}^*$ contains three lines of a pencil having carrier P. As $\mathcal{K}$ can be extended to a dual $(k+1)$-arc by any of L_1 and L_2, the point P must be the intersection of L_1 and L_2. Let $L_0 \in \mathcal{K}$ be the third line through P. Consider the dual $(k+1)$-arc $\mathcal{K}' = \mathcal{K} \cup \{L_1\}$ and let C_t be the curve associated to it according as Theorem 8.10.

If $L \in \mathcal{K} \backslash \{L_0\}$, then the point $L \cap L_2$ is a dual tangent of $\mathcal{K}'$, so L_2 intersects C_t in at least $k-1$ points. Since the point P is on two lines of $\mathcal{K}'$ (L_0 and L_1), the line L_2 cannot be a component of C_t. Thus, by the Fundamental Theorem of Algebra, we have $k - 1 \leq t = q + 2 - (k + 1)$, contradicting our assumption on k.

In the same way, we can prove that $\mathcal{L}$ does not contain three concurrent lines. We have to repeat the previous argument for the set $\mathcal{K} \cup \{R\}$ with $R \in \mathcal{L}$.

Case 2, q odd. The proof is almost the same as in the even case; we use the same notation, too. Suppose that the lines L_1, $L_2 \in \mathcal{L}$ and $L_0 \in \mathcal{K}$ have a point in common, say P. Let C_{2t} be the curve of degree $2t$ associated to $\mathcal{K}$ according to Theorem 8.10. Then, by Proposition 8.12, both L_1 and L_2 are double components of C_{2t}. This means that the intersection multiplicity of C_{2t} and L_0 at P is at least 4, but there is only one element of $\mathcal{K}$ covering P, the line L_0, giving a contradiction. Finally, we can repeat the end of the proof of Part 1 and obtain that no three lines of $\mathcal{L}$ are concurrent. □

It is worth noting that the bound of this theorem is tight if q is even. The example of Tallini Scafati (see Theorem 6.22) is a $(q + 2)/2$-arc that can be extended to a complete $(q + 4)/2$-arc in $q/2$ different ways, by adding one of the points on the line at infinity.

The proofs of the next results need deep theorems, for example the Hasse-Weil bound about the number of points, defined over $\mathrm{GF}(q)$, of an algebraic curve. These are beyond the scope of this book, so we only present some of the results.

Theorem 8.14. *In* $\mathrm{PG}(2, q)$ *the following bounds are known:*

(1) (Segre [153]) $m_2'(2, q) \leq q - \sqrt{q} + 1$, *if q is even,*

(2) (Segre [153]) $m_2'(2, q) \leq q - \sqrt{q}/4 + 25/16$, *if q is odd,*

(3) (Voloch [178]) $m_2'(2, p) \leq 44p/45 + 2$, $p \geq 7$ *prime.*

The next theorem summarizes the best known estimates so far.

Theorem 8.15. *In* $\mathrm{PG}(2, q)$ *the following hold:*

(1) (Voloch [179]) $m_2'(2, q) \leq q + 1 - \sqrt{pq}/4 + 29p/16 + 1$, *if $q = p^h$, $p > 2$ prime and $h \geq 3$ odd,*

(2) (Voloch [179]) $m_2'(2, q) \leq q + 2 - \sqrt{2q}$, *if $q = 2^h$ and $h \geq 3$ odd,*

(3) (Hirschfeld, Korchmáros [89]) $m_2'(2, q) \leq q - \sqrt{q}/2 + 3$, *if $q = p^h$, $p \geq 3$ prime, $q \geq 19^2$, $q \neq 5^5$, and also if $p = 3$, h even and $q \neq 3^6$,*

(4) (Hirschfeld, Korchmáros [89]) *if $q = 2^h$ and $h > 4$ even, then the size of a complete arc in* $\mathrm{PG}(2, q)$ *is $q + 2$, $q - \sqrt{q} + 1$, or at most $q - 2\sqrt{q} + 6$,*

(5) (Ball, Lavrauw [13]) $m_2'(2, q) \leq q - \sqrt{q} + 2 + \sqrt{q}/p$, *if q is an odd square.*

Ball and *Lavrauw* [13] recently also showed that $m_2'(2,q) < q - \sqrt{q} + 7/2$ if q is a prime. This is better than Voloch's bound for small values of q and can be used to describe complete $(q-1)$-arcs. Instead of curves associated to dual arcs, they introduced the *tangent functions*. If $\mathcal{K}$ is a $(q+1-t)$-arc, then every point $P \in \mathcal{K}$ is on t tangents. Let $f_P = \prod_{i=1}^{t} L_i$ where $L_i(X_0, X_1, X_2) = 0$ is the homogeneous equation of the i-th tangent to $\mathcal{K}$ at P. Then f_P is defined up to a scalar factor. Fix the homogeneous coordinates of a base point, say $E = (e_0 : e_1 : e_2)$, on $\mathcal{K}$ and scale the homogeneous coordinates for all other points $P \in \mathcal{K}$ such that $f_P(e_0, e_1, e_2) = (-1)^{t+1} f_E(p_0, p_1, p_2)$. Their results are based on the following reformulation of Segre's Lemma of Tangents (Lemma 6.17). The proof can be found in the original paper [13].

Theorem 8.16 (Ball, Lavrauw [13]). *Let $\mathcal{K}$ be a $(q+1-t)$-arc in* PG$(2,q)$. *Then for any pair of points $X, Y \in \mathcal{K}$:*

$$f_X(y_0, y_1, y_2) = (-1)^{t+1} f_Y(x_0, x_1, x_2).$$

One should also note that all the proofs except the ones by Ball and Lavrauw use Segre's results. This means that the curves are defined in the dual plane and the deep theorems from algebraic geometry are used for those curves, while in the case of Ball and Lavrauw special curves vanishing on the points of the arc are constructed.

Finally, we present an elementary proof that gives almost the same result as Theorem 8.14, Part (1). It contains no algebraic geometry, only the purely combinatorial bound on the size of a (k,n)-arc due to Barlotti (Proposition 7.3).

Theorem 8.17 (J. Thas). *In* PG$(2,q)$, *q even,*

$$m_2'(2,q) \leq q - \sqrt{q + \frac{1}{4}} + \frac{3}{2}.$$

Proof. Apply the dual point of view again. Let $\mathcal{K}$ be a complete dual k-arc and let C_t be the curve associated to $\mathcal{K}$ according to Theorem 8.10. By Proposition 8.12, C_t has no linear component defined over GF(q). This means that each line of PG$(2,q)$ meets C_t in at most t points. Hence if M denotes the number of GF(q)-rational points of C_t, then C_t is an (M, t')-arc with a suitable $t' \leq t$. Thus, by Proposition 7.3, we have $M \leq (t'-1)q + t' \leq (t-1)q + t$. On the other hand, the number of dual tangents to $\mathcal{K}$ is tk, so

$$t(q + 2 - t) = tk \leq (t-1)q + t,$$

hence $q \leq t^2 - t$. From this quadratic inequality we get $t \geq (1 + \sqrt{4q+1})/2$, which proves the statement after rearranging. $\qquad\square$

If q is an even square, then this bound of J. Thas gives $m_2'(2,q) \leq q - \sqrt{q} + 1$, that is the same as the bound of Segre. Comparing this bound and the construction given in Theorem 6.24, we get the exact size of the second largest complete arc.

Corollary 8.18. *In* $\mathrm{PG}(2, q)$, *if* $q > 4$ *is an even square, then*

$$m_2'(2, q) = q - \sqrt{q} + 1.$$

Let us also see an application of algebraic curves to blocking sets. As a preparation, a more general notion of intersection multiplicity (defined earlier for the intersection of a curve and a line) is needed. We shall formulate everything for affine curves; the results can be translated relatively easily to projective ones (see [90]).

We will also use the definition and standard properties of algebraic curves. Even if the curve is defined over $\mathrm{GF}(q)$, it will automatically be considered over the algebraic closure of $\mathrm{GF}(q)$. In many cases we shall be interested in points of the curve defined over $\mathrm{GF}(q)$. Such points are called $\mathrm{GF}(q)$-*rational.* An affine point is $\mathrm{GF}(q)$-rational if its coordinates belong to $\mathrm{GF}(q)$, while a projective point is $\mathrm{GF}(q)$-rational if the coordinate ratios are in $\mathrm{GF}(q)$. Here we summarize the defining properties of *intersection multiplicity* of the curves F and G at a point P, which will be denoted as $I(P; F \cap G)$. These properties will be used frequently later.

(I1) $I(P; F \cap G) < \infty$ if and only if P is not on a common component of F and G,

(I2) $I(P; F \cap G) = 0$ if and only if $P \notin F \cap G$,

(I3) $I(P; F \cap G) = I(P; G \cap F)$,

(I4) $I(P; FG \cap H) = I(P; F \cap H) + I(P; G \cap H)$,

(I5) $I(P; F \cap G) = I(P; F \cap (G + AF))$ for any $A \in \mathbb{K}[\mathbf{X}]$,

(I6) $I(P; F \cap G) \geq m_P(F) m_P(G)$, where $m_P(F)$ denotes the multiplicity of P as a point of F (and $m_P(G)$ is the same for G).

First of all recall that in the first part of this chapter the intersection multiplicity of a curve and a line was already defined using the multiplicity of a root of a one-variable polynomial. The *multiplicity* of a point is the minimum multiplicity of a line through the point with the curve. A point of the curve is called a *simple point* if the multiplicity of the point is 1. This is equivalent to saying that the partial derivatives of the polynomial defining the curve do not vanish at the point. If a point is not simple, then it is called *singular*. We shall not need details about singular points; they are just used in the definition above.

Most of the above requirements are plausible. (I3), (I4) and (I5) essentially mean that the intersection multiplicity only depends on the ideal generated by F and G (see Theorem 3.1 in [90]). (I6) is not very natural, but *Scherk* [148] showed that it is equivalent to

(I6') $I(P; L_1 \cap L_2) = 1$, if $L_1 \neq L_2$ are two lines and $P = L_1 \cap L_2$, which is again quite natural. Let us remind the reader that there is a unique intersection multiplicity which is invariant under projective transformations and satisfies all these requirements (see Theorems 3.8 and 3.9 in [90]). Thus the above requirements can be regarded as a definition of intersection multiplicity. Also note that these properties can be used to actually compute the intersection multiplicity of two curves. Let us briefly recall how this works for the origin. Since any point can be transformed to the origin, this approach works for other points, too.

We wish to compute $I(P; F \cap G)$, where $P = (0,0)$. Let $f(X) = F(X, 0)$, $g(X) = G(X, 0)$ and $d(X)$ denote the greatest common divisor of $f(X)$ and $g(X)$. Then we can write $d(X) = a(X)f(X) + b(X)g(X)$ by the Euclidean algorithm. Put $f_1(X) = f(X)/d(X)$, $g_1(X) = g(X)/d(X)$. Then $f_1(X)a(X) + g_1(X)b(X) = 1$, which means that

$$\begin{pmatrix} -f_1(X) & g_1(X) \\ b(X) & a(X) \end{pmatrix} \begin{pmatrix} G(X,Y) \\ F(X,Y) \end{pmatrix} = \begin{pmatrix} H(X,Y) \\ U(X,Y) \end{pmatrix}$$

is an invertible linear transformation. Hence the ideals (F, G) and (H, U) are the same, so $I(P; F \cap G) = I(P; H \cap U)$. Since Y divides $H(X, Y) = -f_1(X)G(X, Y) + g_1(X)F(X, Y)$, we can use property (I4) of the intersection multiplicity. This process can be continued and after some steps we can recognize that a certain intersection multiplicity is 0, using property (I2).

Let us also remark that the multiplicity of the origin as a point of the curve can also be determined easily. Let $f(X, Y)$ be the equation of the curve. Then $f(X, Y) = f_0(X, Y) + f_1(X, Y) + \ldots$, where $f_i(X, Y)$ is a homogeneous polynomial of degree i (or 0). This form of $f(X, Y)$ is clearly unique. If $(0, 0)$ is a point of the curve, then $f_0(X, Y) = 0$. If the origin is a simple point of f, then $f_0 = 0$ and $f_1 \neq 0$. In general, if f_r is the first f_i which is not zero (so $f_r \neq 0$ and $f_i = 0$ for $i = 0, \ldots, r - 1$), then the multiplicity of the origin is r.

As an illustration, the following useful lemma about intersection multiplicities can be proved, which can be found in Seidenberg's book, [155], Lemma 9.2.

Lemma 8.19. *Let P be the origin.*

(1) *If $I(P; F \cap Y) = r$, $I(P; G \cap Y) = s$ and $r \geq s$, then $I(P; F \cap G) \geq s$.*

(2) *If $r > s$ and P is a simple point of F, then $I(P; F \cap G) = s$.*

Proof. We only prove (1), (2) is left as an exercise. In determining the intersection multiplicity with the approach indicated above, then we have $X^s | d(X)$. So when we compute $I(P; H \cap U)$ with $H = YH_1$, we get $I(P; YH_1 \cap U) = I(P; Y \cap U) + \ldots$, where $\ldots \geq 0$. Now put $U = U_1(X) + YU_2(X, Y)$. Then $I(P; Y \cap U) = I(P; Y \cap U_1)$. Since the terms of U that do not contain Y come from similar terms of F and G, and $d(X)$ is divisible by X^s, X^s divides $U_1(X)$. This implies $I(P; Y \cap U_1(X)) \geq I(P; Y \cap X^s) = s$. □

Let us also recall *Bézout's theorem* that we shall use frequently. We omit the proof, which can be found for example in Chapter 3.2 of [90].

Theorem 8.20 (Bézout). *If two projective curves F and G have no common component, then they intersect in exactly $\deg F \cdot \deg G$ points, if common points are counted according to their intersection multiplicity.*

In the rest of this chapter we will assume that the line at infinity is a tangent to our blocking set and use the following notation:

Notation 8.21. *Let $\mathcal{B} = U\{\infty\}$ be a non-trivial blocking set, $U \subset \mathrm{AG}(2,q)$, $|U| = q+k$, $k < q$. Let $H(X,Y) = \sum_{j=0}^{|U|} h_j(Y)X^{|U|-j}$ be the Rédei polynomial of U.*

Definition 8.22. *Let $\mathcal{C}$ be the affine curve of degree k defined by*

$$f(X,Y) = X^k + h_1(Y)X^{k-1} + \ldots + h_k(Y).$$

Multiple components are allowed here.

Note that the polynomial $f(X,Y)$ has degree k indeed. Observe also that the X-degree of f is its total degree. This will be used later. The next theorem summarizes some important properties of the Rédei polynomial and of this curve. The first part of it shows that the curve f has a lot of $\mathrm{GF}(q)$-rational points, the second and third parts help us translate geometric properties of U into properties of f. It continues Lemma 6.35 using the curve $f(X,Y)$.

Theorem 8.23.

(1) *For a fixed (m), $m \in \mathrm{GF}(q)$, the polynomial $(X^q - X)$ divides $H(X,m)$. Moreover, if $k < q - 1$, then $H(X,m)/(X^q - X) = f(X,m)$ for every $m \in \mathrm{GF}(q)$. Hence $f(X,m)$ splits into linear factors over $\mathrm{GF}(q)$ for every m.*

(2) *For a fixed (m), the element b is an r-fold root of $H(X,m)$ if and only if the line with equation $Y = mX + b$ intersects U in exactly r points.*

(3) *If the line with equation $Y = m$ $(m \in \mathrm{GF}(q))$ meets $f(X,Y)$ at (b,m) with multiplicity t, then the line with equation $Y = mX + b$ meets U in exactly $t + 1$ points.*

Proof. We have seen (2) right after the definition of the Rédei polynomial, see Lemma 6.35. The first part of (1) follows from (2) and the well-known fact that $\prod_{u \in GF(q)}(X - u) = X^q - X$. The rest of (1) is obvious. To prove (3), note that if the intersection multiplicity is t, then b is a $(t + 1)$-fold root of $H(X,m)$. Now the assertion follows from (2). □

The facts given in Theorem 8.23 will be used frequently without further reference. The next lemma shows that the linear components of f correspond to points of $\mathcal{B}$ which are not essential. Recall that in Chapter 6 we have seen that a blocking set is minimal if and only if it had at least one tangent line at any of its points. A point of a blocking set is called *essential* if it has a tangent line, and a point P is called *non-essential* if every line passing through P meets the blocking set in at least two points. The fact that non-essential points correspond to linear components is analogous to Proposition 8.12, where points which can be added to an arc were described similarly.

Lemma 8.24 ([162]).

(1) *If the point $P = (a, b) \in \mathcal{B}$ is not essential, then $X + aY - b$ divides $f(X, Y)$ (as polynomials in two variables).*

(2) *Conversely, if $X + aY - b$ divides $f(X, Y)$, then the point (a, b) is in $\mathcal{B}$ and it is not essential.*

Proof. (1): Take any infinite point (y_0). For this y_0, there are at least two points of $\mathcal{B}$ on the line with slope y_0 through P. Hence $(X + ay_0 - b)$ divides $f(X, y_0)$. In other words, the line $L : X + aY - b$ and the curve $\mathcal{C}$ have a common point for $Y = y_0$. This happens for $q > k$ values of y_0, so Bézout's theorem implies that $L \subset \mathcal{C}$.

(2): Conversely, assume that $X + aY - b$ divides $f(X, Y)$. Then, for every (y_0), the line with slope y_0 through (a, b) intersects $\mathcal{B}$ in at least two points. If $(a, b) \notin B$, then $2q \leq |U| = q + k$. This gives a contradiction, hence $(a, b) \in \mathcal{B}$. Since every line with slope y_0, (y_0), contains at least two points of $\mathcal{B}$, and the vertical line contains (∞) besides (a, b), the point (a, b) cannot be essential. $\qquad\square$

The previous lemma simply says that there are no linear components of f if $|\mathcal{B}| \leq 2q$. Note that also in Segre's theory of complete arcs, we have just seen in the first part of this chapter, there is a lemma corresponding to this one (see Corollary 10.3/(ii) and Theorem 10.4/(ii) in [88], and Proposition 8.12 here), and it plays an important role in proving the incompleteness of arcs.

Now we are going to prove a lower bound on the number of GF(q)-rational points of certain components of f. Again, also in Segre's theory, it is crucial that the curves associated to a k-arc have a lot of GF(q)-rational points, but there the lower bound $(q + 2 - k)k$ is immediate.

Lemma 8.25.

(1) *Let h_P denote the horizontal line through the point P. Then the sum of the intersection multiplicities $I(P, f \cap h_P)$ over all GF(q)-rational points of f is exactly qk. If h is a component of f, then the corresponding sum for h is precisely $q \deg h$.*

(2) *Let $h(X, Y)$ be a divisor of $f(X, Y)$ and suppose that it has neither multiple components nor components with zero partial derivative with respect to X. Then the number of GF(q)-rational points of h is at least*

$$q \deg h - \deg h (\deg h - 1). \tag{8.5}$$

Proof. The assertion for f follows from the fact that $f(X, m)$ is a product of k linear factors for every $m \in \mathrm{GF}(q)$. The same is true for every divisor of f. So the number of points, counted with the intersection multiplicity of h and the horizontal line at that point, is exactly qs, where s denotes the X-degree of h, which is the same as its total degree.

To count the number of points without this multiplicity, we have to subtract the number of points where $I(P; h \cap h_P) \geq 2$. These are precisely the points where $h'_X(P) = 0$, so we have to subtract the affine intersections of h and h'_X. Bézout's theorem then gives the result. We have to note that in this counting of common points of h and h'_X, the common points are counted once if $I(P; h \cap h_P)$ is not divisible by p, and the points with intersection multiplicity divisible by p are not counted at all. To see this, we have to recall Lemma 8.19, (1) and it gives our result. $\qquad\square$

These elementary observations already yield interesting results on blocking sets; see some of the exercises.

The aim of the rest of this chapter is to show that minimal blocking sets of size less than $3(q+1)/2$ intersect every line in 1 modulo p points. This immediately implies Blokhuis' result (Theorem 6.38) for blocking sets in $\mathrm{PG}(2, p)$.

Let now $\mathcal{B}$ be a minimal blocking set of $\mathrm{PG}(2, q)$ and, as before, let U be its affine part, $\mathcal{B} = U \cup \{(\infty)\}$. Only blocking sets with size at most $2q - 1$ will be considered.

(∞). Let $U = \{(a_i, b_i) : i = 1, 2, \ldots, q + k\}$ and write up the Rédei polynomial $H(X, Y) = \prod(X + a_i Y - b_i)$. Since $H(X, Y)$ vanishes for all $(x, y) \in \mathrm{GF}(q) \times \mathrm{GF}(q)$, we can write it as

$$H(X, Y) = (X^q - X)f(X, Y) + (Y^q - Y)g(X, Y),$$

where $\deg f, \deg g \leq k$ as polynomials in two variables. Note that f here is the same as the one defined earlier, so it has degree k. If one fixes $Y = y$, then $H(X, y)$ is divisible by $(X^q - X)$ and for an $(x, y) \in \mathrm{GF}(q) \times \mathrm{GF}(q)$: $f(x, y) = 0$ if and only if the line with equation $Y = yX + x$ intersects U in at least two points. If $(0, c) \in U$, then $H(X, Y)$ is divisible by $(X - c)$, so when we substitute a fixed x, then $H(x, Y)$ will be divisible by $Y^q - Y$ and the quotient is $g(x, Y)$. Note that the product $\prod(X - c)$, where $(0, c) \in U$, divides $g(X, Y)$. Hence we also have that whenever a line $Y = mX + b$ intersects U in more than 1 point, then $g(m, b) = 0$, similarly to (3) in Theorem 8.23.

Lemma 8.26. *For* $(x, y) \in \mathrm{GF}(q) \times \mathrm{GF}(q)$: $f(x, y) = 0$ *implies* $g(x, y) = 0$.

Lemma 8.27. *The polynomials f and g cannot have a common factor if $\mathcal{B}$ is minimal.*

Proof. Such a factor must divide $H(X, Y)$, hence it must be divisible by $X + a_i Y - b_i$ for some i. Theorem 8.23 (2) gives that the point (a_i, b_i) can be deleted, giving a contradiction. $\qquad\square$

Therefore, (f, g) is a pair of polynomials (curves) having no common factor (component), but they pass through more or less the same set of GF(q)-rational points (except of the components of the form $X - c$ of $g(X, Y)$ corresponding to points of U on the y-axis).

Lemmas 8.26 and 8.27 can also be used to show that all the components of f have identically zero partial derivative with respect to X if the minimal blocking set is small.

Recall that for any component h of f the total degree of h is the same as its degree in X.

Theorem 8.28. *If $\mathcal{B}$ is minimal, $k < (q+1)/2$ and $h(X, Y)$ is an irreducible polynomial that divides $f(X, Y)$, then $h'_X = 0$.*

Proof. Suppose to the contrary that h is a component with non-zero partial derivative with respect to X. Denote its degree by s. By Lemma 8.25, the number of GF(q)-rational points on h is at least $qs - s(s-1)$, and these points are also on g. By Lemma 8.27, g and h cannot have a common component; thus Bézout's theorem gives

$$qs - s(s - 1) \leq sk.$$

This immediately implies $q + 1 \leq k + s$ and from $s \leq k$ it follows that $k \geq (q + 1)/2$, a contradiction. □

Corollary 8.29 (Blokhuis, Theorem 6.38). *If $q = p$ is a prime, then $|\mathcal{B}| \geq 3(q + 1)/2$ for the size of a non-trivial blocking set.*

Corollary 8.30. *If $\mathcal{B}$ is a blocking set of size less than $3(q+1)/2$, then each line intersects it in 1 modulo p points.*

Proof. Since all the components of f contain only terms of exponent (in X) divisible by p, for any fixed $Y = y$ the polynomial $f(X, y)$ itself is the p-th power of a polynomial. This means that at each point (x, y) the line $Y = y$ intersects $f(X, Y)$ with multiplicity divisible by p, so the line $Y = yX + x$ intersects U in 1 modulo p points. □

Theorem 8.31. *Let $q = p^2$, p prime and $\mathcal{B}$ be a minimal blocking set which is not a Baer subplane. Then $|\mathcal{B}| \geq 3(q + 1)/2$.*

Proof. Suppose to the contrary that $|\mathcal{B}| < 3(q+1)/2$.

First, we show that every line intersects $\mathcal{B}$ in either 1 or $p+1$ points. By Corollary 8.30, each line intersects $\mathcal{B}$ in $rp + 1$ points where r is a suitable integer. If there were a line ℓ intersecting $\mathcal{B}$ in at least $2p + 1$ points, then we could choose a point $P \in \mathcal{B} \setminus \ell$ and we would see at least $1 + (2p+1)p > 2q$ points, giving a contradiction.

Using this extra information, by Lemma 6.48, $\mathcal{B}$ is either a Baer subplane or a unital. The former option is excluded in the theorem, while in the latter case $|\mathcal{B}| = p^3 + 1 > 3(q + 1)/2$, hence it is also excluded. This contradiction proves the theorem. □

Later *Polverino* [143] also characterized small minimal blocking sets in PG(2, p^3) and showed that they are of Rédei type. On the other hand, *Lunardon, Polito* and *Polverino* developed a method to construct small blocking sets.

Consider the projective plane PG(2, q^t) as the one- and two-dimensional subspaces of the three-dimensional vector space $V = V(3, q^t)$. V can also be regarded as a vector space over GF(q), and it will be a $3t$-dimensional vector space. Points are one-dimensional subspaces over GF(q^t), hence they are t-dimensional subspaces over GF(q). Their pairwise intersection is just the zero vector. Turning to the projective space PG($3t-1, q$) these subspaces correspond to disjoint (projective) subspaces of dimension $t-1$. They form a $(t-1)$-spread of PG($3t-1, q$). Two points determine a line, which means that the $(2t-1)$-dimensional subspace generated by two subspaces corresponds to a line. This line contains $q^t + 1$ points in the original plane PG(2, q^t). This means that the $(t-1)$-spread induces a spread on any $(2t-1)$-dimensional spread generated by two elements of the original spread. Spreads having this property are called *geometric* or *normal* spreads. Now consider a t-dimensional subspace Σ_t in PG($3t-1, q$). This defines a subset $\mathcal{B}$ of PG(2, q^t), namely those points, for which the corresponding $(t-1)$-dimensional subspace intersects Σ_t. As Σ_t intersects each $(2t-1)$-dimensional subspace (by dimension arguments), $\mathcal{B}$ is a blocking set. It is clear that $|\mathcal{B}| \leq (q^{t+1} - 1)/(q - 1)$, so $\mathcal{B}$ is a small blocking set. Blocking sets arising from this construction are called *linear blocking sets*. They were first considered by *Lunardon* [116], who showed that blocking sets of Rédei type can be obtained by this construction. Later *Polito* and *Polverino* [142] showed that for $t \geq 4$ there are linear blocking sets which are not of Rédei type.

An alternative way of constructing these blocking sets was pointed out by *Lunardon, Polito* and *Polverino* [117]. Start from a subgeometry PG(t, q) of PG(t, q^t) and project it onto a plane from a subspace of dimension $(t-3)$. The projection will be a blocking set in the plane PG(2, q^t). To see this, consider the vector space representation over GF(q) again. In this representation, the subgeometry will be a $(t + 1)$-dimensional vector space, while a hyperplane will be of codimension $t + 1$. Hence the subgeometry meets every hyperplane. This property does not change after projection, hence the projection of the subgeometry onto a plane will intersect every line (hyperplane) of that plane.

To illustrate the projection procedure, start from the subgeometry PG(3, p) inside PG(3, p^3) and project it from a point onto a plane. There are two essentially different ways to choose the point: it can lie on a line of the subgeometry or not. In the former case we get a blocking set of size $p^3 + p^2 + 1$, in the latter case a blocking set of size $p^3 + p^2 + p + 1$. The projected blocking set is of Rédei type in both cases.

It is surprising that linear blocking sets and the above projection argument were only discovered in the mid nineties. Since then, the theory of linear sets in general, has emerged rapidly. We just refer to the survey paper by *Lunardon* and *Polverino* [118].

In the general case we have the following theorem. The proof of the lower bound uses lacunary polynomials and it is omitted. The upper bound is combinatorial, and the proof is left as an exercise.

Theorem 8.32. *Let $p^e \neq 4, 8$, $\mathcal{B}$ be a minimal blocking set with $|\mathcal{B}| \leq 3(q+1)/2$, where each line intersects $\mathcal{B}$ in 1 modulo p^e points. Then*

$$q + 1 + p^e \left\lceil \frac{q/p^e + 1}{p^e + 1} \right\rceil \leq |\mathcal{B}| \leq \frac{1 + (p^e + 1)(q + 1) - \sqrt{\Delta}}{2},$$

where $\Delta = (1 + (p^e + 1)(q + 1))^2 - 4(p^e + 1)(q^2 + q + 1)$.

The latest and strongest result on minimal blocking sets with $|\mathcal{B}| \leq 3(q+1)/2$ is due to *Sziklai* [163]. He extended Rédei polynomials using homogeneous coordinates (so for subsets of $\mathrm{PG}(2, q)$) and associated three homogeneous polynomials satisfying a simple relation. With this method he got closer to showing that every such small minimal blocking set is linear and proved the following.

Theorem 8.33 (Sziklai). *Let $\mathcal{B}$ be a minimal blocking set with $|\mathcal{B}| \leq 3(q+1)/2$, $q = p^h$, where each line intersects $\mathcal{B}$ in 1 modulo p^e points. Then the lines intersecting $\mathcal{B}$ in $p^e + 1$ points intersect it in the points of a subline over $\mathrm{GF}(p^e)$. In particular, e divides h.*

Exercises

8.1. Construct a set of $2q$ points in $\mathrm{PG}(2, q)$, two of them on each line through the point $(0 : 0 : 1)$, such that the 6 points on any triple of lines are on a conic, but there is no conic containing all of the $2q$ points. Generalize this to t points on each line.

8.2. Determine the intersection multiplicity of the affine curves $Y - X^2$ and $Y^2 - X^4 + X^3$ at the origin $(0, 0)$.

8.3. Prove Lemma 8.19, Part (2).

8.4. Consider $F = X^2 + XY + Y^3 + X^4 + Y^4$. Determine the multiplicity of the origin and the tangents here.

8.5. If the smallest blocking set of $\mathrm{AG}(2, q)$ has size at most r, then in $\mathrm{PG}(2, q)$ there is a blocking set of size at most $r + 2$ which contains two different minimal blocking sets, and conversely.

8.6. Prove the affine blocking set theorem by *Jamison* [97], *Brouwer–Schrijver* [34]: for blocking sets in $\mathrm{AG}(2, q)$ we have $|\mathcal{B}| \geq 2q - 1$.

8.7. Prove that any blocking set of $\mathrm{PG}(2, q)$, of size at most $2q$ contains a *unique* minimal blocking set.

8.8. Let $\mathcal{B}$ be a blocking set, P be an essential point in $\mathcal{B}$. Then there are at least $t_P \geq 2q - |\mathcal{B}|$ tangents to $\mathcal{B}$ through P.

8.9. Let $\mathcal{B}$ be a minimal blocking set of $PG(2, q)$ containing (∞). Then there are at most k^2 lines not through (∞) that meet $\mathcal{B}$ in at least two points.

8.10. Use the previous two exercises to prove Blokhuis' theorem: in $PG(2, p)$, p prime, the size of a non-trivial blocking set is at least $3(p+1)/2$.

8.11. Use Bézout's theorem for the curves f and g to show that there are at most k^2 lines not through (∞) that meet $\mathcal{B}$ in at least two points (see Exercise 8.9).

8.12. Let $p^e \neq 4, 8$, $\mathcal{B}$ be a minimal blocking set with $|\mathcal{B}| \leq 3(q+1)/2$, where each line intersects $\mathcal{B}$ in 1 modulo p^e points. Prove that

$$|\mathcal{B}| \leq \frac{1 + (p^e + 1)(q + 1) - \sqrt{\Delta}}{2},$$

where $\Delta = (1 + (p^e + 1)(q + 1))^2 - 4(p^e + 1)(q^2 + q + 1)$.

8.13. Using Exercise 8.8. prove that one can assume that the line $X = 0$ contains no point of U, in the final setting.

8.14. Show the same as in Exercise 8.13., using lacunary polynomials.

8.15. Show that $f(X, Y)$ and $g(X, Y)$ have the same set of affine points over $GF(q)$ if $X = 0$ is a tangent to $\mathcal{B}$.

8.16. What has to be modified if the line at infinity is not a tangent to $\mathcal{B}$ but we still assume that (∞) is a point of $\mathcal{B}$? Follow the the proof of the 1 modulo p theorem in this case. Note that the Rédei polynomial and hence the original $f(X, Y)$ has to be multiplied with $\prod(Y - m)$, where (m) runs through the infinite points of $\mathcal{B}$ (different from (∞)).

8.17. In particular, formulate and prove the analogous results to Lemmas 8.26 and 8.24 and Theorem 8.23 for g instead of f,

9

Arcs, caps, unitals and blocking sets in higher dimensional spaces

In the main part of this chapter n-dimensional analogues of planar arcs are investigated. Particular attention is paid to the 3-dimensional case. There are two possible generalizations of the planar property "no three points are collinear", because this can be interpreted as "no $n + 1$ points lie on a hyperplane", too. First we consider the word-by-word generalization.

Definition 9.1. *A set of points in* $\mathrm{PG}(n, q)$ *is called a* k*-cap if it contains* k *points and no three of its points are collinear. A* k*-cap is* complete *if it is not contained in any* $(k + 1)$*-cap.*

As in the planar case, if $\mathcal{K}$ *is a* k*-cap and* ℓ *is a line in* $\mathrm{PG}(n, q)$*, then* ℓ *is called*

- *a* secant *to* $\mathcal{K}$ *if* $|\mathcal{K} \cap \ell| = 2$,

- *a* tangent *to* $\mathcal{K}$ *if* $|\mathcal{K} \cap \ell| = 1$,

- *an* external line *to* $\mathcal{K}$ *if* $|\mathcal{K} \cap \ell| = 0$.

The size of the largest cap in $\mathrm{PG}(n, q)$ is denoted by $m_2(n, q)$, while $m_2'(n, q)$ denotes the size of the second largest complete cap. Our goal is to give estimates on $m_2(n, q)$ and $m_2'(n, q)$.

First, observe that the intersection of a cap and a subspace is a cap in the subspace. This obvious property roughly gives that $m_2(n + 1, q) \leq (q + 1)m_2(n, q)$. The next theorems refine this trivial bound if $n = 2$.

Proposition 9.2. *If* q *is odd, then* $m_2(3, q) = q^2 + 1$.

Proof. Let $\mathcal{K}$ be a cap in $\mathrm{PG}(3, q)$ and $P, R \in \mathcal{K}$ be two points. Since q is odd, by Bose's theorem (Theorem 6.6), we know that each plane containing the line PR meets $\mathcal{K}$ in at most $q - 1$ further points. There are $q + 1$ planes through PR, hence $|\mathcal{K}| \leq (q + 1)(q - 1) + 2 = q^2 + 1$.

On the other hand, an elliptic quadric is a cap and it contains $q^2 + 1$ points. $\square$

If q is even, then this argument gives only $m_2(3,q) \leq q^2 + q + 2$, because in this case the bound in Theorem 6.6 is $m_2(2,q) = q + 2$. This bound is sharp if $q = 2$, because in this case the complement of any hyperplane is an 8-cap in $\mathrm{PG}(3,2)$. Aside from this trivial case, the same upper bound is valid for q even, although the proof is much more complicated.

Theorem 9.3 (Bose, Qvist). *In* $\mathrm{PG}(3,q)$ *if* $q > 2$ *even, then* $m_2(3,q) = q^2 + 1$.

Proof. First we prove that $m_2(3,q) < q^2 + q + 2$. Suppose to the contrary that $\mathcal{K}$ is a $(q^2 + q + 2)$-cap. As there are $q^2 + q + 1$ lines through each point, this implies that $\mathcal{K}$ has no tangent at all. Hence each plane section of $\mathcal{K}$ contains either 0 or $q + 2$ points of $\mathcal{K}$. Let $R \notin \mathcal{K}$ be a point and consider the $q^2 + q + 1$ lines through R. There are $(q^2 + q + 2)/2$ secants among them, hence there are $q^2 + q + 1 - (q^2 + q + 2)/2 = (q^2 + q)/2$ external lines through R. Let e be an external line and take the $q + 1$ planes through e. Each of them meets $\mathcal{K}$ in 0 or $q + 2$ points, so $q + 2$ divides $|\mathcal{K}| = q^2 + q + 2$. Hence $q + 2 | (q^2 + q + 2) - (q + 2)(q - 1) = 4$, contradicting our assumption $q > 2$.

Now, let $\mathcal{K}$ be a complete cap and suppose that $|\mathcal{K}| \geq q^2 + 1$. If $P \in \mathcal{K}$ is a point, then $|\mathcal{K}| < q^2 + q + 2$ implies that there exist some tangents to $\mathcal{K}$ at P; let t be one of them. Then each of the $q + 1$ planes through t contains at most $q + 1$ points of $\mathcal{K}$, because t is a tangent to the corresponding planar arc at P, too. At least one plane contains exactly $q + 1$ points of $\mathcal{K}$, because otherwise $|\mathcal{K}| \leq 1 + (q + 1)(q - 1) = q^2$ would hold. Let Π be a plane for which $|\Pi \cap \mathcal{K}| = q + 1$, and let $\mathcal{A} = \Pi \cap \mathcal{K}$. Then $\mathcal{A}$ is an oval. As q is even, by Theorem 6.10, $\mathcal{A}$ has a nucleus, say N. There exists at least one secant to $\mathcal{K}$ through N, otherwise $\mathcal{K} \cup \{N\}$ would be a cap, contradicting the completeness of $\mathcal{K}$. Let s be a secant to $\mathcal{K}$ through N and let Π' be any plane containing s. Then the line $\Pi \cap \Pi'$ is a tangent to $\mathcal{A}$, hence it is a tangent to $\mathcal{K}$, too. Thus $|\mathcal{K} \cap \Pi'| \leq q + 1$. There are $q + 1$ planes through s, hence we can estimate the size of $\mathcal{K}$ in the same way as we did in the q odd case, and we get $|\mathcal{K}| \leq 2 + (q + 1)(q - 1) = q^2 + 1$. If equality holds, then each plane through s contains exactly $q + 1$ points of $\mathcal{K}$.

An elliptic quadric is a cap and it contains $q^2 + 1$ points. $\square$

Definition 9.4. *A* $(q^2 + 1)$-*cap in* $\mathrm{PG}(3,q)$ *is called an* ovoid.

If $\mathcal{O}$ *is an ovoid and* Π *is a plane in* $\mathrm{PG}(3,q)$, *then* Π *is a* tangent *to* $\mathcal{O}$ *if* $|\mathcal{O} \cap \Pi| = 1$.

Proposition 9.5. *Let* $\mathcal{O}$ *be an ovoid in* $\mathrm{PG}(3,q)$. *There are* $q^2(q^2 + 1)/2$ *secant lines,* $q^2(q^2 + 1)/2$ *external lines and* $(q^2 + 1)(q + 1)$ *tangents to* $\mathcal{O}$.

Proof. As no three points of $\mathcal{O}$ are collinear, the number of secants is $\binom{|\mathcal{O}|}{2} = \binom{q^2+1}{2} = q^2(q^2 + 1)/2$.

If $P \in \mathcal{O}$ is any point, then there are q^2 secants through P, hence the number of tangents to $\mathcal{O}$ at P is $q + 1$, because the total number of lines through P is $q^2 + q + 1$. So altogether there are $(q^2 + 1)(q + 1)$ tangents to $\mathcal{O}$.

There are $(q^2+1)(q^2+q+1)$ lines in $\mathrm{PG}(3,q)$. If a line is neither a secant nor a tangent then it must be an external line. Thus the number of external lines is

$$(q^2+1)(q^2+q+1) - \left(\frac{q^2(q^2+1)}{2} + (q^2+1)(q+1) \right) = \frac{q^2(q^2+1)}{2}.$$

$\square$

Proposition 9.6. *Let $\mathcal{O}$ be an ovoid in $\mathrm{PG}(3,q)$, $q > 2$, and $P \in \mathcal{O}$ be a point. Then the $q+1$ tangent lines to $\mathcal{O}$ through P are coplanar.*

There is a unique tangent plane to $\mathcal{O}$ at each of its points. If a plane is not a tangent to $\mathcal{O}$, then it meets $\mathcal{O}$ in an oval.

Proof. We have already proved (see the proof of Propositions 9.2 and 9.3) that each plane through a secant line contains exactly $q+1$ points of $\mathcal{O}$.

Now let t be a tangent line to $\mathcal{O}$ at P and consider the $q+1$ planes through t. These planes give a partition of the set of points $\mathcal{O} \setminus \{P\}$, and each of them contains either 0 or q points of the set. Hence there are q planes containing q points and one plane has empty intersection with $\mathcal{O} \setminus \{P\}$. This plane is the unique tangent plane to $\mathcal{O}$ at P. The tangent lines at P are precisely the lines of this plane. Thus the total number of tangent planes to $\mathcal{O}$ is q^2+1.

No three points of $\mathcal{O}$ are collinear, hence the number of planes meeting $\mathcal{O}$ in an oval is $\binom{q^2+1}{3} / \binom{q+1}{3} = q^3 + q$. So each of the $q^3 + q^2 + q + 1$ planes of $\mathrm{PG}(3,q)$ is either a tangent to $\mathcal{O}$ or meets it in an oval. $\square$

Proposition 9.7. *Let $\mathcal{O}$ be an ovoid in $\mathrm{PG}(3,q)$. Then there are at most two tangent planes to $\mathcal{O}$ through any line.*

Proof. It is enough to prove that if ℓ is the intersection of two tangent planes, then there is no more tangent plane through ℓ. The line ℓ is obviously an external line to $\mathcal{O}$, so the $q+1$ planes through ℓ give a partition of the points of $\mathcal{O}$. Let t denote the number of tangent planes through ℓ. As each plane is either a tangent to $\mathcal{O}$ or meets it in an oval, we get

$$q^2 + 1 = 1 \cdot t + (q+1) \cdot (q+1-t),$$

so $t = 2$. $\square$

The next theorem is a natural generalization of the famous result of Segre (see Theorem 6.19). We follow Barlotti's proof [15] closely.

Theorem 9.8 (Barlotti, Panella). *If q is odd, then each ovoid of $\mathrm{PG}(3,q)$ is an elliptic quadric.*

Proof. We follow Barlotti's proof closely. Let $\mathcal{O}$ be an ovoid in $\mathrm{PG}(3,q)$. It follows from Proposition 9.6 and from Segre's theorem that each plane section of $\mathcal{O}$ is either a single point, or a conic. Let α_1 be a plane for which $\alpha_1 \cap \mathcal{O} = \mathcal{C}_1$ is a conic, and let $A_1, A_2, \ldots, A_5 \in \mathcal{C}_1$ be five points. Take another plane, say α_2, through the line $A_1 A_2$, and let $\alpha_2 \cap \mathcal{O} = \mathcal{C}_2$. Choose three points $B_1, B_2, B_3 \in \mathcal{C}_2 \setminus \{A_1, A_2\}$. Let τ_i be the tangent plane to $\mathcal{O}$ at A_i for $i = 1, 2$. Then the line $t_i^{(1)} = \alpha_1 \cap \tau_i$ is the unique tangent line to $\mathcal{C}_1$ at A_i, and the line $t_i^{(2)} = \alpha_2 \cap \tau_i$ is the unique tangent line to $\mathcal{C}_2$ at A_i. Let $C \in \mathcal{O} \setminus (\alpha_1 \cup \alpha_2)$ be a point, α_3 be the plane through C and the line $A_1 A_2$ and, finally, let $\mathcal{C}_3$ be the conic $\mathcal{O} \cap \alpha_3$. Then the line $t_i^{(3)} = \alpha_3 \cap \tau_i$ is the tangent line to $\mathcal{C}_3$ at A_i for $i = 1, 2$.

As the dimension of the space of quadratic surfaces in $\mathrm{PG}(3, q)$ is $\binom{3+2}{2} - 1 = 9$, there exists a quadric $\mathcal{Q}$ through the nine points $A_1, A_2, A_3, A_4, A_5, B_1, B_2, B_3$ and C. Then $\mathcal{C}_1 \subset \mathcal{Q}$, because $\mathcal{Q} \cap \alpha_1$ contains the points $A_1, \ldots, A_5$ and five points uniquely determine a conic. In the same way, starting from the points A_1, A_2, B_1, B_2, B_3, we get that $\mathcal{C}_2 \subset \mathcal{Q}$. Let $\mathcal{C}^* = \mathcal{Q} \cap \alpha_3$. We claim that $\mathcal{C}^* = \mathcal{C}_3$. First, observe that the tangent plane to $\mathcal{Q}$ at A_i is τ_i for $i = 1, 2$, because $\mathcal{C}_1, \mathcal{C}_2 \subset \mathcal{Q}$ implies that $t_i^{(j)}$ $(i, j = 1, 2)$ is a tangent line to $\mathcal{Q}$ at A_i. This means that the conic $\mathcal{C}^*$ contains the three points A_1, A_2 and C, and its tangent line at A_i is $t_i^{(3)}$. These five conditions uniquely determine a conic, hence $\mathcal{C}^* = \mathcal{C}_3$.

Now we are ready to prove that $\mathcal{Q} = \mathcal{O}$. Suppose to the contrary that there exists a point $P \in \mathcal{O} \setminus \mathcal{Q}$. Let α be a plane through the line PA_1 such that $A_2 \notin \alpha$ and none of the lines $t_1^{(j)}$, $j = 1, 2, 3$, is contained in α. If $q \geq 5$, then the number of planes through PA_1 is at least six, thus at least two of them fulfill all the conditions. For $j = 1, 2, 3$ the intersection $\alpha \cap \mathcal{C}_j$ is a single point, say D_j, and $D_j \neq A_1$. This means that the conic $\alpha \cap \mathcal{O}$ contains the four points D_1, D_2, D_3 and A_1, and its tangent line at A_1 is $\alpha \cap \tau_1$. These five conditions uniquely determine a conic, but the conic $\alpha \cap \mathcal{Q}$ also fulfills them, hence $\alpha \cap \mathcal{O} = \alpha \cap \mathcal{Q}$. This means that $P \in \mathcal{Q}$, which is a contradiction.

So the quadratic surface $\mathcal{Q}$ contains a $(q^2 + 1)$-cap. A degenerate surface is contained in the union of two planes or it is a quadratic cone (see Exercise 4.4), and any hyperbolic quadric can be covered by the union of $q + 1$ lines, so the maximum size of a cap contained in them is at most $2q + 2 < q^2 + 1$. Thus $\mathcal{Q}$ is an elliptic quadric.

For $q = 3$ direct calculations prove the statement. $\square$

Theorem 9.9 (Barlotti). *If $q \geq 7$ is odd, then $m_2'(3, q) \leq q^2 - q + 6$.*

Proof. The proof is similar to the previous one; the only difficulty encountered is the existence of tangent planes. The proof is again from [15].

Let $\mathcal{K}$ be a cap of size $q^2 - \epsilon$ with $0 \leq \epsilon \leq q - 7$ and Π be any plane. As before, it is clear that $|\mathcal{K} \cap \Pi| \leq q + 1$, and in case of equality the points of the plane section form a conic. Counting the number of points on planes through a secant, we get the following.

(*) If a line ℓ is a secant to $\mathcal{K}$, then there are at least $q - \epsilon \geq 7$ planes through ℓ which meet $\mathcal{K}$ in a conic. Every plane through ℓ meets $\mathcal{K}$ in at least $q - \epsilon$ points.

Let A and B be two distinct points of $\mathcal{K}$, and for $i = 1, 2$, let α_i be a plane through AB such that $\alpha_i \cap \mathcal{K} = \mathcal{C}_i$ is a conic. Let $A, B \neq C \in \mathcal{C}_1$ and

$A, B \neq D \in \mathcal{C}_2$ be two points, and let α_3 be a plane through C and D such that it meets $\mathcal{K}$ in a conic $\mathcal{C}_3$, and it contains none of the points A and B, the tangent line to $\mathcal{C}_1$ at C and the tangent line to $\mathcal{C}_2$ at D. Then the line $\alpha_1 \cap \alpha_3$ contains C but it is not the tangent to $\mathcal{C}_1$ at C, hence it contains another point, say E, of $\mathcal{C}_1$. In the same way, we get that the line $\alpha_2 \cap \alpha_3$ contains a point $D \neq F \in \mathcal{C}_2$. Finally, let G, H and J be further points in $\mathcal{C}_1, \mathcal{C}_2$ and $\mathcal{C}_3$, respectively.

Let $\mathcal{Q}$ be the quadric through the nine points A, B, C, D, E, F, G, H and J. As $A, B, C, E, G \in \mathcal{C}_1$, we get $\mathcal{C}_1 \subset \mathcal{Q}$, and in the same way, we also have $\mathcal{C}_2, \mathcal{C}_3 \subset \mathcal{Q}$.

Let $M \in \mathcal{C}_3$ be a further point and take a plane through the line AM which contains none of the points B, C, D, E and F, and meets $\mathcal{K}$ in a conic $\mathcal{C}_4$. Then $\mathcal{C}_4$ contains A and M, and it intersects each of $\mathcal{C}_1, \mathcal{C}_2$ and $\mathcal{C}_3$. All of these points are in $\mathcal{Q}$, hence $\mathcal{C}_4 \subset \mathcal{Q}$. In the same way, starting from a point $N \neq M \in \mathcal{C}_3$ we get a conic $\mathcal{C}_5 \subset \mathcal{K}$ through A and N with $\mathcal{C}_5 \subset \mathcal{Q}$.

Finally, let $X \in \mathcal{K} \setminus (\mathcal{C}_1 \cup \mathcal{C}_2 \cup \mathcal{C}_4 \cup \mathcal{C}_5)$ be an arbitrary point. Each of these four conics contains the point A, hence there are at most $\binom{4}{2} = 6$ planes through A meeting two of these conics in the same point. Hence it follows from (*) that there exists a plane τ through the line AX which meets $\mathcal{K}$ in a conic $\mathcal{C}_6$ such that $\mathcal{C}_6$ intersects each of the conics $\mathcal{C}_1, \mathcal{C}_2, \mathcal{C}_4$ and $\mathcal{C}_5$ in distinct points. Thus τ meets $\mathcal{Q}$ in at least five points, too. Thus $\mathcal{C}_5 \subset \mathcal{Q}$, hence $X \in \mathcal{Q}$, so $\mathcal{K} \subset \mathcal{Q}$.

As $q \geq 7$ implies $q^2 - q + 7 > 2q + 2$, the argument at the end of the proof of the previous theorem gives that $\mathcal{Q}$ is an elliptic quadric. $\square$

Upper bounds on $m_2'(3, q)$ for $q = 3$ and 5 were also proved by *Barlotti* [16]. We present these results without proof.

Proposition 9.10 (Barlotti). *In* $\mathrm{PG}(3, q)$, *if* $q = 3$ *or* 5, *then any* q^2-*cap is a subset of an elliptic quadric. Hence* $m_2'(3, 3) \leq 8$ *and* $m_2'(3, 5) \leq 24$.

The surprise in these proofs is that they do not need anything about the size of the second largest planar complete arc. The next estimate depends on $m_2'(2, q)$. For the proof see [132].

Theorem 9.11. *Let* $M = \max\{m_2'(2, q), (5q + 19)/6\}$. *If* q *is odd, then*

$$m_2'(3, q) < qM + \frac{3}{4}\left(q + \frac{10}{3} - M\right)^2 - q - 1.$$

Let us turn to the case when q is even. In particular, if $q = 2$, then the exact value of $m_2(n, 2)$ is known and the maximal arcs are characterized, too.

Theorem 9.12 (Bose). *In* $\mathrm{PG}(n, 2)$ *the equality* $m_2(n, 2) = 2^n$ *holds. If* $\mathcal{K}$ *is a* 2^n-*cap, then* $\mathcal{K}$ *is the complement of a hyperplane.*

Proof. Let $\mathcal{K}$ be a cap in $\mathrm{PG}(n,2)$ and let $P \in \mathcal{K}$ be a point. There are $2^n - 1$ lines through P and each of them contains at most one more point of $\mathcal{K}$, hence $|\mathcal{K}| \leq 1 + (2^n - 1) = 2^n$.

If equality holds, then $\mathcal{K}$ has no tangent line at all. Let $\mathcal{H}$ be the complement of $\mathcal{K}$. Then any external line to $\mathcal{K}$ meets $\mathcal{H}$ in 3 points, while any secant to $\mathcal{K}$ meets $\mathcal{H}$ in 1 point. In other words, if $P, R \in \mathcal{H}$ are two distinct points, then the third point of the line PR also belongs to $\mathcal{H}$, hence $\mathcal{H}$ is a subspace. As each line of $\mathrm{PG}(n,2)$ meets $\mathcal{H}$, so the dimension of this subspace must be $n - 1$. So $\mathcal{K}$ is the complement of a hyperplane. $\qquad\square$

The next theorems describe the most important geometric properties of ovoids in $\mathrm{PG}(3,q)$ when $q > 2$ even.

Theorem 9.13. *Let $\mathcal{O}$ be an ovoid in $\mathrm{PG}(3,q)$, $q > 2$ even, and let $N \notin \mathcal{O}$ be a point. Then*

- *there are exactly $q + 1$ tangent lines to $\mathcal{O}$ through N and these lines are coplanar;*

- *there are exactly $q + 1$ tangent planes to $\mathcal{O}$ through N.*

Proof. Let h be a secant line to $\mathcal{O}$ through N. Then each plane through h contains at least two points of $\mathcal{O}$, hence it meets $\mathcal{O}$ in an oval. Let $\mathcal{C}_1$ and $\mathcal{C}_2$ be two of these ovals, and for $i = 1, 2$, let t_i be the tangent to $\mathcal{C}_i$ through N (as q is even, by Theorem 6.10, the tangent lines to an oval cover the plane, thus t_i exists). Let Π be the plane containing the two intersecting lines t_1 and t_2. Then $\Pi \cap \mathcal{O}$ is an oval $\mathcal{C}$ and the lines t_1 and t_2 are tangents to $\mathcal{C}$. Hence, by Theorem 9.3, the intersection of these two tangents, the point N, is the nucleus of $\mathcal{C}$. As the line h is a secant through N, this means that Π does not contain the line h. So the lines of the pencil in Π with carrier N are tangents to $\mathcal{C}$, hence these lines are also tangents to $\mathcal{O}$. We claim that there are no more tangents to $\mathcal{O}$ through N. Suppose that $\ell \not\subset \Pi$ is also a tangent. Let Π' be the plane containing h and ℓ, and let $\mathcal{C}'$ be the oval $\Pi' \cap \mathcal{O}$. Then the line $\Pi \cap \Pi'$ is also a tangent to $\mathcal{C}'$ through N, thus N is the nucleus of $\mathcal{C}'$, too. But this is a contradiction, because $h \subset \Pi'$ is a secant to $\mathcal{C}'$ through N. This proves the first part of the statement.

Then the proof of the second part is obvious. Let $P_1, P_2, \ldots, P_{q+1}$ be the $q + 1$ points of $\mathcal{O}$ for which the lines NP_i are tangents to $\mathcal{O}$. The unique tangent plane to $\mathcal{O}$ at P_i contains N for $i = 1, 2, \ldots, q + 1$, and there is no more tangent plane through N, because there is no more tangent line through N. $\qquad\square$

The previous theorem suggests a natural mapping between $\mathrm{PG}(3,q)$ and its dual space via an ovoid $\mathcal{O}$. If q is odd, then, by Theorem 9.8, $\mathcal{O}$ is an elliptic quadric, hence it coincides with the set of self-conjugate points of an ordinary polarity. If q is even, then we get another type of polarity. Consider a plane Π. If it is a tangent to $\mathcal{O}$, then let its pole be the point of contact, $\Pi \cap \mathcal{O}$. If

Π meets $\mathcal{O}$ in a $(q+1)$-arc, say $\mathcal{K}$, then let the pole of Π be the nucleus of $\mathcal{K}$. We omit the long proof of the following important theorem. It can be found in [150].

Theorem 9.14 (Segre). *The mapping defined above is a null polarity for any ovoid of* $\mathrm{PG}(3,q)$.

Let us remark that for q even, all of the null polarities defined by different ovoids of $\mathrm{PG}(3,q)$ are projectively equivalent. The natural question arises: are there ovoids in $\mathrm{PG}(3,q)$, q even, other than quadratic surfaces? The answer is affirmative. If q is an odd pover of 2, then another class of ovoids is known. These sets of points are called *Suzuki–Tits ovoids*. Here we only give their definition; a detailed description of these objects can be found in Chapter 10.

Theorem 9.15 (Suzuki, Tits). *Let* $q = 2^{2e+1}$ *and let* σ *be the authomorphism of* $\mathrm{GF}(q)$ *defined as* $\sigma : x \mapsto x^{2^{e+1}}$. *Then* $(x^\sigma)^\sigma = x^2$, *so the square of the authomorphism is the same as taking the square in the field. The set of points*

$$\mathcal{S} = \{(t^\sigma + st + s^{\sigma+2} : 1 : s : t) : s, t \in \mathrm{GF}(q)\} \cup \{(1:0:0:0)\} \qquad (9.1)$$

is an ovoid.

The group of projectivities fixing $\mathcal{S}$ is the simple group $Sz(q)$, called the *Suzuki group*. It has order $q^2(q^2+1)(q-1)$ and it acts 2-transitively on the points of $\mathcal{S}$. The existence problem for other ovoids is still open, but there is no more known example.

If q is even, then finding estimates on $m_2'(3,q)$ is more complicated than for q odd, because the ovals are not necessarily conics. We present the best known results without proof.

Theorem 9.16 (J. Thas [169]). *If* $q \geq 8$ *is even, then*

- $m_2'(3,q) < q^2 - (\sqrt{5}-1)q + 5$,

- *for* $q \geq 2048$, $m_2'(3,q) < q^2 - 2q + 3\sqrt{q} + 2$.

In case of $q = 4$ *Hirschfeld* and *J. Thas* [92] proved that $m_2'(3,4) = 14$.

The same natural question arises as in the planar case: are there caps other than subsets of ovoids? The answer is affirmative. The next construction, due to *Segre*, is similar to the planar one given in Theorem 6.21.

Proposition 9.17. *Let* $\mathcal{E}$ *be an elliptic quadric in* $\mathrm{PG}(3,q)$, $q > 3$, *and let* $P \notin \mathcal{E}$ *be a point. Consider the secants and tangents to* $\mathcal{E}$ *through* P *and define the set of points* $\mathcal{K}$ *as follow: If* s *is a secant to* $\mathcal{E}$ *through* P *and* $s \cap \mathcal{E} = \{S_1, S_2\}$ *then put one of the two points* S_1 *and* S_2 *into* $\mathcal{K}$. *If* t *is a tangent to* $\mathcal{E}$ *through* P *and* $s \cap \mathcal{E} = \{T\}$ *then put the point* T *into* $\mathcal{K}$; *finally put the point* P *into* $\mathcal{K}$, *too. Then* $\mathcal{K}$ *is a* $\frac{q^2+q+4}{2}$-*cap and it is not a subset of any ovoid.*

Proof. Obviously no three points of $\mathcal{K}$ are collinear, so $\mathcal{K}$ is a cap. The number of tangents to $\mathcal{E}$ through P is $q+1$, while the the number of secants to $\mathcal{E}$ through P is $q(q-1)$, hence the cardinality of $\mathcal{K}$ is $(q+1)+q(q-1)/2+1 = \frac{q^2+q+4}{2}$ by definition.

If q is odd, then each ovoid is an elliptic quadric and we know that two elliptic quadrics share at most $2(q+1)$ points. But $\mathcal{K} \not\subset \mathcal{E}$ and it contains $(q^2+q+2)/2 > 2(q+1)$ points of $\mathcal{E}$, hence it cannot be a subset of any ovoid.

If q is even, then, by Theorem 9.13, the point P and the points of contact on the tangents to $\mathcal{E}$ through P form a hyperoval. So $\mathcal{K}$ contains $q+2$ coplanar points, hence Theorem 9.6 implies that it cannot be an ovoid. $\square$

Another natural question is to estimate the size of a complete cap in $\mathrm{PG}(3,q)$ from below. The 3-dimensional version of the estimate of Lunelli–Sce (Theorem 6.5) gives that the size of a complete k-cap satisfies the inequality

$$\binom{k}{2}(q-1) + k \geq q^3 + q^2 + q + 1,$$

hence roughly $k \geq \sqrt{2}q$. The next construction of Segre reaches this order of magnitude for q even.

Theorem 9.18 (Segre). *In* $\mathrm{PG}(3,q)$, q *even, there exist complete caps with size at most* $3q+2$.

Proof. Let $\mathrm{PG}(3,q) = \mathrm{AG}(3,q) \cup H_\infty$ where H_∞, the hyperplane at infinity, has equation $X_0 = 0$. Consider the conics

$$\mathcal{C}_1 = \{(1:x:x^2:0) : x \in \mathrm{GF}(q)\} \cup \{(0:0:1:0)\}$$

and

$$\mathcal{C}_2 = \{(1:u:u^2:1) : u \in \mathrm{GF}(q)\} \cup \{(0:0:1:0)\}.$$

They have the same nucleus, the point $N = (0:1:0:0)$. The affine points of $\mathcal{C}_1$ are in the plane having equation $Z = 0$, while the affine points of $\mathcal{C}_2$ are in the plane having equation $Z = 1$. We claim that if $A = (a,b,c)$ is an affine point with $0 \neq c \neq 1$, then there exists a line joining an affine point of $\mathcal{C}_1$ and an affine point of $\mathcal{C}_2$ through A. Let us project $\mathcal{C}_2$ from A to the plane $Z = 0$. This means that we are looking for a coefficient λ for which the third coordinate of the affine combination $\lambda U + (1-\lambda)A$ of the points A and $U = (u,u^2,1)$ is 0. Elementary calculation gives that $\lambda = c/(1+c)$ (remember that $-1 = 1$), hence $1 - \lambda = 1/(1+c)$, so the image of $\mathcal{C}_2$ is the set of points

$$\left\{\left(\frac{uc+a}{1+c}, \frac{u^2c+b}{1+c}, 0\right) : u \in \mathrm{GF}(q)\right\}.$$

These points are on the parabola with equation

$$Y = \frac{1+c}{c}X^2 + \frac{a^2+bc}{c(1+c)}.$$

As $0 \neq (1+c)/c \neq 1$, this parabola shares exactly one point with the parabola $Y = X^2$. Indeed, when we are looking for the points of intersection, then we get $X^2 = (a^2 + bc)/(1 + c)$, and it has a unique solution because q is even. Thus A lies on a (unique) secant of the affine part of the set $\mathcal{C}_1 \cup \mathcal{C}_2$. If A is on the plane $Z = 0$ or $Z = 1$, then we get a secant through A because $\mathcal{C}_1 \cup \{N\}$ is a complete arc in the corresponding projective plane.

The situation is completely different in H_∞. If a is fixed, then the lines joining the pairs of points $(x, x^2, 0)$ and $((x + a), (x + a)^2, 1)$ meet H_∞ in the same point, namely in the point $(0 : a : a^2 : 1)$. These points are on a conic $\mathcal{C}$ in H_∞, the nucleus of $\mathcal{C}$ is also the point N, and $\mathcal{C}$ contains the point $M = (0 : 0 : 1 : 0)$. Let $\mathcal{K}$ be a complete arc in H_∞ which contains the points N and M but disjoint from the set $\{(0 : a : a^2 + 1 : 1) : a \in \mathrm{GF}(q)\}$. For example, the set of points

$$\{(0 : a : a^2 + 1 : 1) : a \in \mathrm{GF}(q)\} \cup \{N, M\}$$

has this property. Then the secants of $\mathcal{K} \cup \mathcal{C}_1 \cup \mathcal{C}_2$ cover the whole space, hence this set is a complete cap and its size is at most $3q + 2$. $\qquad\square$

Now let us look for a general upper bound on the size of a cap in $\mathrm{PG}(n, q)$. In particular, if $q = 2$, then this is a strong estimate. First we prove a lemma about the number of tangents to a cap through a given point. This is similar to the statement in the proof of the even case of Bose's theorem. If $\mathcal{K}$ is a k-cap, then there are $t = q^{n-1} + \ldots + q + 2 - k$ tangents to $\mathcal{K}$ through any of its points.

Lemma 9.19. *Let $\mathcal{K}$ be a k-cap in $\mathrm{PG}(n, q)$, q even. Suppose that there is a secant line to $\mathcal{K}$ through the point $R \notin \mathcal{K}$. Then the number of tangents to $\mathcal{K}$ through the point R is at most $t = q^{n-1} + \ldots + q + 2 - k$.*

Proof. If $n = 2$, then we have already proved the statement in Corollary 8.11. Suppose that $n \geq 3$, let r be a secant to $\mathcal{K}$ through R and let P_1 and P_2 be the points of intersection of $\mathcal{K}$ and r. Consider the planes through the line r. For $i = 1, \ldots, q^{n-2} + \ldots + q + 1$ let Π_i denote these planes, $\mathcal{K}_i$ denote the planar arcs $\mathcal{K} \cap \Pi_i$ and, finally, $t_i(R)$ denote the number of tangents to $\mathcal{K}_i$ through R. Let $t_i - q + 2 - |\mathcal{K}_i|$. As in the plane Π_i the line r is a secant to $\mathcal{K}_i$, Corollary 8.10 gives $t_i(R) \leq t_i$. Summing these inequalities for $i = 1, \ldots, q^{n-2} + \ldots + q + 1$ we get

$$\sum_{i=1}^{q^{n-2}+\ldots+q+1} t_i(R) \leq \sum_{i=1}^{q^{n-2}+\ldots+q+1} (q - (|\mathcal{K}_i| - 2)) = q^{n-1} + \ldots + q - (k - 2) = t,$$

because $2 + \sum_{i=1}^{q^{n-2}+\ldots+q+1}(|\mathcal{K}_i| - 2) = k$. This proves the lemma. $\qquad\square$

Proposition 9.20. *Let $\mathcal{K}$ be a k-cap in $\mathrm{PG}(n, q)$, q even. If $n \geq 2$ and $k \leq q^{n-1} + \ldots + q + 1$, then $k \leq (q^n + \ldots + q + 1)/(q + 1)$ also holds. In particular, if $q = 2$, then $m_2'(n, 2) \leq (2^{n+1} - 1)/3$.*

Proof. The condition $k \leq q^{n-1} + \ldots + q + 1$ implies that the number t in Lemma 9.19 is strictly positive. If $P \in \mathcal{K}$ is any point, then there are exactly t tangents to $\mathcal{K}$ at P, while if $R \notin \mathcal{K}$ is any point, then, according to Lemma 9.19, there are at most t tangents to $\mathcal{K}$ through R.

Let us count the point-line pairs (R, ℓ) where $R \notin \mathcal{K}$ and ℓ is a tangent to $\mathcal{K}$ through R. On the one hand, this number is exactly ktq, because there are t tangents to $\mathcal{K}$ through any of its points and each tangent contains q points not in $\mathcal{K}$. On the other hand, by Lemma 9.19, we get that this number is at most $(q^n + \ldots + q + 1 - k)t$. Hence

$$ktq \leq (q^n + \ldots + q + 1 - k)t.$$

As $t > 0$, this implies

$$k \leq \frac{q^n + \ldots + q + 1}{q + 1}.$$

Substituting $q = 2$ to this inequality, we get the bound on $m_2'(n, 2)$. □

In the case $q = 2$ there are even better estimates. The proof of the next result of *Davydov* and *Tombak* [52] is not geometric, and it requires results from combinatorial group theory, so we skip it.

Theorem 9.21 (Davydov, Tombak). *Let $\mathcal{K}$ be a complete k-cap in $\mathrm{PG}(n, 2)$. If $k \geq 2^{n-1} + 1$, then $k = 2^{n-1} + 2^{n-1-g}$ with a suitable $g = 0, 2, 3, \ldots, n-1$. For each of these values there exists a complete k-cap.*

In general if $q > 2$, then very little is known about caps. The next result shows that estimates about the sizes in dimensions 3 and 4 can be automatically extended to estimates about the sizes in any higher dimensions.

Theorem 9.22 (Hill [84]). *If $q > 2$, then the following hold:*

(i) $m_2(n, q) \leq qm_2(n-1, q) - (q+1)$, *if $n \leq 4$, and*

(ii) $m_2(n, q) \leq q^{n-4}m_2(4, q) + 1 - q^{n-4} - 2(q^{n-5} + \ldots + q + 1)$, *if $n \geq 5$.*

It follows from (i) that $m_2(4, q) < q^4$, hence in dimension 4 (and in any higher dimensions) there is no set of points similar to an ovoid. Part (ii) shows that the 4-dimensional case is of particular relevance. In the next theorem we present the best known bound in dimension 4.

Theorem 9.23. *In $\mathrm{PG}(4, q)$ the following hold:*

(i) ([132]) *If q is odd and $N = \max\{m_2'(3, q), (q^2 + 5q + 2)/2\}$, then $m_2(4, q) < qN + 2q^2$.*

(ii) (Storme, J. Thas, Vereecke [160]) *If $M = m_2'(2, q) \geq (5q + 25)/6$, then*

$$m_2(4, q) < (q+1)\left(qM + \frac{3}{4}\left(q + \frac{10}{3} - M\right)^2 - q - 1 - M\right) + M.$$

(iii) (J. Thas [169]). If $q > 8$ even, then $m_2(4, q) < q^3 - q^2 + 2\sqrt{5}q - 8$.

(iv) (J. Thas [170]). If $q \geq 2048$ even, then

$$m_2(4, q) < q^3 - 2q^2 + 3q\sqrt{q} + 8q - 9\sqrt{q} - 6.$$

Bierbrauer and *Edel* [61] proved $m_2(4, 4) = 41$, while *J. Thas* [169] showed $m_2(4, 8) = 479$. Of course, we can combine these results with Hill's recursive bound, see Theorem 9.22.

In higher dimensional spaces the size of the largest known cap is approximately $q^{2n/3}$. It suggests that the bounds of Hill could be significantly improved in $\mathrm{PG}(n, q)$. Only one result is known in this direction. If q is fixed and the dimension is growing, then the size of the largest cap is much more less than Hill's bound.

Theorem 9.24 (Meshulam [127]). *Let* $q = p^h$, *p odd prime. Then*

$$m_2(n, q) \leq \frac{2q^n}{nh} + m_2(n - 1, q).$$

There are results in the affine case, too. *Bierbrauer* and *Edel* [23] proved that in $\mathrm{AG}(n, q^h)$ a cap contains at most $q^{hn}(hn + 1)/(hn)^2$ points. The exact value of $m_2(n, q)$ is known only in two cases. This number is 20 in $\mathrm{PG}(4, 3)$, see [141], and it is 56 in $\mathrm{PG}(5, 3)$, see [83]. In the 4-dimensional case there are several inequivalent 20-caps; one of them lies in the affine space (see Exercise 9.4). This affine cap has an interesting practical application. If we identify the points of $\mathrm{PG}(4, 3)$ and the 81 cards of the popular game SET, then the cap corresponds to 20 cards of the deck such that no three of them form a SET. In the case of the game SET in dimension n Meshulam's bound is roughly $\frac{3^n}{n}$. It was a longstanding conjecture to find an upper bound of the form c^n with $c < 3$. Extending the extremely elegant use of the polynomial method developed by *Croot, Lev* and *Pach* [49], *Ellenberg* and *Gijswijt* [62] proved the breakthrough upper bound $o(2.756^n)$ for the size of a cap in $\mathrm{AG}(n, 3)$. The best lower bound 2.217^n is due to *Edel* [60].

In the next part of this chapter higher dimensional arcs are considered in general, and particular attention is paid to the 3-dimensional case.

Definition 9.25. *In* $\mathrm{PG}(n, q)$ *a set of points containing at least* $n + 1$ *points is called an* arc *if no* $n + 1$ *of its points lie in a hyperplane. An arc is called a* k-arc, *if it contains exactly* k *points. A* k-arc *is* complete *if it is not contained in any* $(k + 1)$-arc.

A point P is addable *to a* k-arc $\mathcal{K}$ *if* $\mathcal{K} \cup \{P\}$ *is a* $(k + 1)$-arc.

The sizes of the largest and the second largest arcs in $\mathrm{PG}(n, q)$ *are denoted by* $m_n(n, q)$ *and by* $m'_n(n, q)$, *respectively.*

Example 9.26. *The set of points*

$$\mathcal{C}_n = \{(1 : t : t^2 : \ldots : t^n) : t \in \mathrm{GF}(q)\} \cup \{(0 : 0 : \ldots : 0 : 1)\}$$

is a $(q + 1)$-arc *in* $\mathrm{PG}(n, q)$.

Proof. Consider a hyperplane Σ_{n-1} and substitute the coordinates of a point of C_n to the equation of Σ_{n-1}. We get an equation of degree $n-1$ or n according to $(0:0:\ldots:0:1) \in \Sigma_{n-1}$ or not. Hence in both cases no $n+1$ points of C_n lie in Σ_{n-1}. $\qquad\square$

Definition 9.27. *A set of points in* $\mathrm{PG}(n,q)$ *is called a* normal rational curve *if it is projectively equivalent to* C_n *defined in the previous example. If* $n=3$, *then a normal rational curve is called a* twisted cubic.

The next simple observation is very useful in the study of arcs.

Lemma 9.28. *Let* $\mathcal{K}$ *be a* k-arc *in* $\mathrm{PG}(n,q)$, $n \geq 3$. *Let* $P \in \mathcal{K}$ *be a point and let* Σ_{n-1} *be a hyperplane which does not contain* P. *If we project* $\mathcal{K}$ *from* P *to* Σ_{n-1}, *then the image,* $\mathcal{K}'$, *is a* $(k-1)$-arc *in* $\mathrm{PG}(n-1,q)$.

Proof. The set of points $\mathcal{K}'$ can be described as the points of intersections of Σ_{n-1} with the lines joining P and the other points of $\mathcal{K}$, so

$$\mathcal{K}' = \{\Sigma_{n-1} \cap PR : P \neq R \in \mathcal{K}\}.$$

Suppose to the contrary that n points of $\mathcal{K}'$, say $Q_1, Q_2, \ldots, Q_n$, lie in a hyperplane of Σ_{n-1}. This hyperplane is an $(n-2)$-dimensional subspace Σ_{n-2} of $\mathrm{PG}(n,q)$. Hence $\langle P, \Sigma_{n-2} \rangle = \Pi_{n-1}$ is a hyperplane of $\mathrm{PG}(n,q)$. The line PQ_i contains a point $R_i \neq P$ of $\mathcal{K}$ for $i = 1, 2, \ldots, n$, thus Π_{n-1} contains all of the points $P, R_1, R_2, \ldots, R_n$. So at least $n+1$ points of $\mathcal{K}$ lie in a hyperplane of $\mathrm{PG}(n,q)$, giving a contradiction. $\qquad\square$

A longstanding conjecture (it was proposed by *Segre* in the 1950's) stated that if $n < q - 2$, then the maximum size of an arc in $\mathrm{PG}(n,q)$ is $q+1$. The arcs are closely connected to parity check matrices of MDS codes (see Theorem 13.13). This is the reason why the conjecture was called *MDS conjecture* in coding theory. The conjecture was (almost completely) proved by *Ball* in 2012 [7]; later on this proof was extended by *Ball* and *De Beule* [11]. For the sake of completeness we present the best known results, but omit the proof because that exceeds the scope of this book. The proof can be found in detail in Ball's book [8], Chapter 7.

Theorem 9.29 (Ball, De Beule). *Let* $\mathcal{K}$ *be a* k-arc *in* $\mathrm{PG}(n,q)$.

- *If* $q = p^h$ *and* $n \leq 2p - 3$, *then* $k \leq q + 1$.

- *If* q *is odd and* $n \leq \frac{1}{4}\sqrt{q} + \frac{9}{4}$, *then* $k \leq q + 1$.

We prove only some partial results about the sizes of arcs and focus on the 3-dimensional case.

First, suppose that q is odd. In this case the famous result of Segre about ovals has a 3-dimensional analogue. Let us note for the sake of historical accuracy, that this result led to Segre's estimate about the size of the second largest complete planar arc.

Theorem 9.30 (Segre). *In* PG$(3, q)$, $q > 7$ *odd, each* $(q+1)$-*arc is a normal rational curve.*

Proof. (Sketch.) Let $\mathcal{K}$ be a complete $(q+1)$-arc in PG(n, q). Project $\mathcal{K}$ from a point $P_1 \in \mathcal{K}$ to a plane Π. By Lemma 9.28, we get a q-arc $\mathcal{K}_1 \subset \Pi$. Theorem 6.19 implies that $\mathcal{K}_1$ is a subset of a conic in Π, hence $\mathcal{K}$ is a subset of a quadratic cone $\mathcal{C}_1$ having vertex P_1. In the same way, if we project $\mathcal{K}$ from another point $P_2 \in \mathcal{K}$ to Π, then we get that $\mathcal{K}$ is a subset of another quadratic cone $\mathcal{C}_2$ having vertex P_2. The line $P_1 P_2$ is a common generator of the cones $\mathcal{C}_1$ and $\mathcal{C}_2$, and we may assume that the two cones have different tangent planes along the line $P_1 P_2$. Let π_i denote the tangent plane to $\mathcal{C}_i$ along the line $P_1 P_2$ for $i = 1, 2$.

We sketch that in this case $\mathcal{C}_1 \cap \mathcal{C}_2$ is the union of the common generator and a normal rational curve which contains the vertices of $\mathcal{C}_1$ and $\mathcal{C}_2$. We can choose the base points of the coordinate system in the following way: Let $P_1 = (0 : 0 : 0 : 1)$, $P_2 = (1 : 0 : 0 : 0)$, π_1 be the plane $X_2 = 0$ and $\pi_1 \cap \mathcal{C}_2$ be the line $X_2 = X_3 = 0$. Let $\pi_2 \cap \mathcal{C}_1$ be the line ℓ. Then ℓ is a generator of $\mathcal{C}_1$ and it contains the point P_1. Let π_3 denote the tangent plane to $\mathcal{C}_1$ along the line ℓ and let $P_3 = (0 : 1 : 0 : 0)$ be the point of intersection of π_3 and the line $X_2 = X_3 = 0$. In the same way, let π_4 denote the tangent plane to $\mathcal{C}_2$ along the line $P_2 P_3$ and let $P_4 = (0 : 0 : 1 : 0)$ be the point of intersection of π_4 and the line $P_2 P_3$. Finally, let $E = (1 : 1 : 1 : 1)$ be a point of $\mathcal{C}_1 \cap \mathcal{C}_2 \setminus P_1 P_2$.

Let $\mathcal{D}_1$ be the intersection of $\mathcal{C}_1$ and the plane $X_3 = 0$. Then $\mathcal{D}_1$ is a conic through the points P_2, P_4 and $(1 : 1 : 1 : 0)$ which is the intersection of the line $P_1 E$ and the plane $X_3 = 0$. The tangent to $\mathcal{D}_1$ at P_2 is the line $P_2 P_3$ and the tangent to $\mathcal{D}_1$ at P_4 is the line $P_3 P_4$. Hence $\mathcal{D}_1$ is the conic

$$\mathcal{D}_1 = \{(1 : t : t^2 : 0) : t \in \mathrm{GF}(q)\} \cup \{(0 : 0 : 1 : 0)\}.$$

The generators of $\mathcal{C}_1$ are the lines joining a point of $\mathcal{D}_1$ with P_1, so the equation of $\mathcal{C}_1$ is $X_0 X_2 = X_1^2$, and

$$\mathcal{C}_1 = \{(1 : t : t^2 : \alpha) : t, \alpha \in \mathrm{GF}(q)\} \cup P_1 P_4.$$

We get in the same way that

$$\mathcal{D}_2 = \{(0 : 1 : s : s^2) : s \in \mathrm{GF}(q)\} \cup \{(0 : 0 : 0 : 1)\},$$

the equation of $\mathcal{C}_2$ is $X_1 X_3 = X_2^2$, and

$$\mathcal{C}_2 = \{(\beta : 1 : s : s^2) : s, \beta \in \mathrm{GF}(q)\} \cup P_1 P_2.$$

Let σ be the plane with equation $X_0 = 0$. Then $\mathcal{C}_1 \cap \sigma$ is the line $P_1 P_4$ and $\mathcal{C}_2 \cap \sigma$ is the conic $\mathcal{D}_2$. Thus σ contains only one point of $\mathcal{C}_1 \cap \mathcal{C}_2$, that is the point P_1. Now, let $R \in (\mathcal{C}_1 \cap \mathcal{C}_2 \setminus P_1 P_2)$ be a point. Then

$$R = (1 : t : t^2 : \alpha) = (\beta : 1 : s : s^2)$$

with $\beta \neq 0$. Hence $R = (1 : \frac{1}{\beta} : \frac{s}{\beta} : \frac{s^2}{\beta})$. This implies $t = \frac{1}{\beta}$, $t^2 = \frac{s}{\beta}$, and $\alpha = \frac{s^2}{\beta}$. So $s = t$ and $\alpha = t^3$, thus $R = (1 : t : t^2 : t^3)$ which proves the theorem.   □

We can apply the projection trick in higher dimensional spaces, too. The size of the projected arc is one less than the size of the original arc. Hence after $n - 2$ steps we get an estimate on the size of an arc in $PG(n, q)$ from the estimate on the size of arcs in $PG(2, q)$. This method works if the dimension of the space is less than the difference of $q + 1$ and $m'_2(2, q)$. In the following theorem we summarize the best available result obtained with this method. The proof is omitted.

Theorem 9.31 (J. Thas [165], Kaneta, Maruta [101]). *Consider* $PG(n, q)$ *Let* q *be odd and consider* $PG(n, q)$. *Then the following hold:*

- *If* $n < (q + 4) - m'_2(2, q)$, *then the size of the largest arc in* $PG(n, q)$ *is* $q + 1$,

- *if* $n < (q + 3) - m'_2(2, q)$, *then the* $(q + 1)$-*arc is a normal rational curve,*

- *if* $n < (q + 3) - m'_2(2, q)$, *then the size of the second largest complete arc satisfies the inequality* $m'_n(n, q) \leq m'_2(2, q) + n - 2$.

Now let us turn to the investigation of arcs in $PG(3, q)$ when q is even. In this case the notion of "tangent to an arc" is a bit more complicated than the notion of "tangent to a cap". In classical differential geometry one defines the tangent line and the osculating (hyper)plane to a curve using derivatives. The next definition is the combinatorial analogue of these notions.

Definition 9.32. *Let* K *be an arc in* $PG(3, q)$, q *even.*
    A line t *is called a* tangent line *to* K *at the point* P *if the following hold:*

(1) $t \cap K = \{P\}$,

(2) *if a plane* Σ *contains* t, *then* $|\Sigma \cap K| \leq 2$.

A plane Π *is an* osculating plane *at the point* $P \in K$ *if it contains at least one tangent line to* K *and* $\Pi \cap K = \{P\}$.

The proofs of the following propositions are straightforward.

Proposition 9.33. *Let* K *be a* k-*arc in* $PG(3, q)$, q *even. Then* K *can be extended to a* $(k + 1)$-*arc by adding a point* Q, *if and only if the line* PQ *is a tangent to* K *for each point* $P \in K$.

Proposition 9.34. *Any arc in* $PG(3, q)$, q *even, contains at most* $q + 3$ *points.*

In particular, if $q = 2$, then this estimate is tight, because $5 = q + 3$ and the 5 base points of the coordinate system form a 5-arc in $PG(3, 2)$. From now on we assume that $q > 2$ and follow the paper by *Casse* [39].

Proposition 9.35 (Casse). *Let $\mathcal{K}$ be a k-arc in $\mathrm{PG}(3, q)$, $q > 2$ even. Suppose that $k > q - \sqrt{q} + 2$. Then through any point $P \in \mathcal{K}$*

(1) *there are $n = q + 3 - k$ tangent lines to $\mathcal{K}$,*

(2) *there are $m = \binom{n}{2}$ osculating planes to $\mathcal{K}$.*

Proof. Let $P \in \mathcal{K}$ be any point and project $\mathcal{K} \setminus \{P\}$ from P to a plane Π. Let $\mathcal{K}'$ be the resulting planar arc. If t is a tangent line to $\mathcal{K}$ at P, then its image under the projection is the point $T = t \cap \Pi$ and T is addable to $\mathcal{K}$. As $k > q - \sqrt{q} + 2$, we get $|\mathcal{K}'| > q - \sqrt{q} + 1$. Hence, by Theorem 6.19, $\mathcal{K}'$ is contained in a hyperoval. Thus there exist exactly $q + 2 - |\mathcal{K}'| = q + 3 - k = n$ addable points which proves (1). The image of an osculating plane under the projection is a line which contains an addable point and no point of $\mathcal{K}'$. Hence it joins two addable points, which proves (2). Let us remark that for $n = 1$ there are no osculating planes. $\qquad\square$

Corollary 9.36 (Casse). *In $\mathrm{PG}(3, q)$, q even, the following hold:*

- *A $(q + 3)$-arc has no tangent line.*

- *A $(q + 2)$-arc has a unique tangent line at each of its points and it has no osculating plane.*

- *A $(q+1)$-arc has exactly two tangent lines and a unique osculating plane at each of its points. The osculating plane contains the two tangent lines.*

Lemma 9.37 (Casse). *Let $\mathcal{K}$ be a $(q + 3)$-arc in $\mathrm{PG}(3, q)$, q even. Then*

- *no plane of $\mathrm{PG}(3, q)$ meets $\mathcal{K}$ in exactly two points,*

- *the 2-secant lines of $\mathcal{K}$ do not cover all points of $\mathrm{PG}(3, q)$.*

Proof. Suppose that a plane Π meets $\mathcal{K}$ in exactly two points, P and R. Consider the q other planes through the line PR. By the Pigeonhole Principle, at least one of them contains at least four points of $\mathcal{K}$ which is a contradiction.

The second part of the statement follows from the fact that the number of 2-secants to $\mathcal{K}$ is $\frac{(q+3)(q+2)}{2}$, hence these lines cover only $(q + 3) + (q - 1)\frac{(q+3)(q+2)}{2}$ points, and this number is less than $q^3 + q^2 + q + 1$ because $q > 2$. $\qquad\square$

The next lemma is the 3-dimensional analogue of Theorem 6.10.

Lemma 9.38 (Casse). *In $\mathrm{PG}(3, q)$, q even, any $(q + 2)$-arc can be extended to a $(q + 3)$-arc.*

Proof. Let $\mathcal{K} = \{P_0, P_1, \ldots, P_{q+1}\}$ be a $(q + 2)$-arc. Let ℓ_i denote the unique tangent line to $\mathcal{K}$ at P_i for $i = 0, 1, 2, \ldots, q + 1$, and let $\mathcal{K}' = \mathcal{K} \setminus \{P_0\}$. Then the line $P_0 P_i$ is a tangent to $\mathcal{K}'$ at P_i for $i = 1, 2, \ldots, q + 1$, so for every $i = 1, 2, \ldots, q + 1$, both of the planes $\langle \ell_i, P_0 P_i \rangle$ and $\langle \ell_0, P_0 P_i \rangle$ are osculating planes to $\mathcal{K}'$ at P_i. By Corollary 9.36, $\mathcal{K}'$ has a unique osculating plane at each of its points, hence ℓ_0 and ℓ_i are coplanar, so $\ell_0 \cap \ell_i \neq \emptyset$.

In the same way we can prove that any two tangent lines to $\mathcal{K}$ intersect each other. Thus the tangent lines are either coplanar or they have a point in common. In the first case $\mathcal{K}$ itself would be in a plane, contradicting the definition of an arc. So each tangent line contains a fixed point, say Q. By Proposition 9.33, this means that Q is addable to $\mathcal{K}$, so $\mathcal{K} \cup \{Q\}$ is a $(q+3)$-arc. $\qquad\square$

Theorem 9.39 (Casse). *If $q > 2$ even, then there are no $(q+3)$-arcs in* $\mathrm{PG}(3,q)$.

Proof. Suppose to the contrary that $\mathcal{K} = \{A_0, A_1, \ldots, A_{q+2}\}$ is a $(q+3)$-arc in $\mathrm{PG}(3,q)$. We distinguish two cases.

Case 1: $q = 2^{2h}$, that is $q \equiv 1 \pmod 3$. Let $A_0 \neq B \neq A_1$ be a point on the line $A_0 A_1$ and let Π be a plane which does not contain B. Project $\mathcal{K}$ from B to Π. The resulting set is a $(q+2,3)$-arc $\mathcal{K}' = \{A_1', A_2', \ldots, A_{q+2}'\}$ in Π, because $A_0' = A_1'$. Consider the 3-secants of $\mathcal{K}'$ in Π. There is no 3-secant through the point $A_0' = A_1'$, while for $i = 2, 3, \ldots, q+2$, each of the lines through A_i', except the line $A_0' A_i'$, is a 3-secant, because, by the first part of Lemma 9.37, no plane meets $\mathcal{K}$ in exactly 2 points. This implies that there are $q/2$ 3-secants through A_i' $(i = 2, 3, \ldots, q+2)$. Count the point-line pairs (P, ℓ) where $P \in \mathcal{K}'$ and ℓ is a 3-secant to $\mathcal{K}'$ through P in two different ways. If s denotes the total number of 3-secants, then we get $3s = 0 + (q+1)q/2$. The 0 term comes from $A_0' = A_1'$. But $q = 2^{2h}$ implies that $q(q+1)$ is not divisible by 3; this contradiction proves the theorem.

Case 2: $q = 2^{2h+1}$, $h > 0$, that is $q \equiv 2 \pmod 3$. Now choose the center B of the projection so that no 2-secant to $\mathcal{K}$ passes through B. Such a point exists by the second part of Lemma 9.37. In this case $\mathcal{K}'$ is a $(q+3,3)$-arc in Π. As no plane meets $\mathcal{K}$ in exactly 2 points, $\mathcal{K}'$ has no 2-secant line at all. This implies that there are $(q+2)/2$ 3-secants through A_i' $(i = 0, 1, \ldots, q+2)$. Let us count again the same (P, ℓ) pairs as in the proof of Case 1. Now we get $3s = (q+3)(q+2)/2$ where s denotes the total number of 3-secants. But $q = 2^{2h+1}$ implies that $(q+3)(q+2)$ is not divisible by 3, so we get a contradiction again. $\qquad\square$

Corollary 9.40. *If $q > 2$ even, then any arc in* $\mathrm{PG}(3,q)$ *contains at most $q+1$ points.*

With a slight modification of the previous proof one can show that if $\mathcal{K}$ is a $(q+1)$-arc in $\mathrm{PG}(3,q)$, then the tangent lines to $\mathcal{K}$ belong to a suitable hyperbolic quadric. Later on *Casse* and *Glynn* [40] completely characterized $(q+1)$-arcs in $\mathrm{PG}(3,q)$. They also proved that each $(q+1)$-arc in $\mathrm{PG}(4,q)$ is a normal rational curve. This means that there is no difference between the odd and even characteristic spaces if the dimension is 4.

Theorem 9.41 (Casse, Glynn). *If $q \geq 8$ even, then any arc in* $\mathrm{PG}(4,q)$ *contains at most $q+1$ points, and each $(q+1)$-arc is a normal rational curve.*

If we would like to give estimates on the sizes of arcs in higher dimensional spaces, we need results about the sizes of the second largest arcs in PG$(3, q)$ and in PG$(4, q)$. If these results are available, we can apply the method of Theorem 9.31 in the case q even, too. This was done by *Bruen, J. Thas* and *Blokhuis* [36]. The best known results are due to *Storme* and *J. Thas* [159].

Theorem 9.42 (Storme, J. Thas). *Let q be even.*

- *If $n \geq 4$ and $q \geq (2n - 11/2)^2$, then the largest arc in PG(n, q) has $(q + 1)$ points. If $n \geq 4$ and $q > (2n - 7/2)^2$, then each $(q + 1)$-arc is a normal rational curve.*

- *If $n \geq 3$, then the size of the second largest arc satisfies the inequality $m'_n(n, q) \leq q - \sqrt{q}/2 + n - 3/4$.*

Finally, we present a similar result when the dimension is close to q.

Theorem 9.43 (J. Thas). *Let $n \geq 2$ and $q \geq n + 3$. Suppose that each $(q + 1)$-arc in PG(n, q) is a normal rational curve. Then each $(q + 1)$-arc in PG$(q - 1 - n, q)$ is also a normal rational curve. In particular, there are $(q + 2)$-arcs in PG$(q - 2, q)$.*

We will return to the connection of Theorems 9.43 and 9.43 in Chapter 13.

Let us remark that for $n \geq q - 1$ the situation is simple. The base points of the coordinate system, $(1 : 0 : \cdots : 0)$, ..., $(0 : 0 : \cdots : 1)$ and $(1 : 1 : \cdots : 1)$, form an $(n + 2)$-arc and it was proved by *Bush* [38] that this arc is complete.

In the next part of the chapter unitals are investigated. These structures have the same combinatorial properties as Hermitian curves (see Theorem 4.42).

Definition 9.44. *Let Π_{q^2} be a projective plane of order q^2. A set of points $\mathcal{U} \subset \Pi_{q^2}$ is called a* unital, *if $\mathcal{U}$ consists of $q^3 + 1$ points, and each line of Π_{q^2} meets $\mathcal{U}$ in 1 or in $q + 1$ points.*

The representation of unitals in PG$(4, q)$ helps us study their geometric properties. First, we show how the points of PG$(2, q^2)$ can be represented in PG$(4, q)$. This is called the *Bruck–Bose representation.* It is a special case of the general method given in Example 5.20.

Consider GF(q^2) as the field extension GF$(q)[i]$ where the minimal polynomial of i is $f(x) = x^2 - \alpha x - \beta$. The two roots of the equation $f(x) = 0$ are i and i^q, hence $i^q = \alpha - i$ and $i^{q+1} = -\beta$. Every element $t \in$ GF(q^2) can be uniquely written as $t = t_1 + it_2$, where $t_1, t_2 \in$ GF(q). Let PG$(2, q^2) = $ AG$(2, q^2) \cup \ell_\infty$ where the equation of ℓ_∞ is $X_0 = 0$. Then the homogeneous coordinates of the points of AG$(2, q^2)$ can be written as $(1 : x : y)$. If $(1 : x : y)$ is an affine point and $x = x_1 + ix_2$, $y = y_1 + iy_2$ with $x_1, x_2, y_1, y_2 \in$ GF(q), then let the corresponding point in PG$(4, q)$ be $(1 : x_1 : x_2 : y_1 : y_2)$. This is the affine point (x_1, x_2, y_1, y_2) if $\mathrm{PG}(4, q) = \mathrm{AG}(4, q) \cup \Sigma_\infty$ and the hyperplane at infinity, Σ_∞, has equation $X_0 = 0$.

If $(0 : a : b)$ is a point at infinity in $\mathrm{PG}(2, q^2)$ then the corresponding object is a line in Σ_∞. This line is an element of the line spread that defines the parallel classes in Example 5.20. The vertical lines of $\mathrm{AG}(2, q^2)$ correspond to the point $(0 : 0 : 1)$ of ℓ_∞. If we take two affine points from a vertical line, then the line joining the corresponding points in $\mathrm{AG}(4, q)$ meets Σ_∞ in a point whose second and third coordinates are zeros. Hence $(0 : 0 : 1)$ corresponds to the line in Σ_∞ defined by $X_0 = X_1 = X_2 = 0$.

If $a \neq 0$, then the homogeneous coordinates of the point at infinity can be written as $(0 : 1 : m)$. This point is on the lines having affine equation $Y = mX + k$. Let $m = m_1 + im_2$ and $k = k_1 + ik_2$ with $m_1, m_2, k_1, k_2 \in \mathrm{GF}(q)$. If an affine point (x, y), where $x = x_1 + ix_2$ and $y = y_1 + iy_2$ with $x_1, x_2, y_1, y_2 \in \mathrm{GF}(q)$, is on this line, then

$$y_1 + iy_2 = (m_1 + im_2)(x_1 + ix_2) + k_1 + ik_2$$
$$= m_1 x_1 + \beta m_2 x_2 + k_1 + i(m_1 x_2 + m_2 x_1 + \alpha m_2 x_2 + k_2).$$

Hence its coordinates satisfy two equations:

$$y_1 = m_1 x_1 + \beta m_2 x_2 + k_1,$$
$$y_2 = m_1 x_2 + m_2 x_1 + \alpha m_2 x_2 + k_2.$$

These are the affine equations of two hyperplanes. The corresponding homogeneous equations define two hyperplanes in $\mathrm{PG}(4, q)$ whose intersection with Σ_∞ is a line defined by

$$X_0 = 0,$$
$$Y_1 = m_1 X_1 + \beta m_2 X_2 + k_1,$$
$$Y_2 = m_1 X_2 + m_2 X_1 \alpha m_2 X_2 + k_2.$$

This line is an element of the line spread of Σ_∞.

A Hermitian curve $\mathcal{H}$ in $\mathrm{PG}(2, q^2) = \mathrm{AG}(2, q^2) \cup \ell_\infty$ is called *parabolic* or *hyperbolic*, according to whether ℓ_∞ is a tangent to, or a secant of $\mathcal{H}$.

Theorem 9.45 (Buekenhout). *Let $\mathcal{H}$ be a parabolic Hermitian curve in $\mathrm{PG}(2, q^2)$. Then it can be represented in $\mathrm{PG}(4, q)$ as a quadratic cone. The vertex of the cone is a single point and its base is a 3-dimensional elliptic quadric.*

Proof. Without loss of generality we may assume that $\mathcal{H}$ arises from the unitary polarity defined by the matrix

$$\begin{pmatrix} 0 & 0 & 1 \\ 0 & 1 & 0 \\ 1 & 0 & 0 \end{pmatrix},$$

Then ℓ_∞ is the tangent to $\mathcal{H}$ at the point $(0:0:1)$ and the affine equation of $\mathcal{H}$ is

$$X^{q+1} + Y^q + Y = 0. \tag{9.2}$$

We transform this to an equation in $\mathrm{AG}(4,q)$ by substituting $X = X_1 + iX_2$ and $Y = Y_1 + iY_2$. Then Equation (9.2) becomes

$$(X_1 + iX_2)(X_1 + (\alpha - i)X_2) + Y_1 + Y_2 i + Y_1 + (\alpha - i)Y_2 = 0,$$

because $i^q = \alpha - i$. Rearranging it we get

$$X_1^2 + \alpha X_1 X_2 - \beta X_2^2 + 2Y_1 + \alpha Y_2 = 0. \tag{9.3}$$

Consider now this equation in $\mathrm{PG}(4,q)$. After dehomogenization it defines a quadric, $\mathcal{Q}$, whose points at infinity satisfy the equation $X_1^2 + \alpha X_1 X_2 - \beta X_2^2 = 0$. The polynomial $g(x) = x^2 + \alpha x - \beta$ is also irreducible over $\mathrm{GF}(q)$, because if q is even, then $g(x) = f(x)$, and when q is odd, then its discriminant is the same as the discriminant of the irreducible polynomial $f(x)$. Thus $\mathcal{Q}$ meets Σ_∞ in the points $X_1 = X_2 = 0$. This is a set of equations of a line s. This line belongs to the spread in the Bruck-Bose representation.

Elementary calculation shows that $\mathcal{Q}$ has a unique singular point, that is $A = (0:0:\alpha:-2:0)$. If a hyperplane Σ_3 does not contain A, then it meets $\mathcal{Q}$ in a quadric having exactly one point at infinity, namely $s \cap \Sigma_3$. Hence this intersection must be an elliptic quadric in Σ_3. $\qquad \square$

In the case of hyperbolic Hermitian curves similar but much longer computations are needed to prove the following result.

Theorem 9.46 (Buekenhout [37], Schield [149]). *Let $\mathcal{H}$ be a hyperbolic Hermitian curve in $\mathrm{PG}(2,q^2)$. Then it can be represented in $\mathrm{PG}(4,q)$ as a nonsingular parabolic quadric.*

We omit the detailed proof. The main steps are the same as in the parabolic case. Now the affine equation of $\mathcal{H}$ can be chosen as

$$X^{q+1} + Y^{q+1} + 1 = 0, \tag{9.4}$$

and the affine equation of the corresponding quadric is

$$X_1^2 + \alpha X_1 X_2 - \beta X_2^2 + Y_1^2 + \alpha Y_1 Y_2 - \beta Y_2^2 + 1 = 0. \tag{9.5}$$

This quadric is non-singular and meets Σ_∞ in a hyperbolic quadric which contains $q + 1$ elements from the line spread of Σ_∞.

These constructions can be reversed. In the parabolic case this leads to non-classical unitals, too.

Theorem 9.47 (Buekenhout). *Let* $\mathrm{PG}(4,q) = \mathrm{AG}(4,q) \cup \Sigma_\infty$ *and* $\mathcal{S}$ *be a line spread in the hyperplane* Σ_∞. *Let* $\mathcal{C}$ *be a cone whose vertex is a single point* A, *whose base is an ovoid* $\mathcal{O}$ *in a 3-dimensional subspace* Σ_3, *and suppose that the intersection of* $\mathcal{C}$ *and* Σ_∞ *is a line that belongs to* $\mathcal{S}$. *Then in the reverse of the Bruck-Bose representation* $\mathcal{C}$ *corresponds to a parabolic unital in* $\mathrm{PG}(2,q^2)$.

Proof. The cone $\mathcal{C}$ contains $1 + q(q^2 + 1) = q^3 + q + 1$ points, but $q + 1$ of its points form a line of $\mathcal{S}$. Hence the reverse of the Bruck-Bose representation produces $q^3 + 1$ points in $\mathrm{PG}(2,q^2)$. Let $\mathcal{U}$ denote this set of points.

The line at infinity of $\mathrm{PG}(2,q^2)$ is a tangent to $\mathcal{U}$, because $\mathcal{C}$ contains exactly one line of $\mathcal{S}$. The other lines of $\mathrm{PG}(2,q^2)$ correspond to affine planes of $\mathrm{PG}(4,q)$. First, let Π be an affine plane that does not contain A. Project Π from A to Σ_3. Its image is a plane that meets $\mathcal{O}$ in 1 or in $q + 1$ points, hence the corresponding line of $\mathrm{PG}(2,q^2)$ meets $\mathcal{U}$ in 1 or in $q + 1$ points. If the affine plane Π contains A, then it corresponds to a line of $\mathrm{PG}(2,q^2)$ if and only if it contains the unique spread element s through A. Hence the intersection $\Pi \cap \mathcal{C}$ is either the single line s, or two lines, one of them being s. Thus the corresponding line of $\mathrm{PG}(2,q^2)$ meets $\mathcal{U}$ in 1 point in the former case and in $q + 1$ points in the latter case. $\qquad\square$

If q is even, then there exist ovoids in $\mathrm{PG}(3,q)$ that are not elliptic quadrics. If $\mathcal{O}$ is a Suzuki-Tits ovoid (see its definition in Theorem 9.15), then the construction above gives a unital different from a Hermitian curve.

The line spread of Σ_∞ also could be arbitrary. Hence the construction produces unitals in all translation planes that can be represented in $\mathrm{PG}(4,q)$. These are the translation planes whose coordinatising quasifield has order q^2 with a kernel of order q.

Finally, we mention that the process can be extended to planes of order q^{2k} instead of order q^2. Then in the proof of Theorem 9.45 the variables x_i, y_i belong to $\mathrm{GF}(q^k)$. So if we would like to use coordinates from $\mathrm{GF}(q)$, then there were k coordinates for each variable and there would be a quadratic equation in each coordinate.

In Theorem 9.47, if $\mathcal{C}$ is a quadratic cone, then it is not obvious whether the constructed unital is a Hermitian curve or not. *Metz* [130] proved that unitals different from Hermitian curves (called non-classical unitals or Hermitian arcs) also arise from quadratic cones.

Theorem 9.48 (Metz). *If* $q > 3$, *then the construction described in Theorem 9.47 could result in non-classical unitals also in the case when* $\mathcal{C}$ *is a quadratic cone.*

Proof. If the line at infinity is a tangent to a Hermitian curve $\mathcal{H}$ at the point $(0 : 0 : 1)$, then the affine equation of the curve can be written as

$$X^{q+1} + a^q X^q + aX + b^q Y^q + bY + c = 0$$

where $0 \neq b \in \mathrm{GF}(q^2)$, $a \in \mathrm{GF}(q^2)$ and $c \in \mathrm{GF}(q)$ are arbitrary elements. Thus the total number of these curves is $(q^2 - 1)q^3$. Each of them corresponds to a quadratic cone in the Bruck-Bose representation. We claim that if $q > 3$, then the number of quadratic cones is larger than $(q^2 - 1)q^3$, which obviously implies the theorem.

Fix the vertex A of the cone on the line $s : X_0 = X_1 = X_2 = 0$. We have $q + 1$ choices for A. Take a 3-dimensional subspace Σ_3 that does not contain s. Now we are looking for the number of elliptic quadrics in Σ_3 having exactly one point at infinity.

Let $\Pi \subset \Sigma_3$ be a plane and $\mathcal{C}$ be an irreducible conic in Π such that $\mathcal{C} \cap \Sigma_\infty = \emptyset$. Then $\mathcal{C}$ can be uniquely extended to an elliptic quadric having exactly one point at infinity. As $\Pi \cap \Sigma_\infty$ is a line, we have to count the number of irreducible conics having empty intersection with a fixed line. According to Corollary 4.70, the number of irreducible conics is $q^5 - q^2$. We also know that there are $q(q-1)/2$ external lines to a conic, hence the number of ordered pairs (conic, external line to it) is $(q^5 - q^2)q(q-1)/2$. On the other hand, the number of lines is $q^2 + q + 1$ and each line is external to the same number of conics, thus the number that we are looking for is

$$(q+1)\frac{(q^5 - q^2)q(q-1)}{2(q^2 + q + 1)} = (q^2 - 1)q^3\frac{q-1}{2}.$$

If $q > 3$, then this number is much greater than the number of Hermitian curves. $\qquad\square$

Definition 9.49. *The parabolic unitals arising from quadratic cones are called Buekenhout–Metz unitals. If q is even, then the unitals arising from cones whose base is a Suzuki–Tits ovoid are called Buekenhout–Tits unitals.*

The Buekenhout–Metz unitals can be represented by planar coordinates, too. We present it only when q is odd.

Proposition 9.50 (Baker, Ebert [5]). *Let q be odd and let $a, b \in \mathrm{GF}(q^2)$ such that $(b^q - b)^2 + 4a^{q+1}$ is a non-squre element in $\mathrm{GF}(q)$. Then the set of points*

$$U_{ab} = \{(1 : x : ax^2 + bx^{q+1} + r) : x \in \mathrm{GF}(q^2), r \in \mathrm{GF}(q)\} \cup \{(0 : 0 : 1)\},$$

is a Buekenhout–Metz unital.

Proof. As q is odd we may assume that $\mathrm{GF}(q^2) = \mathrm{GF}(q)[i]$ where the primitive polynomial of i is $f(x) = x^2 - \beta$. Let $a = a_1 + a_2 i$, $b = b_1 + b_2 i$, $x = x_1 + x_2 i$ and $y = y_1 + y_2 i$, with $a_1, a_2, b_1, b_2, x_1, x_2, y_1, y_2 \in \mathrm{GF}(q)$.

The affine point $(1 : x : y)$ belongs to U_{ab} if and only if $ax^2 + bx^{q+1} - y \in \mathrm{GF}(q)$. This can be written as

$$a_2(x_1^2 + \beta x_2^2) + b_2(x_1^2 - \beta x_2^2) + 2a_1 x_1 x_2 - y_2 = 0.$$

If we use homogeneous coordinates $(x_0 : x_1 : x_2 : y_1 : y_2)$ in $\mathrm{PG}(4, q)$ then these points are on a quadratic cone whose vertex is $(0 : 0 : 0 : 1 : 0)$, because y_1 is missing from the equation. The base of this cone is an elliptic quadric if and only if $(2a_1)^2 - 4(a_2 + b_2)(a_2 - b_2)\beta$ is a non-square element in $\mathrm{GF}(q)$. Elementary calculation shows that this is equivalent to the condition of the statement. $\square$

One can prove that all Buekenhout–Metz unitals can be represented in this way. We should mention that direct calculations in the plane also show that each line meets U_{ab} either in 1 or in $q + 1$ points. The unital U_{ab} is a Hermitian curve if and only if $a = 0$, and it is a union of conics of a pencil if and only if $b = 0$. If q is even, then the similar planar description needs a bit more calculation; the reader can find it e.g., in the book [17]. Finally, we remark that *Barwick* showed that all hyperbolic Buekenhout–Metz unitals are classical. She used a counting argument, and the proof can also be found in [17].

In the last part of this chapter we study blocking sets in higher dimensions. Similarly to the case of arcs, there are different objects that can be regarded as the generalizations of planar blocking sets. The general definition is the following.

Definition 9.51. *A k-blocking set of* $\mathrm{PG}(n, q)$ *is a set of point which intersects every* $(n - k)$-*dimensional subpace. A k-blocking set is called* non-trivial *if it contains no k-subspace. A 1-blocking set will simply be called a* blocking set. *As usual,* minimality *is defined with respect to set theoretical inclusion.*

The above definition implies immediately that the projection of a k-blocking set from a point not in the set onto a hyperplane will be a k-blocking set in that hyperplane (see Exercise 9.18). Similarly, a k-blocking set of a subspace is a k-blocking set of the entire space.

First, we are going to study blocking sets in $\mathrm{PG}(3, q)$ in more detail. In this case we have (1-)blocking sets, which intersect every plane, and 2-blocking sets, which intersect every line. Let us begin with blocking sets. Clearly, a coplanar set of points will intersect every plane if and only if the set intersects every line of the plane containing it. The next proposition shows that the smallest blocking set is a line and the smallest non-trivial blocking set is a non-trivial planar blocking set contained in a plane of $\mathrm{PG}(3, q)$. Note that there are examples of non-trivial blocking sets of size $1 + 3q/2$, (q even), and $3(q + 1)/2$, (q odd); we may use that $b(q) \leq 3(q + 1)/2$.

Proposition 9.52 (Beutelspacher, Bruen, Heim). *Every blocking set $\mathcal{B}$ in* $\mathrm{PG}(3, q)$ *has at least $q + 1$ points. If $|\mathcal{B}| = q + 1$, then it is a line. Let $b(q)$ denote the size of the smallest blocking set in $\mathrm{PG}(2, q)$. If $\mathcal{B}$ is non-trivial, then $|\mathcal{B}| \geq b(q)$. In case of equality $\mathcal{B}$ is a blocking set in a plane.*

Proof. We may assume that $\mathcal{B}$ is minimal. Pick a point $P \notin \mathcal{B}$ which lies on a line joining two points of $\mathcal{B}$. Project $\mathcal{B}$ from P onto a plane. If there is no such point P, then $\mathcal{B}$ is a line. This projection $\mathcal{B}'$ is a blocking set that has one point less than $\mathcal{B}$. Since $\mathcal{B}'$ has at least $q + 1$ points, the first two assertions are immediate. If $|\mathcal{B}| \leq b(q)$, then $|B'| < b(q)$, so $\mathcal{B}'$ contains a line ℓ. The plane $\pi = \langle P, \ell \rangle$ contains at least $q + 2$ points of $\mathcal{B}$. If $B \cap \pi$ is a blocking set, then we are done. If not, there is a line $\ell' \subset \pi$, which is disjoint from $\mathcal{B}$. The planes through ℓ' intersect $\mathcal{B}$, and there are q such planes different from π, hence $|\mathcal{B} \setminus \pi| \geq q$ and $|\mathcal{B}|$ would be much larger than $b(q)$. This gives a contradiction. $\qquad\square$

Now we turn to the case of 2-blocking sets. These are sets intersecting every line in $\mathrm{PG}(3, q)$.

Theorem 9.53 (Beutelspacher, Heim). *Let $M = b(q)$ be the size of the smallest non-trivial minimal blocking set of $\mathrm{PG}(2, q)$. The size of a 2-blocking set B in $\mathrm{PG}(3, q)$ not containing a plane is at least $1 + qM$ with equality if B is a cone over a planar blocking set of size M, or $q = 2$ and B consists of the 10 points outside an elliptic quadric.*

The following proof is due to *Blokhuis*. We are indebted to him for the help with this proof.

Proof. We assume that $\mathcal{B}$ is minimal, so for every point of $\mathcal{B}$ there is a *tangent*, that is a line of $\mathrm{PG}(3, q)$ containing that point only. We also assume that $\mathcal{B}$ is not a cone over a planar blocking set (in particular, $\mathcal{B}$ is not a plane) and want to derive a contradiction.

In the case $q = 2$ we look at the complement of $\mathcal{B}$: this is a set that has a point in every plane, and does not contain a line. It is an exercise to show that the complement of $\mathcal{B}$ is an ovoid, so an elliptic quadric, see Exercise 9.19.

Therefore, from now on we assume $q \geq 3$ (and actually, $q > 3$).

In the proof the following notation will be used: A point of $\mathcal{B}$ will be called a $\mathcal{B}$-*point*, a line contained in $\mathcal{B}$ a $\mathcal{B}$-*line* and a plane containing more than $\frac{1}{2}(q + 5)$ $\mathcal{B}$-lines a $\mathcal{B}$-*plane*. The bound $\frac{1}{2}(q + 5)$ comes from the upper bound $3(q + 1)/2$ on $M = b(q)$. The proof consists of a sequence of six lemmas.

Lemma 9.54. *Every $\mathcal{B}$-point is contained in a $\mathcal{B}$-line.*

Proof. Let t be a tangent through the $\mathcal{B}$-point P. Planes through t either contain at least M points, or at most $M - 1$. In the latter case they contain a line. If only the first case occurs, then

$$|\mathcal{B}| \geq 1 + (q + 1)(M - 1) = 1 + qM + (M - q - 1) > 1 + qM,$$

which is larger than our bound on the size. So, there is a plane containing a $\mathcal{B}$-line ℓ through P. $\qquad\square$

Lemma 9.55. *Every $\mathcal{B}$-point is contained in at least two $\mathcal{B}$-lines.*

Proof. Let P be contained in exactly one $\mathcal{B}$-line, say ℓ. Then the plane π on P containing ℓ and the tangent t has at least $1+q$ points, and all other planes on t have at least $1+(M-1)$ points, so $|\mathcal{B}| \geq 1+q+q(M-1) = 1+qM$ and we have equality everywhere.

Let m be another $\mathcal{B}$-line. As m intersects π in a point of ℓ, all $\mathcal{B}$-lines intersect ℓ. If m also has the property that it is the only $\mathcal{B}$-line on one of its points, then all remaining $\mathcal{B}$-lines intersect both ℓ and m, so they are either contained in the plane $\sigma = \langle \ell, m \rangle$ or they pass through their intersection point, $Q = \ell \cap m$.

Consider a plane π not through Q, and let $\mathcal{B}'$ be the intersection of the $\mathcal{B}$-lines with π. Then $\mathcal{B}'$ is a blocking set in π. Suppose to the contrary that there is a line in π disjoint from $\mathcal{B}'$. Then there is a plane on Q not containing a $\mathcal{B}$-line through Q. This plane intersects σ in a line through Q that is not contained in $\mathcal{B}$, so has a point $R \notin \mathcal{B}$. Now take a line through R in this plane but not through Q. It is not blocked by the original $\mathcal{B}$, giving a contradiction.

So ℓ is the only line with this property. Now let Q be another $\mathcal{B}$-point, and consider the plane $\langle Q, \ell \rangle$. Every $\mathcal{B}$-point in this plane, not on ℓ is on at least two $\mathcal{B}$-lines, so we find at least $q+(q-1)+\cdots+1 = \frac{1}{2}q(q+1)$ $\mathcal{B}$-points in this plane. Since $M \leq \frac{3}{2}(q+1)$, we can have at most two such planes. Now $\mathcal{B}$ is completely contained in the union of two planes, but that means it contains one of the planes. □

Lemma 9.56. *Every $\mathcal{B}$-point P is contained in at least three $\mathcal{B}$-lines that are independent (so not coplanar).*

Proof. Let P be contained in the two $\mathcal{B}$-lines, ℓ_1 and ℓ_2, and suppose P is not contained in a $\mathcal{B}$-line ℓ_3 independent of ℓ_1 and ℓ_2. Let $\pi = \langle \ell_1, \ell_2 \rangle$ and let every other line in this plane have at least a additional $\mathcal{B}$-points (and there is a line with exactly a). Then we see that

$$|\mathcal{B}| \geq 1+a+2q+(q-2)a+q(M-a-1) = 1+qM+(q-a) > 1+qM.$$

□

From now on we assume $q > 3$.

Lemma 9.57. *Every $\mathcal{B}$-line is contained in a $\mathcal{B}$-plane, a plane having more than $(q+3)/2$ additional $\mathcal{B}$-lines.*

Proof. Let ℓ be a $\mathcal{B}$-line, and for every point P on ℓ choose (exactly) two $\mathcal{B}$-lines different from ℓ, not forming a coplanar triple with ℓ. Consider the planes π_i, $1 \leq i \leq q+1$ on ℓ, and let π_i contain a_i of these chosen lines. Then $\pi_i \setminus \ell$ contains at least $q+(q-1)+\cdots+(q-a_i+1)$ $\mathcal{B}$-points. We have $2(q+1)$ lines chosen, so $\sum a_i = 2(q+1)$. Counting $\mathcal{B}$-points on the planes π_i we get

$$1+q+2q(q+1) - \frac{1}{2}\sum a_i(a_i-1) \leq 1+qM \leq 1+\frac{3}{2}q(q+1).$$

which implies $\sum a_i^2 \geq q^2+5q+2$.

If all $a_i \leq (q+3)/2$, then the sum of the a_i^2 is at most $q^2 + 2q + 13 < q^2 + 5q + 2$, so there is an i such that $a_i > (q+3)/2$. $\quad\square$

Lemma 9.58. *There are at most three $\mathcal{B}$-planes, and they cover the entire $\mathcal{B}$.*

Proof. A $\mathcal{B}$-plane contains at least $\frac{3}{8}q(q+6)$ $\mathcal{B}$-points and four such planes would contain at least $\frac{3}{2}q(q+6) - 6q - 3 > 1 + \frac{3}{2}q(q+1)$ $\mathcal{B}$-points. $\quad\square$

Lemma 9.59. *Every $\mathcal{B}$-point is in at least two $\mathcal{B}$-planes.*

Proof. A $\mathcal{B}$-point is contained in at least three independent $\mathcal{B}$-lines. So if a $\mathcal{B}$-point were in a unique $\mathcal{B}$-plane, then there would be a $\mathcal{B}$-line through this point not in any $\mathcal{B}$-plane, giving a contradiction. $\quad\square$

With the last lemma, the proof of the theorem for $q > 3$ is now finished, since a $\mathcal{B}$-plane can contain at most two $\mathcal{B}$-lines (namely the intersections with the other $\mathcal{B}$-planes) and $\frac{1}{2}(q+5) > 2$.

The case $q = 3$ can be done by essentially the same reasoning as above. The proof is left as an exercise (see Exercise 9.20). $\quad\square$

The previous proofs show the typical methods for k-blocking sets as well. A relatively easy general result determines the smallest k-blocking sets in $\mathrm{PG}(n, q)$.

Theorem 9.60 (Bose, Burton). *Every k-blocking set in $\mathrm{PG}(n, q)$ has at least $(q^{k+1} - 1)/(q - 1)$ points. In case of equality the k-blocking set is a subspace of dimension k.*

Proof. The proof goes by induction on n. For $n = k$, the assertion is trivial. If the k-blocking set $\mathcal{B}$ in $\mathrm{PG}(n + 1, q)$ is projected from a point outside to $\mathrm{PG}(n, q)$, then no two points can be projected onto the same point, because we would get a k-blocking set of size strictly less than $(q^{k+1} - 1)/(q - 1)$, contradicting the induction hypothesis. This means that the line through any two points of $\mathcal{B}$ is contained in $\mathcal{B}$, which implies that $\mathcal{B}$ is a subspace. Because of its size, the set has to be a subspace of dimension k. $\quad\square$

Recall that $\mathrm{PG}(t, q)$ is a blocking set in $\mathrm{PG}(t, q^t)$. By projecting it, one can also obtain blocking sets in dimensions smaller than t. *Lunardon* and *Polverino* proved that the linear blocking sets introduced at the end of Chapter 8 can also be obtained by projecting subgeometries, where the projection technique was also explained. This can also be combined with building cones over a blocking set, as we did in the case of Buekenhout's construction of unitals. Finally, one can also copy Buekenhout's construction in the sense that by choosing an appropriate spread in the hyperplane at infinity, one can again obtain k-blocking sets in a smaller dimensional space. Such constructions are discussed in detail by *Mazzocca, Polverino* [125] and *Mazzocca, Polverino, Storme* [126]. A general survey on blocking sets with focus on higher dimensions is [29].

Some of the constuctions can be found here as exercises (see Exercises 9.21, 9.22).

For the sake of completeness, let us see the results in arbitrary dimensions.

Proposition 9.61 (Beutelspacher, Bruen, Heim). *Every blocking set* $\mathcal{B}$ *in* $\mathrm{PG}(n, q)$ *has at least* $q + 1$ *points. If* $|\mathcal{B}| = q + 1$, *then it is a line. Let* $b(q)$ *denote the size of the smallest blocking set in* $\mathrm{PG}(2, q)$. *If* $\mathcal{B}$ *is non-trivial, then* $|\mathcal{B}| \geq b(q)$. *In case of equality* $\mathcal{B}$ *is a blocking set in a plane.*

The analogous result for k-blocking sets is due to Heim [82].

Theorem 9.62 (Heim). *Let* $\mathcal{B}$ *be a non-trivial* k-*blocking set in* $\mathrm{PG}(n, q)$, $n > k$, $q > 2$. *Let* $b(q) = q + 1 + r(q)$ *denote the size of the smallest blocking set in* $\mathrm{PG}(2, q)$. *Then* $|\mathcal{B}| \geq (q^{k+1} - 1)/(q - 1)$. *In case of equality, the set* $\mathcal{B}$ *is a come, whose base is a non-trivial planar blocking set of size* $b(q)$ (*in a* 2-*dimensional subspace* π), *and its vertex is a subspace of dimension* $k - 2$, *disjoint from* π.

In the special case when q is a square (and hence $r(q) = \sqrt{q}$), the above theorem was proved earlier by Beutelspacher [20]. The case $q = 2$ was done by *Govaerts* and *Storme*, [75].

Exercises

9.1. Verify that the definition of an arc is stronger than the definition of a cap in the sense that each arc is a cap, but there exist caps that are not arcs.

9.2. Prove Theorem 9.8 in the case $q = 3$.

9.3. Prove that in $\mathrm{AG}(3, q)$ any cap consists of at most q^2 points.

9.4. (Pellegrino, [141]) Consider $\mathrm{PG}(4, 3)$ as $\mathrm{AG}(4, 3) \cup \mathrm{PG}(3, 3)$ and let $\mathcal{O} = \{P_1, P_2, \ldots, P_{10}\}$ be an ovoid in $\mathrm{PG}(3, 3)$. Let V be a point in $\mathrm{AG}(4, 3)$ and R_i, $S_i \in \mathrm{AG}(4, 3)$ be two points on the line VP_i for $i = 1, 2, \ldots, 10$, such that $R_i \neq V \neq S_i$. Prove that the set $\{R_i, S_i : i = 1, 2, \ldots, 10\}$ is a 20-cap.

9.5. Show that $\mathrm{AG}(4, 3)$ can be divided into the union of three parallel hyperplanes in 40 different ways.

9.6. Suppose that $\mathcal{K}$ is a 21-cap in $\mathrm{AG}(4, 3)$. Let us divide $\mathrm{AG}(4, 3)$ into the union of three parallel hyperplanes $\mathcal{H}_1$, $\mathcal{H}_2$ and $\mathcal{H}_3$ and consider the multiset $S = \{|\mathcal{H}_1 \cap \mathcal{K}|, |\mathcal{H}_2 \cap \mathcal{K}|, |\mathcal{H}_1 \cap \mathcal{K}|\}$.

• Show that S is one of the following:

$$\{9, 9, 3\}, \{9, 8, 4\}, \{9, 7, 5\}, \{9, 6, 6\}, \{8, 8, 5\}, \{8, 7, 6\}, \{7, 7, 7\}.$$

- Take all of the 40 different partitions of $AG(4,3)$ into three parallel hyperplanes and calculate the corrresponding multiset to each partition. Let s_{ijk} denote the number of multisets $\{i,j,k\}$. Show that

$$s_{993} + s_{984} + s_{975} + s_{966} + s_{885} + s_{876} + s_{777} = 40,$$
$$75s_{993} + 70s_{984} + 67s_{975} + 66s_{966} + 66s_{885} + 64s_{876} + 63s_{777} = 40,$$
$$169s_{993} + 144s_{984} + 129s_{975} + 124s_{966} + 122s_{885} + 111s_{876} + 105s_{777} = 40.$$

9.7. Show that the previous three equations imply $12s_{993} = 210$.

9.8. Prove that there is no 21-cap in $AG(4,3)$.

9.9. In $PG(3,q)$ let $P_t = (1 : t : t^2 : t^3)$ and consider the twisted cubic

$$C_3 = \{P_t : t \in GF(q)\} \cup \{(0 : 0 : 0 : 1)\}$$

Show that

- The plane containing three distinct points, P_{t_1}, P_{t_2} and P_{t_3} of C_3, has equation

$$t_1 t_2 t_3 X_0 - (t_1 t_2 + t_2 t_3 + t_3 t_1)X_1 + (t_1 + t_2 + t_3)X_2 - X_3 = 0.$$

- Let Π_t be the plane with equation

$$t^3 X_0 - 3t^2 X_1 + 3t X_2 - X_3 = 0$$

and ℓ_t be the line with Plücker coordinates

$$(1 : 2t : 3t^2 : t^2 : -2t^3 : t^4).$$

Show that ℓ_t is a tangent line to C_3 at P_t, and the plane Π_t is an osculating plane to C_3 at P_t.

- Let Q_1 be the quadratic cone whose vertex is $V_1 = (1 : 0 : 0 : 0)$ and base is the conic

$$C_1 = \{(0 : 1 : t : t^2) : t \in GF(q)\} \cup \{(0 : 0 : 0 : 1)\},$$

and Q_2 be the quadratic cone whose vertex is $V_2 = (0 : 0 : 0 : 1)$ and base is the conic

$$C_2 = \{(1 : t : t^2 : 0) : t \in GF(q)\} \cup \{(1 : 0 : 0 : 0)\}.$$

Prove that $Q_1 \cap Q_2$ is the union of C_3 and the line $V_1 V_2$.

9.10. Prove that any twisted cubic is contained in a hyperbolic quadric.

9.11. Let q be odd and $\delta \in \mathrm{GF}(q)$ be a non-square element. Prove that

$$u^2 + \frac{\delta}{u^2} = v^2 + \frac{\delta}{v^2}$$

implies $u = \pm v$.

9.12. Let q be odd and $\beta \in \mathrm{GF}(q)$ be a fixed non-square element. Consider $\mathrm{GF}(q^2)$ as $\mathrm{GF}(q)[i]$ with $i^2 = \beta$.

Prove that if $\delta = a + ib$, $a, 0 \neq b \in \mathrm{GF}(q)$ is fixed, then

- there are exactly $(q-1)/2$ elements $u \in \mathrm{GF}(q)$ such that $\delta - u$ is a square element in $\mathrm{GF}(q^2)$;

- there are exactly $(q+1)/2$ elements $u \in \mathrm{GF}(q)$ such that $\delta - u$ is a non-square element in $\mathrm{GF}(q^2)$.

9.13. Let q be odd and α be a fixed non-square element of $\mathrm{GF}(q^2)$. For $a \in \mathrm{GF}(q^2)$ let $\mathcal{P}_a$ be the conic in $\mathrm{PG}(2, q^2)$ with equation

$$\mathcal{P}_a : \quad X_2 X_0 = X_1^2 + \alpha a X_0^2.$$

Then $\mathcal{P}_a$ is a parabola in the affine plane $\mathrm{AG}(2, q^2)$.

Let $\{a_1, a_2, \ldots, a_q\}$ be the elements of $\mathrm{GF}(q) \subset \mathrm{GF}(q^2)$ and consider the union of the q conics

$$\mathcal{U} = \bigcup_{i=1}^{q} \mathcal{P}_{a_i}.$$

- Show that the line ℓ_∞ is a tangent to $\mathcal{U}$.

- Show that a tangent to $\mathcal{P}_a$ at any of its affine points is an external line to $\mathcal{P}_b$ if and only if $a - b$ is a non-square element in $\mathrm{GF}(q^2)$.

- Show that if ℓ is a tangent line to any of the parabolas $\mathcal{P}_{a_i}$ then ℓ is a tangent line to $\mathcal{U}$, too.

- Show that if ℓ is a non-tangent line to any of the parabolas $\mathcal{P}_{a_i}$ then ℓ intersects $\mathcal{U}$ in $q + 1$ points.

- Prove that $\mathcal{U}$ is a unital.

- Let $\mathcal{U}_{a0}$ denote the Buekenhout-Metz unital (described in Proposition 9.50) in the case $b = 0$. Show that $\mathcal{U}$ is isomorphic to $\mathcal{U}_{a0}$.

9.14. Let $\mathcal{U}_{0b}$ denote the Buekenhout-Metz unital (described in Proposition 9.50) in the case $a = 0$.

- Show that $\mathcal{U}_{0b}$ is a Hermitian curve.

- Let $c \in \mathrm{GF}(q^2)$ and $f \in \mathrm{GF}(q)$ be arbitrary elements. Prove that each of the transformations given by the matrices

$$T_c = \begin{pmatrix} 1 & c & bc^{q+1} \\ 0 & 1 & (b-b^q)c^q \\ 0 & 0 & 1 \end{pmatrix} \text{ and } L_f = \begin{pmatrix} 1 & 0 & f \\ 0 & 1 & 0 \\ 0 & 0 & 1 \end{pmatrix}$$

  fixes $\mathcal{U}_{0b}$ setwise and fixes the point $(0:0:1)$.

- Show that if a linear transformation fixes $\mathcal{U}_{0b}$ setwise and fixes the points $(0:0:1)$ and $(1:0:0)$, then its matrix is diagonal.

9.15. Prove that the unitary group $\mathrm{PGU}(3,q^2)$ acts double transitively on the points of a Hermitian curve.

9.16. Prove that the subgroup of the unitary group $\mathrm{PGU}(3,q^2)$ fixing two points of a Hermitian curve is isomorphic to the multiplicative group $\mathrm{GF}(q^2)^*$.

9.17. Let $P_0 = (1:t_0:t_0^2:t_0^3)$ be a point on the normal rational curve $\mathcal{C}_3$. Calculate the set of equations of the tangent line to $\mathcal{C}_3$ at P, and the equation of the osculating plane at P.

9.18. Let $\mathcal{B}$ be a k-blocking set in $\mathrm{PG}(n,q)$, $n \geq 3$, Σ be a hyperplane and $P \notin \Sigma$ be a point. Let $\mathcal{B}'$ denote the projection of $\mathcal{B}$ from P to Σ. Prove that $\mathcal{B}'$ is a k-blocking set in Σ.

9.19. Let $\mathcal{D}$ be a set of points in $\mathrm{PG}(3,2)$. Suppose that $\mathcal{D}$ has at least one point in each plane, but it does not contain any line. Prove that $\mathcal{D}$ is an elliptic quadric.

9.20. Prove Theorem 9.53 for $q = 3$.

9.21. Let $\mathcal{O}$ be an ovoid of a 3-space Σ_3 and embed Σ_3 in $\mathrm{PG}(h+2,q)$, $h \geq 2$. Construct the cone $\mathcal{B}$ in $\mathrm{PG}(h+2,q)$ with base $\mathcal{O}$ and vertex V, an $(h-2)$-space disjoint from Σ_3.

  - Show that $\mathcal{B}$ blocks all the planes in $\mathrm{PG}(h+2,q)$.
  - A plane not meeting the vertex V contains either 1 or $q+1$ points from $\mathcal{B}$. Any 3-space disjoint from V meets $\mathcal{B}$ in q^2+1 points.

9.22. Let $\mathcal{S}$ be the $(h-1)$-spread of the hyperplane Σ_∞ at infinity of $\mathrm{PG}(2h,q)$, defining the plane $\mathrm{PG}(2,q^h)$. Construct the cone $\mathcal{B}$ of the previous exercise. Put $\mathcal{B}$ in $\mathrm{PG}(2h,q)$ in such a way that the hyperplane Σ_∞ contains the vertex V and exactly one point T of the ovoid $\mathcal{O}$ (so Σ_∞ meets the 3-space Σ_3 in a tangent plane of the ovoid $\mathcal{O}$). Place $\mathcal{B}$ so that the $(h-1)$-space $\langle V,T\rangle$ is an element σ of the spread $\mathcal{S}$. Then the corresponding set $\mathcal{B}'$ of size $q^{h+1}+1$ in $\mathrm{PG}(2,q^h)$ is a minimal blocking set. (This set contains one point at infinity, namely σ.)

9.23. Try to generalize the previous construction, yielding minimal blocking sets in a projective plane of order p^h, when (instead of the ovoid $\mathcal{O}$ in $PG(3,q)$) we have a relatively large minimal blocking set in some $PG(r,q)$ for a fixed r. (For example, start from a large planar minimal blocking set.)

10

Generalized polygons, Möbius planes

In this chapter some further classes of incidence geometries are investigated. Generalized polygons can be considered as natural generalizations of projective planes. Among them we pay particular attention to generalized quadrangles, because these objects are closely connected to polarities of projective spaces. The Möbius planes belong to circle geometries; they generalize the incidence properties of points and circles in the real spherical plane.

Definition 10.1. *Let* $\mathcal{S} = (\mathcal{P}, \mathcal{L}, I)$ *be finite point-line incidence geometry and* $n \geq 2$ *be an integer.*

A chain of length h *is a sequence of* $h + 1$ *elements* $x_0, x_1, \ldots, x_h \in \mathcal{P} \cup \mathcal{L}$ *such that*

$$x_0 I x_1 I \ldots I x_h.$$

We say that the chain joins x_0 *and* x_h. $\mathcal{S}$ *is* connected *if for any two elements of* $\mathcal{P} \cup \mathcal{L}$ *there exists a chain joining them. The* distance *of two elements* x *and* y *of* $\mathcal{P} \cup \mathcal{L}$, *denoted by* $d(x, y)$, *is the length of the shortest chain joining them.*

$\mathcal{S}$ *is called a generalized* n*-gon if it is connected and satisfies the following axioms.*

- **Gn1.** $d(x, y) \leq n$ *for all* $x, y \in \mathcal{P} \cup \mathcal{L}$.

- **Gn2.** *If* $d(x, y) = k < n$, *then there exists a unique chain of length* k *joining* x *and* y.

- **Gn3.** *For all* $x \in \mathcal{P} \cup \mathcal{L}$ *there exists* $y \in \mathcal{P} \cup \mathcal{L}$ *such that* $d(x, y) = n$.

The notions of *collineation* and *isomorphism* for generalized polygons are defined in the usual way.

We recall a definition from graph theory. The incidence graph was originally defined for configurations (that is for incidence structures where points are incident with at least three lines and lines are incident with at least three points).

Definition 10.2. *The* incidence graph *(sometimes called* Levi graph*) of a point-line incidence geometry is a bipartite graph whose two classes of vertices* correspond to the set of points and the set of lines of the geometry, and two vertices are adjacent if and only if the corresponding point-line pair is incident in the geometry.

It is easy to see that the distance of two elements in a generalized n-gon S is the same as the usual graph theoretical distance of the corresponding vertices in the Levi graph of S. We will study the generalized polygons from a graph theoretical point of view, and we will also provide some of their applications to the cage and the degree/diameter problems in Chapter 12 (see Theorems 12.33 and 12.34).

Let us start with some examples.

Example 10.3. Let N be a proper n-gon in the Euclidean plane. Let $\mathcal{P}$ be the set of vertices and $\mathcal{L}$ be the set of sides of N, respectively. If the incidence is the natural one then $(\mathcal{P}, \mathcal{L}, I)$ is obviously a generalized n-gon.

Example 10.4. Let $S = (\mathcal{P}, \mathcal{L}, I)$ be a generalized n-gon. Its *dual* is defined as $S^* = (\mathcal{L}, \mathcal{P}, I^*)$ where I^* is the same as I. The dual of a generalized n-gon is also a generalized n-gon.

The following simple observations are very useful.

Lemma 10.5. *Let $S = (\mathcal{P}, \mathcal{L}, I)$ be a generalized n-gon. Then the distance of any two points or any two lines of S is even and the distance of a point and a line is odd.*

Proof. Let $x, y \in \mathcal{P} \cup \mathcal{L}$ be two elements and let

$$x \, I \, x_1 \, I \, \ldots \, I \, x_{h-1} \, I \, y$$

be a chain joining them. As exactly one of any two consecutive elements of the chain belongs to $\mathcal{P}$ and the other one belongs to $\mathcal{L}$, the length of the chain is odd if and only if exactly one of x and y is a point and the other one is a line. Thus $d(x, y)$ is odd if and only if exactly one of x and y is a point and the other one is a line. □

Let $S = (\mathcal{P}, \mathcal{L}, I)$ be a generalized n-gon. A set of $2k$ elements, $\{x_1, x_2, \ldots, x_{2k}\} \subset (\mathcal{P} \cup \mathcal{L})$, is called a *proper k-gon* if $x_1 \, I \, x_2 \, I \, \ldots \, I \, x_{2k} \, I \, x_1$.

Lemma 10.6. *Let $S = (\mathcal{P}, \mathcal{L}, I)$ be a generalized n-gon and $2 \leq k < n$ be an integer. Then S does not contain any proper k-gon.*

Proof. Suppose to the contrary that S contains a proper k-gon, $x_1 \, I \, x_2 \, I \, \ldots$ $I \, x_{2k} I \, x_1$. Then $d(x_1, x_{k+1}) \leq k$ because $x_1 \, I \, x_2 \, I \, \ldots \, I \, x_{k+1}$ is a chain of length k joining them. The other part of the proper k-gon, $x_{k+1} \, I \, x_{k+2} \, I \, \ldots \, I \, x_{2k} \, I \, x_1$, is another chain of length k joining the two elements. As $k < n$, this contradicts to axiom Gn2, which proves the lemma. □

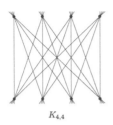
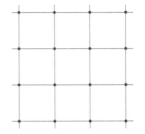

$K_{4,4}$

FIGURE 10.1
Trivial generalized 4-gons

If $n = 2$, then the distance of any point-line pair is an odd number which is at most 2, thus the distance must be 1. This means that every point is incident with every line. Hence generalized 2-gons are trivial structures (their Levi graphs are the complete bipartite graphs).

If $n = 3$, then the distance of any two distinct points is a positive even number which is at most 3, thus the distance must be 2. This means that the points are collinear, and the line joining them is unique, because of axiom Gn2. In the same way we get that any two distinct lines have a unique point of intersection. Thus the generalized 3-gons satisfy axioms P1 and P2 of projective planes. Conversely, any (not necessarily non-degenerate) projective plane containing at least two points and two lines is a generalized 3-gon. The degenerate planes are trivial generalized 3-gons.

There are other more or less trivial generalized n-gons, too. We leave it as an exercise to prove that the following two structures are generalized 4-gons.

Example 10.7. Let $a \geq 2$ and $b \geq 2$ be positive integers and $K_{a,b}$ be the complete bipartite graph with a and b vertices in its classes. Let $\mathcal{P}$ be the set of vertices of $K_{a,b}$ and $\mathcal{L}$ be the set of edges of $K_{a,b}$. Define I such that a point is on a line if and only if the corresponding vertex is an endpoint of the corresponding edge in $K_{a,b}$. Then $\mathcal{S} = (\mathcal{P}, \mathcal{L}, \mathrm{I})$ is a generalized 4-gon. This type of generalized 4-gons is called a *grid* (see Figure 10.1).

Example 10.8. Take a Cartesian coordinate system in the classical Euclidean plane and let $a \geq 2$ and $b \geq 2$ be positive integers. Let $\mathcal{P}$ be the set of lattice points (x, y) whose coordinates are integers and satisfy the inequalities $0 \leq x < a$ and $0 \leq y < b$. Let $\mathcal{L}$ be the set of lines with equation $X = c$ or $Y = d$ where $0 \leq c < a$ and $0 \leq d < b$. If the incidence is the same as in the Euclidean plane, then $\mathcal{S} = (\mathcal{P}, \mathcal{L}, \mathrm{I})$ is a generalized 4-gon.

Let us remark that the previous examples are duals of each other.

We would like to avoid trivial structures. It can be done by adding an extra axiom.

Definition 10.9. *A generalized polygon is called* thick, *if it satisfies the following axiom, too.*

Gn4. *Each line is incident with at least three points and each point is incident with at least three lines.*

Theorem 10.10. *In a thick finite generalized n-gon, each line is incident with the same number of points and each point is incident with the same number of lines. If n is odd, then these two numbers are equal.*

Proof. Let $\mathcal{S} = (\mathcal{P}, \mathcal{L}, \mathrm{I})$ be a finite thick generalized n-gon. For each element $x \in (\mathcal{P} \cup \mathcal{L})$ let

$$N(x) = \{y : y \in (\mathcal{P} \cup \mathcal{L}) , d(x,y) = 1\}$$

be the set of neighbours of x.

First, we show that $|N(x)| = |N(y)|$ if $d(x,y) = n$. Let $N(x) = \{x_1, x_2, \ldots, x_k\}$ and $N(y) = \{y_1, y_2, \ldots, y_m\}$. Then $d(x, y_i) < n$, because, by Lemma 10.5, its parity is different from the parity of $d(x,y)$. On the other hand, $d(x, y_i) \geq n - 1$, because $d(x, y_i) + 1 \geq d(x,y)$ obviously holds. Hence $d(x, y_i) = n - 1$ for all $i = 1, 2, \ldots, m$. Thus, for each i there exists a unique chain, $\mathcal{C}_i : x \mathrm{I} x_{i_1} \mathrm{I} \ldots \mathrm{I} y_i$ of length $n - 1$ joining x and y_i. We claim that for $i \neq j$ the chains $\mathcal{C}_i$ and $\mathcal{C}_j$ have only one element in common, namely x. Suppose to the contrary that z is another common element. Then $d(z, y_i) \leq n - 2$ and $d(z, y_j) \leq n - 2$ imply that $z \mathrm{I} \ldots \mathrm{I} y_i \mathrm{I} y \mathrm{I} y_j \mathrm{I} \ldots \mathrm{I} z$ is a proper k-gon with $k \leq n - 1$, contradicting Lemma 10.6. Hence $i \neq j$ implies $x_{i_1} \neq x_{j_1}$, so $|N(x)| \geq |N(y)|$. In the same way, we can prove that $|N(y)| \geq |N(x)|$, thus $|N(x)| = |N(y)|$ follows from the finiteness of $\mathcal{S}$.

Now, we show that any two points, P_1 and P_2, have the same number of neighbours. If their distance is n, then the assertion follows from the previous paragraph. Suppose that $d(P_1, P_2) = 2k < n$ and let

$$P_1 \mathrm{I} x_1 \ldots \mathrm{I} x_{k-1} \mathrm{I} x_k \mathrm{I} x_{k+1} \mathrm{I} \ldots \mathrm{I} x_{2k-1} \mathrm{I} P_2$$

be the unique chain of length $2k$ joining them. From the thickness of $\mathcal{S}$, it follows that there exists $y \mathrm{I} x_k$ such that $x_{k-1} \neq y \neq x_{k+1}$. The thickness of $\mathcal{S}$ also implies that there is a chain $x_k \mathrm{I} y \mathrm{I} \ldots v$ such that $d(x_k, v) = n - k$. This means that

$$P_1 \mathrm{I} x_1 \ldots \mathrm{I} x_{k-1} \mathrm{I} x_k \mathrm{I} y \mathrm{I} \ldots v$$

and

$$v \mathrm{I} \ldots y \mathrm{I} x_k \mathrm{I} x_{k+1} \mathrm{I} \ldots \mathrm{I} x_{2k-1} \mathrm{I} P_2$$

are chains of length n joining P_1 and v, and P_2 and v, respectively. Hence $d(P_1, v) = d(P_2, v) = n$, because if there were a chain of length $h < n$ joining P_i and v, then $\mathcal{S}$ would contain a proper $(h + n)/2$-gon contradicting axiom Gn2. Thus $|N(P_1)| = |N(v)| = |N(P_2)|$.

In the same way we can prove that any two lines have the same number of neighbours. Finally, if n is odd, then take a point P and a line ℓ such that $d(P, \ell) = n$. Then $|N(P)| = |N(\ell)|$ which proves the last statement of the theorem. $\square$

Definition 10.11. *A finite generalized polygon is called of order (s,t) if each of its lines contains $s + 1$ points and there are $t + 1$ lines through each of its points.*

The proof of the following theorem is beyond the scope of this book. It can be found in [66].

Theorem 10.12 (Feit, Higman). *Finite thick generalized n-gons exist if and only if $n = 2, 3, 4, 6$ and 8.*

Let us consider thick generalized n-gons for small values of n. We have already seen that generalized 2-gons are trivial structures. For the next value, $n = 3$, the characterization is easy.

Theorem 10.13. *Finite thick generalized 3-gons and finite projective planes are the same point-line incidence geometries.*

Proof. Let $\mathcal{S} = (\mathcal{P}, \mathcal{L}, I)$ be a finite thick generalized 3-gon. By Lemma 10.5, the distance of any two distinct points or any two distinct lines of $\mathcal{S}$ is an even number. Because of axiom Gn1, this number is less than or equal to 3, hence it must be 2. If the distance of two elements is 2, then axiom Gn2 gives that the chain of length 2 joining them is unique. Thus there is a unique line joining any two distinct points of $\mathcal{S}$ and there is a unique point of intersection for any two distinct lines of $\mathcal{S}$. So $\mathcal{S}$ satisfies axioms P1 and P2 of projective planes. As $\mathcal{S}$ is thick, axioms P3 and P4 are also satisfied.

Conversely, if $\Pi = (\mathcal{P}, \mathcal{L}, I)$ is a projective plane of order $m \geq 2$, then, by axiom P1, the distance of any two distinct points is 2, by axiom P2, the distance of any two distinct lines is 2 and there is a unique chain of length 2 joining the two elements in both cases. If (P, ℓ) is a non-incident point-line pair, then their distance is 3, thus Π is a generalized 3-gon. Finally, axioms P3 and P4 imply the thickness of Π. $\square$

$\mathcal{S}$ is called a *generalized quadrangle* if $n = 4$. The usual abbreviation for generalized quadrangle is GQ. First, we give an alternate definition for these objects. We leave it as an exercise to prove that the two definitions are equivalent.

Definition 10.14. *Let $s > 1$ and $t > 1$ be integers. A point-line incidence geometry $\mathcal{S} = (\mathcal{P}, \mathcal{L}, I)$ is a generalized quadrangle of order (s,t) if it satisfies the following axioms.*

- **GQ1.** *Each point is incident with $t + 1$ lines and two distinct points are incident with at most one line.*

- **GQ2.** *Each line is incident with $s + 1$ points and two distinct lines are incident with at most one point.*

- **GQ3.** *If $(P, \ell) \subset \mathcal{P} \times \mathcal{L}$ is a non-incident point-line pair, then there is a unique pair $(P', \ell') \subset \mathcal{P} \times \mathcal{L}$ for which $P \, I \, \ell' \, I \, P' \, I \, \ell$.*

First, we describe the three types of so-called *classical generalized quad-rangles*. These structures are closely connected to the polarities of $PG(n, q)$ which were discussed in Chapter 4.

Example 10.15. Let $n = 3, 4$ or 5 and $\mathcal{Q}$ be a non-singular quadric of projective index 1 in $PG(n, q)$. Then the points of $\mathcal{Q}$ and the lines contained in $\mathcal{Q}$ with the incidence inherited from $PG(n, q)$ form a generalized quadrangle. The orders of these quadrangles are the following:

$$
\begin{aligned}
s &= q \text{ and } t = 1, &&\text{if } n = 3, \\
s &= q \text{ and } t = q, &&\text{if } n = 4, \\
s &= q \text{ and } t = q^2, &&\text{if } n = 5.
\end{aligned}
$$

These quadrangles are denoted by $\mathcal{Q}(n, q)$.

The lines of $\mathcal{Q}(n, q)$ are lines in $PG(n, q)$, hence each of them contains $q + 1$ points, so axiom GQ2 is satisfied with $s = q$.

By Theorem 4.59, for each point $P \in \mathcal{Q}$ the lines contained in $\mathcal{Q}$ and through P form a cone whose base, $\mathcal{B}$, is an $(n - 2)$-dimensional non-singular quadric of the same character as $\mathcal{Q}$. Hence if $n = 3, 4$ or 5, then $\mathcal{B} \subset PG(n - 2, q)$ is a hyperbolic, a parabolic or an elliptic quadric, respectively. So $\mathcal{B}$ consists of 2, $q + 1$ or $q^2 + 1$ points according to $n = 3, 4$ or 5, which proves that $\mathcal{Q}(n, q)$ satisfies axiom GQ1.

Finally, let (P, ℓ) be a non-incident point-line pair in $\mathcal{Q}(n, q)$. The lines of $\mathcal{Q}(n, q)$ through P are in the polar hyperplane of P with respect to $\mathcal{Q}$ in $PG(n, q)$. This polar hyperplane does not contain ℓ, hence it intersects ℓ in a unique point, say P', in $PG(n, q)$. As ℓ is on $\mathcal{Q}$, the point P' is also on $\mathcal{Q}$ and the line PP' is contained in $\mathcal{Q}$. Hence this is a line of $\mathcal{Q}(n, q)$. As no more lines of $\mathcal{Q}(n, q)$ through P meet ℓ, we showed that axiom GQ3 is also satisfied by $\mathcal{Q}(n, q)$.

Example 10.16. Let $n = 3$ or 4 and $\mathcal{U}$ be a non-singular Hermitian variety in $PG(n, q^2)$. Then the points of $\mathcal{U}$ and the lines contained in $\mathcal{U}$ with the incidence inherited from $PG(n, q^2)$ form a generalized quadrangle. The orders of these quadrangles are the following:

$$
\begin{aligned}
s &= q^2 \text{ and } t = q, &&\text{if } n = 3, \\
s &= q^2 \text{ and } t = q^3, &&\text{if } n = 4.
\end{aligned}
$$

These quadrangles are denoted by $\mathcal{U}(n, q^2)$.

The lines of $\mathcal{U}(n, q^2)$ are lines in $PG(n, q^2)$, hence each of them contains $q^2 + 1$ points, so axiom GQ2 is satisfied with $s = q^2$.

By Theorem 4.45, for each point $P \in \mathcal{U}$ the lines contained in $\mathcal{U}$ and through P form a cone whose base, $\mathcal{B}$, is an $(n - 2)$-dimensional non-singular Hermitian variety. So, by Theorem 4.33, $\mathcal{B}$ consists of $q + 1$ or $q^3 + 1$ points according as $n = 3$ or 4. This proves that $\mathcal{Q}(n, q)$ satisfies axiom GQ1 with $t = q$ if $n = 3$, and with $t = q^3$ if $n = 4$.

Finally, let (P, ℓ) be a non-incident point-line pair in $\mathcal{U}(n, q^2)$. The lines of $\mathcal{U}(n, q^2)$ through P are in the polar hyperplane of P with respect to $\mathcal{U}$ in $PG(n, q^2)$. This polar hyperplane does not contain ℓ, hence it intersects ℓ in a unique point, say P', in $PG(n, q^2)$. As ℓ is on $\mathcal{U}$, the point P' is also on $\mathcal{U}$ and the line PP' is contained in $\mathcal{U}$. Hence this is a line also in $\mathcal{U}(n, q^2)$. As no more lines of $\mathcal{U}(n, q^2)$ through P meet ℓ, we showed that axiom GQ3 is also satisfied by $\mathcal{U}(n, q^2)$.

Example 10.17. In $PG(3, q)$ all points of the space and the self-conjugate lines of a null polarity with the incidence inherited from $PG(3, q)$ form a generalized quadrangle of order (q, q). This quadrangle is denoted by $\mathcal{W}(q)$.

The lines of $\mathcal{W}(q)$ are lines in $PG(3, q)$, hence each of them contains $q + 1$ points, so axiom GQ2 is satisfied with $s = q$.

If α is a null polarity of $PG(3, q)$, then each point is self-conjugate. Let P be an arbitrary point. If ℓ is a line through the points P and R, then ℓ^{α} is the intersection of the planes P^{α} and R^{α}. Hence ℓ is self-conjugate if and only if $P \in R^{\alpha}$ (and $R \in P^{\alpha}$). As $P \in P^{\alpha}$ also holds, this means that the self-conjugate lines through P are the lines of the pencil with carrier P in the plane P^{α}. Each pencil contains $q + 1$ lines hence $\mathcal{W}(q)$ satisfies axiom GQ1 with $t = q$.

Let (P, ℓ) be a non-incident point-line pair in $\mathcal{W}(q)$. Then ℓ is a self-conjugate line of α not containing P, hence ℓ is not contained in the plane P^{α}. So ℓ intersects P^{α} in a unique point, say P' in $PG(3, q)$. Then PP' is the unique self-conjugate line through P which meets ℓ, hence axiom GQ3 is satisfied by $\mathcal{W}(q)$.

There is only one known type of pairs (s, t) for which s and t are not powers of the same prime and there exist thick generalized quadrangles of order (s, t). For each prime power q, *Ahrens* and *Szekeres* [1] constructed a generalized quadrangle of order $(q - 1, q + 1)$. For q even, the same quadrangles were found independently by *Hall* [78]. We present the latter, simple construction.

Example 10.18. Let π be a plane in $PG(3, q)$, q even, and $\mathcal{H}$ be a hyperoval in π. Let $\mathcal{S} = (\mathcal{P}, \mathcal{L}, \mathrm{I})$ be the point-line incidence geometry whose set of points, $\mathcal{P}$, consists of the points of the affine space $AG(3, q) - PG(3, q) \setminus \pi$, the set of lines, $\mathcal{L}$, consists of those lines in $AG(3, q)$ whose point at infinity belongs to $\mathcal{H}$ and the incidence is the same as in $AG(3, q)$. Then $\mathcal{S}$ is a generalized quadrangle of order $(q - 1, q + 1)$. This generalized quadrangle is denoted by $\mathcal{T}_2^*(\mathcal{H})$.

As $\mathcal{H}$ contains $q + 2$ points and each of them can be joined to any point of $PG(3, q) \setminus \pi$ by a unique line in $PG(3, q)$, axiom GQ1 is satisfied with $t = q + 1$. Axiom GQ2 is also satisfied with $s = q - 1$, because each line of $AG(3, q)$ contains q affine points.

Let (P, ℓ) be a non-incident point-line pair in $\mathcal{T}_2^*(\mathcal{H})$. Then P and ℓ uniquely define a plane τ in $PG(3, q)$. The line $\pi \cap \tau$ contains the point $R = \ell \cap \mathcal{H}$, hence, as a hyperoval has no tangent line, it intersects $\mathcal{H}$ in two points. Let $\tau \cap \mathcal{H} = \{R, T\}$ and ℓ' be the line joining P and T in $\mathrm{PG}(3, q)$. Then ℓ' is also a line in $\mathcal{T}_2^*(\mathcal{H})$, and obviously $P \mathrm{I} \ell' \mathrm{I} T$. The lines ℓ and ℓ' are coplanar in $\mathrm{PG}(3, q)$, hence they have a unique point of intersection, say P'. The point P' is in $\mathrm{AG}(3, q)$, because ℓ and ℓ' meet π in distinct points, R and T. Thus P' is a point of $\mathcal{T}_2^*(\mathcal{H})$, so $P \mathrm{I} \ell' \mathrm{I} P' \mathrm{I} \ell$. The unicity of the pair (P', ℓ') follows, because the other lines of $\mathcal{T}_2^*(\mathcal{H})$ through P are not in the plane τ, hence none of them can meet ℓ.

Example 10.19. Let $n = 3$ or $n = 4$, π be a hyperplane in $\mathrm{PG}(n, q)$ and $\mathcal{O} \subset \pi$ be an oval if $n = 3$ and an ovoid if $n = 4$. Let $\mathcal{S} = (\mathcal{P}, \mathcal{L}, \mathrm{I})$ be the point-line incidence geometry whose set of points consists of the following three types of objects:

(i) the points of $\mathrm{PG}(n, q) \setminus \pi = \mathrm{AG}(n, q)$,

(ii) the hyperplanes of $\mathrm{PG}(n, q)$ which are tangent to $\mathcal{O}$,

(iii) a new symbol, denoted by (∞).

The set of lines $\mathcal{L}$ consists of the the following two types of objects:

(a) the lines of $\mathrm{PG}(n, q)$ which are not contained in π and meet $\mathcal{O}$ (necessarily in a unique point),

(b) the points of $\mathcal{O}$.

Incidence is defined in the following way: a point of type (i) is incident with a line of type (a) if and only if the corresponding point-line pair is incident in $\mathrm{PG}(n, q)$, and a point of type (i) is never incident with a line of type (b). A point of type (ii) is incident with all lines of type (a) contained in it, and the unique line of type (b) which corresponds to the point of contact of the corresponding hyperplane and $\mathcal{O}$. Finally, the point of type (iii) is incident with no line of type (a) and all lines of type (b). These generalized quadrangles are denoted by $\mathcal{T}_{n-1}(\mathcal{O})$.

If $n = 3$, then the order of $\mathcal{T}_2(\mathcal{O})$ is (q, q), while if $n = 4$ then the order of $\mathcal{T}_3(\mathcal{O})$ is (q, q^2). Using the geometric properties of ovals and ovoids (see in Chapters 6 and 9) it is easy to prove that $\mathcal{T}_{n-1}(\mathcal{O})$ satisfies axioms GQ1-GQ3. We leave the proofs as exercise.

Consider the points of type (i) of $\mathcal{T}_2(\mathcal{O})$. These points are points in the affine space $\mathrm{AG}(3, q) \subset \mathrm{PG}(3, q)$, too. A translation of $\mathrm{AG}(3, q)$ fixes the hyperplane at infinity of $\mathrm{PG}(3, q)$ pointwise, thus the group G of translations of $\mathrm{AG}(3, q)$ can be considered as a collineation group of $\mathcal{T}_2(\mathcal{O})$. Then G acts transitively on the points of type (i) and any element of G fixes each point of types (ii) and (iii) and each line of type (b). If we fix a base point O in $\mathrm{AG}(3, q)$ then the elements of G can be identified with the points of $\mathrm{AG}(3, q)$ in the natural way: $g \in G$ corresponds to O^g, that is the image of O under the collineation g. In this correspondence the image of the set of points of a line or a plane through O is the set of transformations of a subgroup of order q or q^2, respectively. The incidence relation of $\mathcal{T}_2(\mathcal{O})$ can be easily described by the containment relation of the cosets of these subgroups. It was discovered by *Kantor* [102] that this process also works in the opposite direction. If one finds a suitable group with "good" subgroups, then one can construct a generalized quadrangle.

Example 10.20. Let $1 < s, t$ be integers and G be a group of order $s^2 t$. Suppose that G admits two sets of subgroups $\{H_1, H_2, \ldots, H_{t+1}\}$ and $\{H_1^*, H_2^*, \ldots, H_{t+1}^*\}$ satisfying the following conditions for all $i = 1, 2, \ldots, t+1$:

1. the order of H_i is s, the order of H_i^* is st;

2. $H_i \subset H_i^*$;

3. the intersection $H_i H_j \cap H_k$ consists of only the identity element of G if $i \neq k \neq j$.

Let $\mathcal{S} = (\mathcal{P}, \mathcal{L}, \mathrm{I})$ be the point-line incidence geometry whose set of points consists of the the following three types of objects:

(*i*) the elements of G;

(*ii*) the $(t+1)s$ cosets of type $H_i^* g$, where $g \in G$;

(*iii*) a new symbol, denoted by (∞).

The set of lines, $\mathcal{L}$, consists of the following two types of objects:

(*a*) the $(t+1)st$ cosets of type $H_i g$ where $g \in G$;

(*b*) $t+1$ new symbols, denoted by $[H_1], [H_2], \ldots, [H_{t+1}]$.

Incidence is defined in the following way: A point g of type (*i*) is incident with a line of type (*a*) if and only if the line corresponds to a coset of form $H_i g$. A point $H_i^* g$ of type (*ii*) is incident with a line $H_i h$ of type (*a*) if and only if $H_i h \subset H_i^* g$, and with the line $[H_i]$ of type (*b*). Finally, the point of type (*iii*) is incident with no line of type (*a*) and with all lines of type (*b*).

A group and its subgroups satisfying the three conditions of the previous example are called a 4-*gonal family*.

The existence problem of generalized quadrangles with given order (s, t) is open in general. In particular, it is also open for small parameters; for example it is unknown whether a generalized quadrangle of order $(4, 11)$ does or does not exist. Each of the known generalized quadrangles has one of the following orders:

- $(s, 1)$ with $s \geq 1$;

- $(1, t)$ with $t \geq 1$;

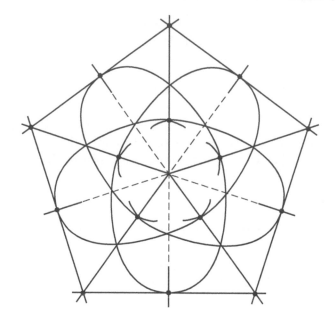

FIGURE 10.2
The generalized quadrangle of order $(2, 2)$

- (q, q) where q is a power of a prime;

- (q, q^2) or (q^2, q), where q is a power of a prime;

- (q^2, q^3) or (q^3, q^2), where q is a power of a prime;

- $(q - 1, q + 1)$ or $(q + 1, q - 1)$, where q is a power of a prime.

We have already seen examples belonging to each class on this list. It is easy to see that quadrangles of order $(s, 1)$ and $(1, t)$ are unique up to isomorphism. We leave it as an exercise to prove that the generalized quadrangle of order $(2, 2)$ (see Figure 10.2) is also unique; it is isomorphic to $\mathcal{W}(2)$.

Some necessary conditions of existence are known. Before proving them, we show some basic combinatorial properties of generalized quadrangles.

Lemma 10.21. *Suppose that the order of the generalized quadrangle S is (s, t). Then S contains $v = (s + 1)(st + 1)$ points and $b = (t + 1)(st + 1)$ lines.*

Proof. Let ℓ be a fixed line in S. By axiom GQ2, there are $s + 1$ points on ℓ and, by axiom GQ1, there are t lines different from ℓ through each of these points. Each of these lines contains s points not on ℓ. So there are $(s + 1)ts$ points not on ℓ and any two of these points are distinct because of axiom GQ3. As there are $s + 1$ points on ℓ, we get that S contains altogether $(s + 1) + (s + 1)ts = (s + 1)(st + 1)$ points.

By the principle of duality, we get the statement about the number of lines of $\mathcal{S}$. $\qquad\square$

Let $\mathcal{S} = (\mathcal{P}, \mathcal{L}, \mathrm{I})$ be a generalized quadrangle and P and Q be two, not necessarily distinct points of $\mathcal{S}$. We say that P and Q are collinear, and denote it as $P \sim Q$, if there exists a line of $\mathcal{S}$ which is incident with both P and Q. If the two points are not collinear then we denote them as $P \not\sim Q$. We introduce some notation: if P is a point, then $P^\perp = \{Q \in \mathcal{P} : P \sim Q\}$; if $\mathcal{X} = \{P_i : i \in I\}$ is a set of points, then $\mathcal{X}^\perp = \cap_{i \in I} P_i^\perp$.

Lemma 10.22. *Let P and P' be two distinct points in a generalized quadrangle $\mathcal{S}$ of order (s,t). Then*

$$|\{P, P'\}^\perp| = \begin{cases} s+1, & \text{if } P \sim P', \\ t+1, & \text{if } P \not\sim P'. \end{cases}$$

Proof. If $P \sim P'$, then $\{P, P'\}^\perp$ obviously contains the $s+1$ points on the line PP'. By axiom GQ3, $\mathcal{S}$ does not contain any proper triangle, hence there is no other point in $\{P, P'\}^\perp$.

If $P \not\sim P'$ and ℓ is a line through P, then the distance of ℓ and P' is 2, so again by axiom GQ3, there is a unique point on ℓ which is collinear with P'. As there are $t+1$ distinct lines through P and no two of them meet in another point, we get $|\{P, P'\}^\perp| = t+1$. $\qquad\square$

Theorem 10.23. *If there exists a generalized quadrangle of order (s,t), then*

$$s + t \mid st(s+1)(t+1).$$

Proof. Suppose that $\mathcal{S}$ is a generalized quadrangle of order (s,t). By Lemma 10.21, $\mathcal{S}$ contains $v = (s+1)(st+1)$ points. Let $P_1, P_2, \ldots, P_v$ denote the points of $\mathcal{S}$ and $A = (a_{ij})$ denote the $v \times v$ matrix over $\mathbb{R}$ defined as

$$a_{ij} = \begin{cases} 1, & \text{if } P_i \sim P_j \text{ and } i \neq j, \\ 0, & \text{otherwise.} \end{cases}$$

The $0-1$ matrix A is called an adjacency matrix of $\mathcal{S}$. Let $A^2 = (c_{ij})$. Then c_{ij} equals the number of points collinear with both P_i and P_j but distinct from both of them, hence

$$c_{ij} = \begin{cases} (t+1)s, & \text{if } i = j; \\ s-1, & \text{if } i \neq j \text{ and } P_i \sim P_j; \\ t+1, & \text{if } P_i \not\sim P_j. \end{cases}$$

Thus the adjacency matrix satisfies the equation

$$A^2 = (s-t-2)A + (t+1)(s-1)I + (t+1)J \qquad (10.1)$$

where I and J are the $v \times v$ identity and all 1 matrices, respectively.

Rearranging Equation (10.1) we get

$$A^2 - (s - t - 2)A - (t+1)(s-1)I = (t+1)J.$$

It is well-known that the matrix J has two eigenvalues, v with multiplicity 1 and 0 with multiplicity $(v - 1)$. The eigenvector belonging to v is $\mathbf{1} = (1, 1, \ldots, 1)$. As each point of $\mathcal{S}$ is collinear with $(t+1)s$ points, we get $\mathbf{1}A = (t+1)s\mathbf{1}$, so $(t+1)s$ is an eigenvalue of A. Since

$$\left((t+1)s\right)^2 - (s-t-2)(t+1)s - (t+1)(s-1) = (t+1)v,$$

this eigenvalue of A corresponds to the eigenvalue v of J, its multiplicity is 1. All other eigenvalues of A correspond to the eigenvalue 0 of J, so they are roots of the equation

$$\lambda^2 - (s-t-2)\lambda - (t+1)(s-1) = 0.$$

The two roots of this quadratic equation are $\lambda_1 = s - 1$ and $\lambda_2 = -t - 1$. Let m_i denote the multiplicity of λ_i for $i = 1, 2$. Then

$$1 + m_1 + m_2 = v$$

and

$$(t+1)s + m_1(s-1) + m_2(-t-1) = \operatorname{tr}A = 0.$$

Solving this set of equations we get

$$m_1 = \frac{st(s+1)(t+1)}{s+t}, \qquad m_2 = \frac{(st+1)s^2}{s+t}.$$

Each multiplicity must be an integer, thus $s+t$ divides $st(s+1)(t+1)$ proving the statement.

Let us remark that the other divisibility condition, $(s+t) \mid (st+1)s^2$, is equivalent with the one we just proved. $\square$

Theorem 10.24 (Higman's inequality). *If there exists a generalized quadrangle of order (s,t) and $s > 1$ and $t > 1$, then $s \leq t^2$ and $t \leq s^2$.*

Proof. By duality, it is enough to prove that $t \leq s^2$. Let $\mathcal{S}$ be a generalized quadrangle of order (s, t) and P and P' be two non-collinear points in $\mathcal{S}$. Let the set of points $\mathcal{R}$ be definied as

$$\mathcal{R} = \{T \in \mathcal{P} : T \nsim P \text{ and } T \nsim P'\}.$$

Then, by Lemmas 10.21 and 10.22,

$$|\mathcal{R}| = (s+1)(st+1) - 2 - 2(t+1)s + t + 1 = d.$$

Let $R_1, R_2, \ldots, R_d$ denote the points in $\mathcal{R}$. For $i = 1, 2, \ldots, d$, define the numbers m_i as

$$m_i = |\{T \in \{P, P'\}^{\perp} : T \sim R_i\}|.$$

The standard double countings of the ordered pairs $(R_i, T) \in \mathcal{R} \times \{P, P'\}^\perp$ for which $T \sim R_i$, and the ordered triples $(R_i, T, T') \in \mathcal{R} \times \{P, P'\}^\perp \times \{P, P'\}^\perp$ for which $T \neq T'$ and $T \sim R_i \sim T'$, give

$$\sum_{i=1}^{d} m_i = (t+1)(t-1)s \quad \text{and} \quad \sum_{i=1}^{d} m_i(m_i-1) = (t+1)t(t-1).$$

Summing these two equations we get

$$\sum_{i=1}^{d} m_i^2 = (t+1)(t-1)(s+t).$$

According to the inequality between the arithmetic and quadratic means these imply

$$d \sum_{i=1}^{d} m_i^2 \geq \left(\sum_{i=1}^{d} m_i \right)^2,$$

hence

$$d(t+1)(t-1)(s+t) \geq (t+1)^2(t-1)^2 s^2.$$

After rearranging it

$$t(s-1)(s^2-t) \geq 0$$

follows, which proves the theorem because $s > 1$. $\qquad\square$

A stronger version of Higman's inequality can be proven if we combine the previous two theorems.

Theorem 10.25. *If $s > 1$, $t > 1$, $s \neq t^2$, $t \neq s^2$ and there exists a generalized quadrangle of order (s,t), then $s = t^2 - t$, $s = t^2 - t - 1$ or $s \leq t^2 - 2t$, and $t = s^2 - s$, $t = s^2 - s - 1$ or $t \leq s^2 - 2s$.*

Proof. Again, by duality, it is enough to prove the statements about t and s^2. Because of the Higman inequality there exists a positive integer e such that $t = s^2 - e$. Then, by Theorem 10.23, $s + s^2 - e$ divides $s(s^2 - e)(s+1)(s^2 - e + 1)$. Elementary calculation shows that

$$s(s^2 - e)(s+1)(s^2 - e + 1) \equiv (s^2 - e)(s^2 + s)(s^2 - e + 1)$$
$$\equiv (-s)e(-s+1)$$
$$\equiv e(e - 2s) \pmod{s + s^2 - e},$$

hence

$$s + s^2 - e \mid e(e - 2s). \tag{10.2}$$

If $e \geq 2s$, then we are done. If $e < 2s$, then the divisibility condition (10.2) implies

$$s + s^2 - e \leq e(2s - e) = -e^2 + 2se.$$

The solution of this quadratic inequality gives $s \leq e \leq s+1$. As e is an integer, the proof is complete. $\qquad\square$

Regarding generalized quadrangles with small parameters and considering the case $s \leq t$, Theorems 10.23 and 10.25 give several non-existence results.

Corollary 10.26. *If a generalized quadrangle of order (s, t) exists and*

- $s = 2$, *then* $t \in \{2, 4\}$,

- $s = 3$, *then* $t \in \{3, 5, 6, 9\}$,

- $s = 4$, *then* $t \in \{4, 6, 8, 11, 12, 16\}$,

- $s = 5$, *then* $t \in \{5, 7, 10, 15, 19, 20, 25\}$.

This corollary gives only necessary but not sufficient conditions for the existence. For example, it is known that there is no generalized quadrangle of order $(3, 6)$. For more details, we refer the reader to [140, Chapter 6].

There exist several isomorphisms among the known GQ's. In the next paragraphs we consider them and we also prove some properties of $\mathcal{W}(q)$ using the Klein correspondence (see Chapter 5).

Theorem 10.27. *The dual of $\mathcal{W}(q)$ is isomorphic to $\mathcal{Q}(4, q)$.*

Proof. The lines of $\mathcal{W}(q)$ form a general linear complex, hence their image under the Klein correspondence is the intersection of $\mathcal{H}_5$ with a non-tangent hyperplane. As the hyperplane is isomorphic to $\mathrm{PG}(4, q)$, the intersection is a 4-dimensional non-singular quadric $\mathcal{P}_4$. The lines of $\mathcal{W}(q)$ which are incident with a given point, form a pencil in $\mathrm{PG}(3, q)$, hence their image is a line on $\mathcal{P}_4$. As the points of $\mathcal{P}_4$ and lines contained in $\mathcal{P}_4$ form $\mathcal{Q}(4, q)$, the theorem is proved. □

The third known class of GQ's of order (q, q) is $\mathcal{T}_2(\mathcal{O})$. The proof of the following theorem can be found in [140].

Theorem 10.28. *$\mathcal{T}_2(\mathcal{O})$ is isomorphic to $\mathcal{Q}(4, q)$ if and only if $\mathcal{O}$ is a conic.*

The proofs of the corresponding theorems about GQ's of order (q, q^2) can also be found in [140].

Theorem 10.29.

- *The dual of $\mathcal{U}(3, q^2)$ is isomorphic to $\mathcal{Q}(5, q)$.*

- *$\mathcal{T}_3(\mathcal{O})$ is isomorphic to $\mathcal{Q}(5, q)$ if and only if $\mathcal{O}$ is an elliptic quadric.*

A polarity of a GQ of order (s, s) can be defined in the natural way.

Definition 10.30. *Let $\mathcal{S}$ be a GQ of order (s, s) and $\mathcal{S}^*$ be its dual GQ. A collineation $\varphi : \mathcal{S} \to \mathcal{S}^*$ interchanges the points and lines of $\mathcal{S}$, hence it can be considered as a collineation $\mathcal{S}^* \to \mathcal{S}$, too. If φ^2 considering in this form is the identity, then φ is called a polarity. A point or a line x of $\mathcal{S}$ is self-conjugate if $x \, \mathrm{I} \, x^\varphi$.*

Theorem 10.31. *Let $q = 2^h$. Then $\mathcal{W}(q)$ is self-dual. If h is odd, then $\mathcal{W}(q)$ admits a polarity.*

Proof. We may assume without loss of generality that $\mathcal{W}(q)$ is defined by the null polarity of $\mathrm{PG}(3, q)$ given by the matrix

$$A = \begin{pmatrix} 0 & 0 & 1 & 0 \\ 0 & 0 & 0 & 1 \\ 1 & 0 & 0 & 0 \\ 0 & 1 & 0 & 0 \end{pmatrix}.$$

Then, by Proposition 5.9 and Definition 5.8, a line in $\mathrm{PG}(3, q)$ with Plücker coordinates $(p_{01} : p_{02} : p_{03} : p_{23} : p_{31} : p_{12})$ is a line of $\mathcal{W}(q)$ if and only if $p_{02} + p_{31} = 0$. Applying Corollary 5.3 this gives $p_{01}p_{23} + p_{02}^2 + p_{03}p_{12} = 0$. Hence the image of the set of lines of $\mathcal{W}(q)$ under the Klein correspondence is the parabolic quadric $\mathcal{P}_4$ with equation

$$X_0 X_3 + X_1^2 + X_2 X_4 = 0.$$

The points and lines of this quadric correspond to the points and lines of $\mathcal{Q}(4, q)$.

Consider the mapping

$$\tau : \mathcal{P}_4 \to \mathrm{PG}(3, q), \quad (x_0 : x_1 : x_2 : x_3 : x_4) \mapsto (x_0 : x_2 : x_3 : x_4).$$

By Theorem 4.53, the point $(0 : 1 : 0 : 0 : 0)$ is the nucleus of $\mathcal{P}_4$, thus τ is a bijection. The line joining the points $Y = (y_0 : y_1 : y_2 : y_3 : y_4)$ and $Z = (z_0 : z_1 : z_2 : z_3 : z_4)$ of $\mathcal{P}_4$, is contained in $\mathcal{P}_4$ if and only if the two points are conjugates with respect to $\mathcal{P}_4$. Thus Y and Z are collinear in $\mathcal{Q}(4, q)$ if and only if $G(\mathbf{y}, \mathbf{z}) = 0$. As q is even,

$$\begin{aligned} G(\mathbf{y}, \mathbf{z}) &= (y_1 + z_1)^2 + (y_0 + z_0)(y_3 + z_3) + (y_2 + z_2)(y_4 + z_4) \\ &\quad - (y_1^2 + y_0 y_3 + y_2 y_4) - (z_1^2 + z_0 z_3 + z_2 z_4) \\ &= y_0 z_3 + y_3 z_0 + y_2 z_4 + y_4 z_2. \end{aligned}$$

So $G(\mathbf{y}, \mathbf{z}) = 0$ if and only if Y^τ and Z^τ are conjugates with respect to the null polarity defined by A, hence if and only if Y^τ and Z^τ are collinear points in $\mathcal{W}(q)$. This means that τ gives an isomorphism between $\mathcal{Q}(4, q)$ and $\mathcal{W}(q)$. By Theorem 10.27, the dual of $\mathcal{W}(q)$ is isomorphic to $\mathcal{Q}(4, q)$ for all q, hence $\mathcal{W}(q)$ is self-dual if q is even.

Now, suppose that h is odd. Let ρ be the isomorphism given in Theorem 10.27. We calculate the action of ρ^2 on the points of $\mathcal{W}(q)$ using coordinates. Let $Y = (y_0 : y_2 : y_3 : y_4)$ be an arbitrary point of $\mathcal{W}(q)$. First, suppose that $y_4 \neq 0$. Consider the points $V = (y_4 : y_3 : 0 : 0)$ and $W = (0 : y_0 : y_4 : 0)$. Then $\mathbf{y} A \mathbf{v}^{\mathrm{T}} = \mathbf{y} A \mathbf{v}^{\mathrm{T}} = 0$, so YV and YW are lines of $\mathcal{W}(q)$. Their Plücker coordinates are $(y_0 y_3 + y_2 y_4 : y_3 y_4 : y_4^2 : 0 : y_3 y_4 : y_3^2)$ and $(y_0^2 : y_0 y_4 : 0 : y_4^2 : y_0 y_4 : y_0 y_3 + y_2 y_4)$, respectively. We have $y_0 y_3 + y_2 y_4 = y_1^2$, so the τ images of these lines are the points $(y_1^2 : y_4^2 : 0 : y_3^2)$ and $(y_0^2 : 0 : y_4^2 : y_1^2)$. Thus Y^ρ has Plücker coordinates

$$(y_0^2 y_4^2 : y_1^2 y_4^2 : y_1^4 + y_0^2 y_3^2 : y_3^2 y_4^2 : y_1^2 y_4^2 : y_4^4).$$

Applying $y_1^4 + y_0^2 y_3^2 = y_2^2 y_4^2$ and using the homogenity, the coordinates of the line can be written as

$$(y_0^2 : y_1^2 : y_2^2 : y_3^2 : y_1^2 : y_4^2).$$

This means that the τ-image of Y, that is the same as Y^{ρ^2}, is the point $(y_0^2 : y_2^2 : y_3^2 : y_4^2)$. This point is different from Y, hence ρ itself is not a polarity, but after a slight modification it becomes a polarity. As h is odd, $\sigma : x \mapsto x^{2^{(h+1)/2}}$ is an automorphism of $\mathrm{GF}(2^h)$. This automorphism obviously commutes with both τ and ρ. Hence if $\pi = \rho\sigma$, then $Y^{\pi^2} = Y$ for all points whose last homogeneous coordinate is not 0. This implies that π^2 fixes each point of an affine space, hence it fixes each point of the projective closure of this affine space, too. Thus π is a polarity of $\mathcal{W}(q)$. $\qquad\square$

There is a general principle formulated by Tits about the abstract automorphisms of structures embedded into projective spaces: the automorphisms are those collineations of the space that preserve the structure setwise. In the case of $q = 2^h$, h even, this implies at once that $\mathcal{W}(q)$ is not self-dual, because in $\mathrm{PG}(3,q)$ there is no collineation whose action is the same as taking squares coordinatewise for each point.

For the sake of completeness we mention the q odd case, but we omit the proof (see in [140]) of the following statement.

Theorem 10.32. *If q is odd, then $\mathcal{W}(q)$ is not self-dual.*

The notion of a line spread of a projective space can be extended to GQ's in the natural way. We introduce its dual object, too.

Definition 10.33. *Let S be a generalized quadrangle. A* spread *of S is a set $\mathcal{R}$ of its lines such that each point of S is incident with a unique line of $\mathcal{R}$.*

An ovoid *of S is a set $\mathcal{O}$ of its points such that each line of S is incident with a unique point of $\mathcal{O}$.*

Proposition 10.34. *Let S be a GQ of order (s, t). Then each ovoid of S consists of $st + 1$ points and each spread of S contains $st + 1$ lines.*

Proof. There are $t + 1$ lines through each point of S. Hence if an ovoid consists of k points, then, by Lemma 10.21, $(t + 1)(st + 1) = k(t + 1)$, so $k = st + 1$. The number of lines in a spread follows by duality. $\qquad\square$

Let us remark that the above proposition gives only necessary conditions; there are GQ's that do not contain any spread or ovoid. The existence problem of these objects is a hard question in general. The next theorem shows that one can construct an ovoid if a GQ admits a polarity.

Theorem 10.35. *Let* $\mathcal{S}$ *be a GQ of order* (t, t). *If* $\mathcal{S}$ *admits a polarity* π, *then the self-conjugate points of* π *form an ovoid.*

Proof. We claim that each line of $\mathcal{S}$ contains exactly one self-conjugate point of π. Let ℓ be a line and A denote the point ℓ^π.

First, suppose that A is on ℓ. If another self-conjugate point, say B, were on ℓ, then $B \operatorname{I} A^\pi$ would imply $A \operatorname{I} B^\pi$, hence B^π would also be the line $AB = \ell$, giving a contradiction. Thus ℓ contains exactly one self-conjugate point in this case.

If A is not on ℓ, then there is a unique point, say B, on ℓ which is collinear with A. As A and B are collinear, the lines $A^\pi = \ell$ and $B^\pi = \ell_1$ intersect each other. On the other hand, $B \operatorname{I} \ell$ implies $B^\pi \operatorname{I} \ell^\pi$, thus ℓ_1 contains the point A. As there is a unique line through A which intersects ℓ, we get $\ell_1 = AB$, so B is self-conjugate, hence ℓ contains at least one self-conjugate point in this case, too. Now, suppose that ℓ contains another self-conjugate point, say C, and let $C^\pi = \ell_2$. The points B and C are collinear, hence the lines ℓ_1 and ℓ_2 intersect in a point D. This means that B, C and D form a triangle, which is a contradiction again. $\qquad\square$

Corollary 10.36. *Let* $q = 2^h$. *If h is odd, then* $\mathcal{W}(q)$ *has an ovoid.*

Proof. By Theorem 10.31, $\mathcal{W}(q)$ admits a polarity whose self-conjugate points form an ovoid. $\qquad\square$

We have seen in Chapter 9 that for q even each ovoid in $\mathrm{PG}(3, q)$ induces a null polarity (Theorem 9.14). The null polarities are projectively equivalent and each of them defines $\mathcal{W}(q)$. Let $\mathcal{O}$ be an ovoid in $\mathrm{PG}(3, q)$ which defines $\mathcal{W}(q)$. Then the pole of a tangent plane to $\mathcal{O}$ is its point of contact, thus the set of points of $\mathcal{O}$ is an ovoid in $\mathcal{W}(q)$, too. The converse is also true.

Theorem 10.37. *Let* $\mathcal{O}$ *be an ovoid in* $\mathcal{W}(q)$. *Then the set of points of* $\mathcal{O}$ *is an ovoid in* $\mathrm{PG}(3, q)$, *too.*

Proof. Suppose to the contrary that in $\mathrm{PG}(3, q)$ a line ℓ meets the set of points of $\mathcal{O}$ in $m > 2$ points. For $i = 1, 2, \ldots, q+1$, let σ_i denote the planes through ℓ, R_i denote the pole of σ_i in the null polarity which defines $\mathcal{W}(q)$ and $\mathcal{P}_i$ denote the pencil in σ_i with carrier R_i. This means that the lines of $\mathcal{W}(q)$ are those lines of $\mathrm{PG}(3, q)$ which belong to one of the pencils $\mathcal{P}_i$. Each of these lines meets $\mathcal{O}$ in exactly 1 point, because $\mathcal{O}$ is an ovoid in $\mathcal{W}(q)$, so σ_i meets $\mathcal{O}$ in at most $q + 1$ points for $i = 1, 2, \ldots, q+1$. As ℓ contains m points of $\mathcal{O}$ we get

$$|\mathcal{O}| \le m + (q + 1)(q + 1 - m) \le (q + 1)^2 - 3q < q^2 + 1,$$

giving a contradiction. $\qquad\square$

In Theorem 10.31 we explicitly calculated a polarity of $\mathcal{W}(q)$ if $q = 2^h$ and h is odd. A long, but straightforward calculation gives the homogeneous coordinates of the self-conjugate points in PG$(3, q)$. By Theorem 10.37, this is an ovoid in PG$(3, q)$, too. This class of ovoids is called the *Suzuki-Tits ovoids*, and we gave their analytic description in Theorem 9.15.

Finally, we present some results about the existence of ovoids and spreads in GQ's. We omit the proofs; they can be found in the book of *Payne* and *J. Thas* [140].

Theorem 10.38.

- $\mathcal{Q}(4, q)$ *has ovoids for all* q*, and it has spreads if and only if* q *is even.*

- $\mathcal{Q}(5, q)$ *has spreads but it has no ovoid.*

- $\mathcal{U}(4, q^2)$ *has no ovoid.*

- $\mathcal{T}_2(\mathcal{O})$ *always has an ovoid.*

- $\mathcal{T}_3(\mathcal{O})$ *always has spreads but it has no ovoid.*

Most of the known constructions for generalized hexagons and octagons are based on group theoretical results. Due to the page limitations of our book, we ignore their study. For the detailed description of these objects we refer the reader to the book of *Van Maldeghem* [177]. We only mention an example which was introduced by *Tits* [172], and *Payne* [137] gave its geometric description as follows.

Example 10.39. Let q be odd and $\mathcal{Q}_6$ be a non-singular parabolic quadric in PG$(6, q)$ with equation $X_0 X_4 + X_1 X_5 + X_2 X_6 - X_3^2 = 0$. Then the *split Cayley hexagon* $H(q)$ is the following:

- the points of $H(q)$ are all points of $\mathcal{Q}_6$;

- the lines of $H(q)$ are the lines on $\mathcal{Q}_6$ whose Grassmannian coordinates $(p_{01}, p_{02}, ..., p_{56})$ satisfy the six relations $p_{12} = p_{34}$, $p_{56} = p_{03}$, $p_{45} = p_{23}$, $p_{01} = p_{36}$, $p_{02} = -p_{35}$ and $p_{46} = -p_{13}$;

- the incidence is the same as in PG$(6, q)$.

Up to duality $H(q)$ is the only known generalized hexagon of order (q, q).

In the rest of this chapter Möbius planes are considered. These structures generalize the incidence properties of points and circles of classical spherical geometry.

Definition 10.40. *An incidence geometry* $\mathcal{M} = (\mathcal{P}, \mathcal{C}, \mathrm{I})$ *is called a* Möbius plane *if it satisfies the following three axioms:*

- **M1.** *For any three distinct elements of* $\mathcal{P}$ *there is precisely one element of* $\mathcal{C}$*, which is in relation* I *with all of them.*

- **M2.** *For each ordered triple* $(P, P', c) \subset \mathcal{P} \times \mathcal{P} \times \mathcal{C}$ *satisfying* $P\,\mathrm{I}\,c$ *and* P' *is not in relation* I *with* c, *there is precisely one element* $c' \in \mathcal{C}$ *such that* $P\,\mathrm{I}\,c'\,\mathrm{I}\,P'$ *and there is no* $P \neq Q \in \mathcal{P}$ *for which* $c\,\mathrm{I}\,Q\,\mathrm{I}\,c'$.

- **M3.** *There exist four distinct elements of* $\mathcal{P}$ *such that no element of* $\mathcal{C}$ *is in relation* I *with each of them.*

As in the case of abstract projective and affine planes, the usual geometric terminology will also be used in the case of abstract Möbius planes, so we shall talk about the circle through three points, the points of intersection of two circles and we will say that two circles are tangent to each other if they have exactly one point in common. Points will be denoted by upper case Latin letters, circles by lower case Latin letters. For example, axiom M2 means the following: if c is a circle through the point P and the point P' is not on c, then there is a unique tangent circle to c at P which contains P'.

The notions of *automorphism* and *isomorphism* for Möbius planes are defined in the usual way.

In the 3-dimensional classical Euclidean space the points of a sphere $\mathcal{S}$ and the circles on $\mathcal{S}$ form a Möbius plane. Our next example is the natural generalization of this model.

Example 10.41. Let $\mathcal{O}$ be an ovoid in $\mathrm{PG}(3, q)$. Let $\mathcal{P}$ be the set of points of $\mathcal{O}$, $\mathcal{C}$ be the set of planes of $\mathrm{PG}(3, q)$ meeting $\mathcal{O}$ in $q + 1$ points, and define the incidence relation I as the set theoretical containment relation. Then $\mathcal{M} = (\mathcal{P}, \mathcal{C}, \mathrm{I})$ is a Möbius plane.

These planes are called *egglike Möbius planes*.

By definition, no three points of $\mathcal{O}$ are collinear. Hence there is a unique plane in $\mathrm{PG}(3, q)$ through any three distinct points of $\mathcal{O}$. If a plane contains more than one point of $\mathcal{O}$, then it intersects $\mathcal{O}$ in a $(q + 1)$-arc, thus axiom M1 is satisfied.

If $c \in \mathcal{C}$ is a circle, then let Π_c denote the corresponding plane in $\mathrm{PG}(3, q)$, while if $P \in \mathcal{P}$ is a point, then let Σ_P denote the tangent plane to $\mathcal{O}$ at P in $\mathrm{PG}(3, q)$. Suppose that an ordered triple $(P, P', c) \subset \mathcal{P} \times \mathcal{P} \times \mathcal{C}$ with $P\,\mathrm{I}\,c$ and P' is not in relation I with c is given. Let ℓ denote the line $\Sigma_P \cap \Pi_c$ in $\mathrm{PG}(3, q)$. A circle c' contains P in $\mathcal{M}$ if and only if in $\mathrm{PG}(3, q)$ the line $\ell_{c'} := \Sigma_P \cap \Pi_{c'}$ contains P. We claim that c' is a tangent circle to c at P in $\mathcal{M}$ if and only if $\ell = \ell_{c'}$ in $\mathrm{PG}(3, q)$. If $\ell = \ell_{c'}$, then $(\Pi_c \cap \Pi_{c'}) \subset \Sigma_P$, hence $\mathcal{O}$ contains exactly one point of the intersection in $\mathrm{PG}(3, q)$, so the two circles are tangent to each other in $\mathcal{M}$. If ℓ and $\ell_{c'}$ are distinct lines then the line $\Pi_c \cap \Pi_{c'}$ is a secant to $\mathcal{O}$ in $\mathrm{PG}(3, q)$, because it contains P but it is not contained in Σ_P. This means that if R is the other point in the set $\Pi_c \cap \Pi_{c'} \cap \mathcal{O}$, then both circles contain R in $\mathcal{M}$. As P' is not on the line ℓ, there is a unique plane $\Pi' = \langle P', \ell \rangle$ in $\mathrm{PG}(3, q)$, and $\Pi' \cap \mathcal{O}$ defines the unique circle satisfying axiom M2 in $\mathcal{M}$.

Finally, axiom M3 obviously holds in this model.

The notion of an ovoid can be extended to arbitrary projective spaces in the following way.

Definition 10.42. *A set of points $\mathcal{O}$ is an* ovoid *in* $\mathrm{PG}(3, \mathbb{K})$ *if*

- *no three points of $\mathcal{O}$ are collinear,*

- *for each point $P \in \mathcal{O}$ the set of tangent lines to $\mathcal{O}$ at P coincide with the set of all elements of a pencil with carrier P.*

The plane formed by the tangent lines to $\mathcal{O}$ at P is called the tangent plane to $\mathcal{O}$ at P.

It is easy to see that the construction described in Example 10.41 produces an infinite Möbius plane if $\mathcal{O}$ is an ovoid in a projective space over an infinite field $\mathbb{K}$. Thus there exist infinite egglike planes, too.

Our next example gives an algebraic construction.

Example 10.43. Let $\mathbb{K} = \mathrm{GF}(q)$, $\mathbb{F} = \mathrm{GF}(q^2)$ and $\mathcal{E} = \mathrm{PG}(1, q^2)$ be the projective line over $\mathbb{F}$. Then the group $G = \mathrm{PGL}(2, q^2)$ acts on the set of points of $\mathcal{E}$. This action can be described in the following way. The homogeneous coordinates of the points of $\mathcal{E}$ can be chosen such that the set of points is

$$\{(1 : x) : \ x \in \mathbb{F}\} \cup \{(0 : 1)\}.$$

Thus the points can be identified with the elements of the set

$$\mathcal{P} = \mathbb{F} \cup \{\infty\}.$$

If a point has homogeneous coordinates $(x_0 : x_1)$ and $A = (a_{ij})$ is a 2×2 non-singular matrix, then the image of the point has homogeneous coordinates $(x_0 : x_1)A$. Hence the formula

$$x \mapsto \frac{a_{01} + a_{11}x}{a_{00} + a_{10}x}$$

gives the action of A on the set $\mathcal{P}$ using the calculation rules $x + \infty = \infty$, $x \cdot \infty = \infty$ if $x \neq 0$ and $0 \cdot \infty = 1$. For an arbitrary element $\sigma \in G$, let

$$(\mathbb{K} \cup \{\infty\})^\sigma : \ = \{k^\sigma : \ k \in \mathbb{K}\} \cup \{\infty^\sigma\},$$

and

$$\mathcal{C} : \ = \{(\mathbb{K} \cup \{\infty\})^\sigma : \ \sigma \in G\}.$$

Let us define the incidence relation I as the set theoretical containment relation. Then the triple $(\mathcal{P}, \mathcal{C}, \mathrm{I})$ is a Möbius plane.

Let H denote the subgroup $\mathrm{PGL}(2, q) \subset G$. Then two elements σ_1 and σ_2 of G define the same circle if and only if they are in the same coset of H. By Theorem 4.16, $\mathrm{PGL}(2, q^2)$ acts sharply 3-transitively on the points of $\mathcal{E}$ and

H acts sharply 3-transitively on the points of $PG(1, q)$. On the one hand, this means that each circle is uniquely determined by the images of the points 0, 1 and ∞, so axiom M1 is satisfied. On the other hand, the group is transitive on the incident point-circle pairs, hence it is enough to prove axiom M2 in the case when the circle c is the set $\mathbb{K} \cup \{\infty\}$ and P is the point ∞. In this case P' corresponds to an element $x' \in (\mathbb{F} \setminus \mathbb{K})$. The additive group of $\mathbb{K}$, denoted by $\mathbb{K}_+$, is a subgroup of the additive group of $\mathbb{F}$, thus x' belongs to a unique coset of $\mathbb{K}_+$. Hence there is a unique $t \in (\mathbb{F} \setminus \mathbb{K})$ such that $x' \in \mathbb{K}_+ + t$. Let $\sigma \in G$ be defined as

$$x \mapsto x + t.$$

We claim that $c' = (\mathbb{K} \cup \{\infty\})^\sigma$ is the unique circle satisfying the requirements of axiom M2. By definition, $c' = \{\mathbb{K}_+ + t\} \cup \{\infty\}$, so it is a tangent circle to c at ∞. Let $c_1 = (\mathbb{K} \cup \{\infty\})^{\sigma_1}$ be any circle through ∞. Because of the 3-transitivity of H, we may assume that $\infty^{\sigma_1} = \infty$. Let $0^{\sigma_1} = s$ and $1^{\sigma_1} = r$. Then we can calculate the entries of the corresponding matrix and we get

$$x^{\sigma_1} = s + (r - s)x.$$

Hence if $x_1 \neq x_2$ elements of $\mathbb{K}$ and $x_1^{\sigma_1}$ and $x_2^{\sigma_1}$ are in the same coset, then $r - s \in \mathbb{K}$. So $\mathbb{K}^{\sigma_1}$ either coincides with a coset of $\mathbb{K}_+$ or intersects each of the q cosets in exactly 1 element. In the former case, as P' is on c', the coset must be $\mathbb{K}_+ + t$ which implies $c' = c$; in the latter case, $\mathbb{K}^{\sigma_1}$ contains exactly 1 element of $\mathbb{K}$, hence c' intersects c not only in ∞, but in one more point, so it is not a tangent circle to c.

Finally, $q \geq 2$ implies axiom M3 at once.

The following observation plays a crucial role in the proof of the basic combinatorial properties of finite Möbius planes.

Lemma 10.44. *Let P be a fixed point of the Möbius plane $\mathcal{M} = (\mathcal{P}, \mathcal{C}, I)$. Let $\mathcal{C}_P \subset \mathcal{C}$ denote the set of circles through P. Then $\mathcal{M}_P = (\mathcal{P} \setminus \{P\}, \mathcal{C}_P, I)$ is an affine plane.*

This affine plane is called the derived plane *of $\mathcal{M}$ with respect to P.*

Proof. We claim that $\mathcal{M}_P$ satisfies axioms A1–A4 of Definition 1.19.

If A and B are two distinct points in $\mathcal{P} \setminus \{P\}$, then, by axiom M1, there is a unique circle $c \in \mathcal{C}$ through the three points A, B and P. As c contains P, we have $c \in \mathcal{C}_P$. Thus $\mathcal{M}_P$ satisfies axiom A1.

If a circle $c \in \mathcal{C}_P$ and a point A not on c are given, then $P I c$ thus, by axiom M2, there is a unique circle c' through A which is a tangent to c at the point P. Hence $c \cap c' \cap (\mathcal{P} \setminus \{P\}) = \emptyset$, so $\mathcal{M}_P$ satisfies axiom A2, too.

Axioms A3 and A4 obviously follow from axiom M3. $\square$

Theorem 10.45. *If there is a circle which is incident with $n + 1$ points in a finite Möbius plane $\mathcal{M}$, then*

- *$\mathcal{M}$ contains $n^2 + 1$ points,*

- *every circle of $\mathcal{M}$ is incident with $n+1$ points,*

- *every point of $\mathcal{M}$ is incident with $n(n+1)$ circles,*

- *$\mathcal{M}$ contains $n^3 + n$ circles.*

The number n is called the order *of the Möbius plane.*

Proof. Let c be a circle of the Möbius plane $\mathcal{M} = (\mathcal{P}, \mathcal{C}, I)$ containing exactly $n+1$ points. Then for each point $P I c$, the derived affine plane $\mathcal{M}_P$ has order n. Thus, by Theorem 1.21, there are $n(n+1)$ circles through each point $P I c$, and if a circle c' has non-empty intersection with c, then c' contains $n+1$ points, because any line of an affine plane of order n contains n points.

As $\mathcal{M}$ contains one more point than $\mathcal{M}_P$ and, by Theorem 1.21, $\mathcal{M}_P$ has n^2 points, we get that $\mathcal{M}$ contains $n^2 + 1$ points.

Now, let d be a circle which has empty intersection with c. Take two points, $R I d$ and $P I c$, and let c' be any circle through P and R. As c' has $n+1$ points, the derived affine plane $\mathcal{M}_R$ has order n which implies that d contains $n+1$ points, too. This proves the second statement of the theorem.

If R is an arbitrary point not on c, and we choose c' through the points R and $P I c$, then c' contains $n+1$ points. Thus the derived affine plane $\mathcal{M}_R$ has order n which implies that there are $n(n+1)$ circles through R, proving the third statement of the theorem.

Finally, double counting of incident point-circle pairs gives

$$(n^2 + 1) \cdot \left(n(n+1)\right) = x \cdot (n+1),$$

where x denotes the total number of circles of the plane. This gives $x = n^3 + n$ completing the proof. □

Corollary 10.46. *Let P and R be two distinct points and c be a circle through P in a Möbius plane $\mathcal{M}$ of order n. Then*

1. *there are $n+1$ circles through P and R;*

2. *there are n mutually tangent circles through P such that one of them is c.*

These $n+1$ or n circles give a partition of the points of $\mathcal{M}$ distinct from P and R, or P, respectively.

In the classical Euclidean plane there are three types of pencils of circles: hyperbolic, parabolic and elliptic. Each of these circle sets can be defined in Möbius planes, but the definition of a pencil of elliptic type is much more complicated. As we do not need this type later in this chapter for our characterization theorems, we only give the definitions of the first and second types.

Definition 10.47. *The set of $n+1$ circles through two distinct points P and R of a Möbius plane of order n is called a* hyperbolic pencil (or bundle) *of circles with carriers P and R.*

The set of n mutually tangent circles through a fixed point P of a Möbius plane of order n is called a parabolic pencil *of circles with carrier P.*

The next statement is a direct consequence of axioms M1 and M2.

Proposition 10.48. *Let c_1 and c_2 be two circles with non-empty intersection in a Möbius plane. Then there is a unique pencil of circles containing both c_1 and c_2. The pencil is hyperbolic if $c_1 \cap c_2$ consists of two points, and it is parabolic if $c_1 \cap c_2$ contains a single point.*

An egglike Möbius plane $\mathcal{M}$ arising from an ovoid $\mathcal{O}$ in $\mathrm{PG}(3,q)$ has order q. Let ℓ be a line in $\mathrm{PG}(3,q)$ and $\{\Pi_1, \Pi_2, \ldots, \Pi_{q+1}\}$ be the set of $q+1$ planes through ℓ. It is obvious from the geometric properties of $\mathcal{O}$ that if ℓ is a secant to $\mathcal{O}$ and $\ell \cap \mathcal{O} = \{P, R\}$ then the set of circles $\{\Pi_i \cap \mathcal{O} : i = 1, 2, \ldots, q+1\}$ is a hyperbolic pencil of circles with carriers P and R, while if ℓ is a tangent line to $\mathcal{O}$ at P and Π_1 is the tangent plane to $\mathcal{O}$ at P, then $\{\Pi_i \cap \mathcal{O} : i = 2, 3, \ldots, q+1\}$ is a parabolic pencil of circles with carrier P. If ℓ is an external line to $\mathcal{O}$ and Π_1 and Π_2 are the two tangent planes to $\mathcal{O}$ through ℓ, then the set of $q-1$ circles $\{\Pi_i \cap \mathcal{O} : i = 3, 4, \ldots, q+1\}$ gives a partition of all but two points of $\mathcal{M}$. These circles give an elliptic pencil of circles determined by ℓ and the two distinguished points are the carriers.

There are two famous configurational theorems in classical circle geometries. To formulate them we need the following definition.

Definition 10.49. *A* 4-chain of circles *is an ordered quadruple (c_1, c_2, c_3, c_4) of circles such that no three of them have a point in common, but $c_i \cap c_{i+i} \neq \emptyset$ when the subscripts are taken modulo 4.*

Let (c_1, c_2, c_3, c_4) be a 4-chain of circles and for $i = 1, 2, 3, 4$ let $c_i \cap c_{i+i} = \{A_i, B_i\}$. Then A_1, A_2, A_3 and A_4 are four distinct points, but $A_i = B_i$ could happen for some i.

Theorem 10.50 (Bundle Theorem). *For any 4-chain of circles (c_1, c_2, c_3, c_4) the two pencils of circles determined by $c_1 \cap c_2$ and $c_3 \cap c_4$ have a common circle if and only if the two pencils of circles determined by $c_1 \cap c_4$ and $c_2 \cap c_3$ have a common circle (see Figure 10.3).*

Theorem 10.51 (Theorem of Miquel). *For any 4-chain of circles (c_1, c_2, c_3, c_4) the points A_1, A_2, A_3, A_4 are cocircular if and only if the points B_1, B_2, B_3, B_4 are cocircular (see Figure 10.4).*

A Möbius plane is called *Miquelian* if it satisfies Miquel's theorem. These two theorems play a role in the study of Möbius planes similar to that of theorems of Desargues and Pappus in the theory of projective planes.

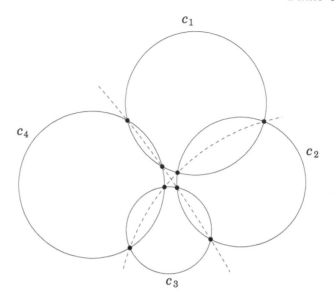

FIGURE 10.3
Bundle Theorem

Theorem 10.52. *A Möbius plane is egglike if and only if it satisfies the Bundle Theorem.*

Theorem 10.53. *A Möbius plane is Miquelian if and only if it is isomorphic to a plane described in Example* 10.41 *with $\mathcal{O}$ being an elliptic quadric.*

The proofs of these characterization theorems are beyond the scope of this book. Theorem 10.53 and the part of Theorem 10.52 stating that every egglike plane satisfies the Bundle Theorem, were proved by *van der Waerden* and *Smid* [175], while the opposite direction of Theorem 10.52 is due to *Kahn* [99], [100]. As a consequence of these theorems we get that Miquel's theorem implies the Bundle Theorem. The existence of Suzuki-Tits ovoids shows that there are non-Miquelian Möbius planes, thus the Bundle Theorem does not imply Miquel's theorem, so they have the same logical relation as theorems of Pappus and Desargues have, see Theorem 3.34.

It is worth mentioning that each known finite Möbius plane is egglike, but there exist infinite, non-egglike planes, see [64]. Finally, we present the important characterization result of *Dembowski* and *Hughes* about Möbius planes of even order. The proof can be found in [56].

Theorem 10.54. *Every finite Möbius plane of even order n is egglike, and so satisfies the Bundle Theorem and n is a power of* 2.

An interesting, and in general unsolved, problem is the extendability of an affine plane to a Möbius plane by adding one point. It can be considered as

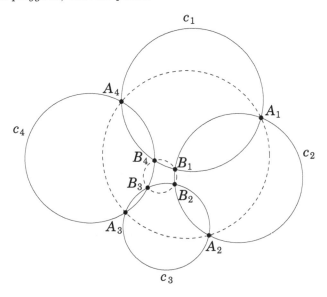

FIGURE 10.4
Theorem of Miquel

the reverse of the derivation process described in Lemma 10.44. If the affine plane is the Desarguesian plane $AG(2, \mathbb{K})$, then one can define the circles using the cartesian coordinates and one can extend the lines to circles by adding the extra point (∞). So the construction is easy in this case. It is also easy to prove that these planes are isomorphic to the Miquelian Möbius planes over $\mathbb{K}$. On the other hand, if Π is a non-Desarguesian affine plane, then nothing is known about the extendability. There is no example and there is no theorem about the impossibility of the extension.

It is known that the derivation process starting from a Miquelian plane and any of its points gives a Desarguesian affine plane. In the opposite direction, an important result of *J. Thas* [168] shows that if the order of a finite Möbius plane $\mathcal{M}$ is not equal to $11, 23$ or 59, and at least one derived affine plane of $\mathcal{M}$ is Desarguesian, then $\mathcal{M}$ is Miquelian. For more results about Möbius planes we refer the reader to [55], Chapter 6.

Exercises

10.1. Show that the Levi graph of a generalized n-gon is a graph of diameter n and girth $2n$.

10.2. Show that a generalized n-gon exists if and only if a graph of diameter n and girth $2n$ exists.

10.3. Prove that Definitions 10.1 and 10.14 are equivalent for thick finite generalized quadrangles.

10.4. Show that the GQ of order $(2, 2)$ is unique up to isomorphism.

10.5. Let $\mathcal{S}$ be a thick generalized n-gon of order (s, t). Show that

- if n is even, then $\mathcal{S}$ contains $(s + 1)\frac{(st)^{n/2}-1}{st-1}$ points and $(t + 1)\frac{(st)^{n/2}-1}{st-1}$ lines;

- if n is odd, then $\mathcal{S}$ contains $1 + (s^2 + s)\frac{s^{n-1}-1}{s^2-1}$ points and the same number of lines.

10.6. Let $\mathcal{S}$ be a finite thick generalized n-gon, $A = (a_{ij})$ be the incidence matrix of $\mathcal{S}$ (the rows of A correspond to lines of $\mathcal{S}$, the columns of A correspond to points of $\mathcal{S}$, and $a_{ij} = 1$ if and only if $P_j \mathrm{I} \ell_i$). Let $M = A^{\mathrm{T}} A$ and consider the matrix $M^k = (m_{ij,k})$. Show that $m_{ij,k}$ is the number of chains of length $2k$ joining P_i and P_j.

10.7. Let $\mathcal{P}$ be any set of five elements and $\mathcal{C}$ be the set of all 3-element subsets of $\mathcal{P}$. Let us define the incidence relation I as the set theoretical containment relation. Show that the triple $(\mathcal{P}, \mathcal{C}, \mathrm{I})$ is a Möbius plane.

10.8. Prove that there is a unique Möbius plane of order 2 and 3.

11

Hyperovals

In this chapter $(q+2)$-arcs in $\mathrm{PG}(2, q)$ are considered. It was proven in Chapter 6 that hyperovals exist only in planes of even order, hence we assume that q is even (some theorems of this chapter are also valid when q is odd; we will indicate them). If q is even, then the characteristic of $\mathrm{GF}(q)$ is 2. This means that $a + a = 0$ for all $a \in \mathrm{GF}(q)$, hence $-1 = 1$, so $a - b = a + b$ also holds. We will use these identities without any further notice for field elements and for polynomials over $\mathrm{GF}(q)$, too.

As we have already seen in Theorem 6.9, the classical example of a hyperoval is the union of a conic and its nucleus. This motivates the following definition.

Definition 11.1. *A hyperoval in* $\mathrm{PG}(2, q)$ *is called* regular *if it is the union of a conic and its nucleus. A hyperoval is called* irregular *if it is not regular.*

The first natural question is the existence of irregular hyperovals. Are there hyperovals other than the regular ones? The answer "no" would be the Theorem of Segre for q even, but, as we will show, the correct answer is "yes". If $q > 8$ even, then $\mathrm{PG}(2, q)$ always contains irregular hyperovals.

Because linear transformations map hyperovals to hyperovals, by Theorem 4.16, we may assume without loss of generality that the hyperoval under investigation contains the four base points of the coordinate system. The next theorem gives a useful algebraic representation of these hyperovals.

Theorem 11.2. *Let* $\mathcal{H}$ *be a hyperoval containing the points* $(1 : 0 . 0)$, $(0 : 1 : 0)$, $(0 : 0 : 1)$ *and* $(1 : 1 : 1)$. *Then there exists a polynomial* $f \in \mathrm{GF}(q)[x]$ *which permutes the elements of* $\mathrm{GF}(q)$ *such that* $f(0) = 0$ *and* $f(1) = 1$, *and*

$$\mathcal{H} = \{(1 : t : f(t)) : t \in \mathrm{GF}(q)\} \cup \{(0 : 1 : 0), (0 : 0 : 1)\}.$$

Proof. Let ℓ_∞ be the line with equation $X_0 = 0$ and $\mathrm{AG}(2, q) = \mathrm{PG}(2, q) \setminus \ell_\infty$. As $\mathcal{H}$ has no tangent line and it contains the point $(0 : 0 : 1)$, each vertical line of the affine plane must contain exactly one affine point of $\mathcal{H}$. Thus the affine part of $\mathcal{H}$ can be considered as a graph of a function $f \colon \mathrm{GF}(q) \to \mathrm{GF}(q)$. Each function $f \colon \mathrm{GF}(q) \to \mathrm{GF}(q)$ can be given by a polynomial, so we may assume without loss of generality that $f \in \mathrm{GF}(q)[x]$. The point at infinity of the horizontal lines of the affine plane also belongs to the affine part of $\mathcal{H}$, hence each horizontal line contains exactly one affine point of $\mathcal{H}$, too. This means that f defines a permutation on the elements of $\mathrm{GF}(q)$. Finally, $f(0) = 0$ and $f(1) = 1$ follow from the fact that the points $(1:0:0)$ and $(1:1:1)$ belong to $\mathcal{H}$. $\qquad\square$

Definition 11.3. *A polynomial $f \in \mathrm{GF}(q)[x]$ is called a* permutation polynomial, *if it permutes the elements of $\mathrm{GF}(q)$, so $f(x) \neq f(y)$ if $x \neq y$.*

A polynomial $f \in \mathrm{GF}(q)[x]$ is called an o-polynomial, if applying the process described in Theorem 11.2, the resulting set of points,

$$\mathcal{D}(f) = \{(1:t:f(t)) : t \in \mathrm{GF}(q)\} \cup \{(0:1:0),(0:0:1)\}$$

is a hyperoval. In particular, if f is an o-polynomial then $f(0) = 0$ and $f(1) = 1$.

Permutation polynomials exist over arbitrary finite fields, while o-polynomials exist only over finite fields of characteristic 2. Let us remark that different o-polynomials could produce the same hyperoval; for example each of the polynomials $f(x) = x^2$, $f(x) = x^{q-2}$ and $f(x) = x^{q/2}$ defines the same regular hyperoval (see Exercise 11.1).

There is an algebraic criterion due to *Dickson* which helps us decide whether a polynomial over an arbitrary finite field is a permutation polynomial or not. Before formulating it we recall two classical formulae about the expression of power sums in terms of symmetric polynomials and vice versa.

Theorem 11.4 (Newton–Waring–Girard Power Sum Formulae). *Let $X_1, X_2, \ldots, X_n$ be variables and $\mathbb{K}$ be an arbitrary field. In the polynomial ring $\mathbb{K}[X_1, X_2, \ldots, X_n]$ let*

$$\sigma_k(X_1, X_2, \ldots, X_n) = \sum_{1 \leq j_1 < j_2 < \cdots < j_k \leq n} X_{j_1} X_{j_2} \cdot \cdots \cdot X_{j_k}$$

and

$$p_k(X_1, X_2, \ldots, X_n) = \sum_{i=1}^{n} X_i^k$$

be the k-th symmetric polynomial and the k-th power sum, respectively. Then for all $n \geq 1$ and $k \geq 1$

$$p_k(X_1, X_2, \ldots, X_n) = (-1)^{k-1} k \sigma_k(X_1, X_2, \ldots, X_n)$$
$$+ \sum_{i=1}^{k-1} (-1)^{k-1+i} \sigma_{k-i}(X_1, X_2, \ldots, X_n) p_i(X_1, X_2, \ldots, X_n)$$

where $\sigma_k(X_1, X_2, \ldots, X_n) = 0$ if $k > n$.

Corollary 11.5. *The symmetric polynomials can be expressed in terms of the power sums in the following way:*

$$k\sigma_k(X_1, X_2, \ldots, X_n) = \sum_{i=1}^{k} (-1)^{i-1}\sigma_{k-i}(X_1, X_2, \ldots, X_n)p_i(X_1, X_2, \ldots, X_n).$$

Corollary 11.6. *In a finite field* GF(q) *the sum of the k-th powers of the elements is the following:*

$$\sum_{u \in \mathrm{GF}(q)} u^k = \begin{cases} 0, & \text{if } k \not\equiv 0 \pmod{q-1}, \\ -1, & \text{if } k \equiv 0 \pmod{q-1}. \end{cases}$$

Proof. The elements of GF(q), $u_1, u_2, \ldots, u_q$, are the roots of the equation $X^q - X = 0$. Thus, by Viéta's formulae

$$\sigma_k(u_1, u_2, \ldots, u_q) = \begin{cases} 0, & \text{if } k \neq q-1, \\ -1, & \text{if } k = q-1. \end{cases}$$

By substituting these to the formulae of Theorem 11.4 we get the statement.

□

The next statement is the "sufficiency" part of Dicksons' criterion. It works for arbitrary prime powers q.

Lemma 11.7. *Let* $q = p^n$, *p prime. If a polynomial* $f \in \mathrm{GF}(q)[x]$ *satisfies the conditions*

$$\sum_{x \in \mathrm{GF}(q)} f(x)^k = 0 \tag{11.1}$$

for $k = 1, 2, \ldots, q-2$, *then one of the following hold:*

(i) *f is a permutation polynomial and* $\sum_{x \in \mathrm{GF}(q)} f(x)^{q-1} = -1$,

(ii) $\sum_{x \in \mathrm{GF}(q)} f(x)^{q-1} = 0$ *and in the range of f the number of appearance of each element is divisible by p.*

Proof. Consider the polynomial

$$F(U) = \prod_{x \in \mathrm{GF}(q)} (U - f(x)).$$

Let $s_k(f)$ be a shorthand notation for the k-th symmetric polynomial on q variables evaluated at $f(x_1), f(x_2), \ldots, f(x_q)$ where $x_1, x_2, \ldots, x_q$ denote the elements of GF(q), that is,

$$s_k(f) = \sigma_k(f(x_1), f(x_2), \ldots, f(x_q)).$$

With this notation

$$F(U) = \sum_{i=0}^{q} (-1)^i s_k(f) U^{q-i}.$$

Then, by Corollary 11.5, Equations (11.1) imply that $s_k(f) = 0$ if $k \leq q - 2$ and p does not divide k. Hence, by Viéta's formulae,

$$F(U) = U^q + (-1)^p s_p(f) U^{q-p} + \cdots + s_{q-p}(f) U^p + (-1)^{q-1} s_{q-1}(f) U + (-1)^q s_q(f). \tag{11.2}$$

If $f(x) \neq 0$ for all $x \in \mathrm{GF}(q)$, then

$$\sum_{x \in \mathrm{GF}(q)} f(x)^{q-1} = \sum_{x \in \mathrm{GF}(q)} 1 = 0,$$

hence Corollary 11.5 implies $s_{q-1}(f) = 0$, too. Thus Equation (11.2) can be written as $F(U) = G(U^p)$ with a suitable polynomial G. As G is fully reducible over $\mathrm{GF}(q)$, its roots are p-th powers, so we get case (ii).

If $f(x) = 0$ for at least one $x \in \mathrm{GF}(q)$, then $s_q(f) = 0$. This implies that $F'(U) = s_{q-1}(f)$. If $s_{q-1}(f) = 0$, then we get back the previous case. If $s_{q-1}(f) \neq 0$, then F has no multiple root. On the one hand, this gives that f is a permutation polynomial. On the other hand, this means that $F(U) = U^q - U$, thus, by Corollary 11.6, $s_{q-1}(f) = -1$ also holds, so we get case (i). $\square$

Theorem 11.8 (Dickson's criterion). *Let $q = p^n$, p prime. Let $f \in \mathrm{GF}(q)[x]$ be a polynomial with degree at most $q - 1$. Then f is a permutation polynomial if and only if the following hold:*

(i) *for r odd and $r < q - 1$, the degree of $f(x)^r$ reduced modulo $(x^q - x)$ is at most $q - 2$,*

(ii) *the equation $f(t) = 0$ has exactly one solution in $\mathrm{GF}(q)$.*

Proof. First, suppose that $f(x) = \sum_{i=0}^{q-1} a_i x^i$ is a permutation polynomial. Then condition (ii) obviously holds. We claim that condition (i) is also satisfied. As each element of $\mathrm{GF}(q)$ appears exactly once in the range of $f(x)$, the elements of the set $\{f(x)^r : x \in \mathrm{GF}(q)\}$ are exactly the r-th powers of the elements of $\mathrm{GF}(q)$. Thus, by Corollary 11.6, we have

$$\sum_{x \in \mathrm{GF}(q)} f(x)^r = \sum_{x \in \mathrm{GF}(q)} x^r = 0$$

for $r = 0, 1, \ldots, q - 2$. Let $f(x)^r$ reduced modulo $(x^q - x)$ be the polynomial $\sum_{i=0}^{q-1} b_i x^i$. Then for $r < q - 1$ we have

$$0 = \sum_{x \in \mathrm{GF}(q)} f(x)^r = \sum_{x \in \mathrm{GF}(q)} \sum_{i=0}^{q-1} b_i x^i = \sum_{i=0}^{q-1} \left(b_i \sum_{x \in \mathrm{GF}(q)} x^i \right)$$

$$= b_{q-1} \sum_{x \in \mathrm{GF}(q)} x^{q-1} = -b_{q-1},$$

hence $b_{q-1} = 0$ and this proves condition (i).

Now suppose that $f(x)$ satisfies conditions (i) and (ii). We claim that $\sum_{x \in \mathrm{GF}(q)} f(x)^r = 0$ for all $r < q - 1$. If r is not divisible by p, then it follows from condition (i), while if r is divisible by p, then it follows from the fact that taking the p-th power is an automorphism of $\mathrm{GF}(q)$, hence $\sum_{x \in \mathrm{GF}(q)} f(x)^r = \sum_{x \in \mathrm{GF}(q)} f(x)^{r/p}$. So we can apply Lemma 11.7. Condition (ii) excludes case (ii) of the lemma, so case (i) must hold which proves the theorem. $\square$

Using permutation polynomials it can easily be formulated when a polynomial f is an o-polynomial.

Theorem 11.9. *If $q > 2$ even, then $f \in \mathrm{GF}(q)[x]$ is an o-polynomial if and only if the following hold:*

(i) *f is a permutation polynomial with $f(0) = 0$ and $f(1) = 1$,*

(ii) *for each $s \in \mathrm{GF}(q)$, the polynomial*

$$f_s(x) = \frac{f(x+s) + f(s)}{x}, \quad \text{where } f(0) = 0$$

is also a permutation polynomial and $f_s(0) = 0$.

Proof. We have already proved condition (i); it is a part of Theorem 11.2. As $f_s(x)$ gives the slope of the line joining the points $(s, f(s))$ and $(x+s, f(x+s))$ and condition (i) guarantees that these slopes are different for different values of x. Thus no three points of the affine part of $\mathcal{D}(f)$ are collinear. $\square$

If the polynomial is a monomial $f(x) = x^r$, then the conditions are even simpler.

Corollary 11.10. *If $q > 2$ even, then the monomial $f(x) = x^r \in \mathrm{GF}(q)[x]$ is an o-polynomial if and only if the following hold:*

(i) *$(r, q - 1) = 1$;*

(ii) *$(r - 1, q - 1) = 1$;*

(iii) *$((x+1)^r + 1)/x$ is also a permutation polynomial.*

Proof. Condition (i) is equivalent to $f(x)$ being a permutation polynomial.

Condition (ii) is satisfied if and only if x^{r-1} is a permutation polynomial. This is condition (ii) for Theorem 11.9 in the case $s = 0$.

Finally, condition (iii) corresponds to condition (ii) of Theorem 11.9 when $s \neq 0$ because

$$\frac{(x+s)^r + s^r}{x} = \frac{(\frac{x}{s}+1)^r + 1}{\frac{x}{s}} s^{r-1},$$

and according to condition (iii) the polynomial on the right-hand side is a permutation polynomial. $\square$

Let us look at some more examples when the monomial $f(x) = x^r$ defines a hyperoval. These hyperovals are called *monomial hyperovals*.

Proposition 11.11. *Let $q = 2^n$ and $r = 2^h$. Then $f(x) = x^r \in \mathrm{GF}(q)[x]$ is an o-polynomial if and only if n and h are coprime.*

Proof. Condition (i) of Corollary 11.10 is obviously satisfied. If $r = 2^h$, then $(x+1)^r + 1 = x^r$, hence in this case either both, or none of conditions (iii) and (ii) of Corollary 11.10 hold. Elementary calculation shows that $(2^n - 1, 2^h - 1) = 2^{(n,h)} - 1$, so the conditions are satisfied if and only if n and h are coprime.   □

We can characterize the monomial hyperovals geometrically, too. In order to do this we need the following definition.

Definition 11.12. *Let $f(x) \in \mathrm{GF}(q)[x]$ be an o-polynomial. The hyperoval*

$$\mathcal{D}(f) = \{(1 : t : f(t)) : t \in \mathrm{GF}(q)\} \cup \{(0 : 1 : 0), (0 : 0 : 1)\}$$

defined by f is called a translation hyperoval, *if for all $a \in \mathrm{GF}(q)$ the translation of $\mathrm{AG}(2, q)$ mapping the point $(0, 0)$ into $(a, f(a))$ fixes $\mathcal{D}(f)$ setwise. This condition means that $f(u + v) = f(u) + f(v)$ for all $u, v \in \mathrm{GF}(q)$.*

Translation hyperovals were characterized independently by *Hirschfeld* [86] and *Payne* [138]. We present their result without proof.

Theorem 11.13 (Payne, Hirschfeld). *Let $q = 2^n$ and $r = 2^h$. Then $\mathcal{D}(f)$ is a translation hyperoval in $\mathrm{PG}(2, q)$ if and only if f is a monomial, $f(x) = x^r$, and $(h, n) = 1$.*

It could happen that for distinct values, $r_1 \neq r_2$, the monomials x^{r_1} and x^{r_2} define projectively equivalent hyperovals. This situation can be controlled. In particular, it is known when a translation hyperoval is regular. We cite the following two theorems of *Hirschfeld* [86] without proofs.

Theorem 11.14. *Let $1 < r_1 \leq r_2 < q - 1$ be integers. The hyperovals $\mathcal{D}(x^{r_1})$ and $\mathcal{D}(x^{r_2})$ in $\mathrm{PG}(2, q)$ are projectively equivalent in the following cases:*

- *if $r_1 r_2 \equiv 1 \pmod{q - 1}$,*

- *if $(r_1 - 1)(r_2 - 1) \equiv 1 \pmod{q - 1}$,*

- *if $r_1 + r_2 = q$.*

Corollary 11.15. *The translation hyperoval $\mathcal{D}(x^r)$ is a regular hyperoval in $\mathrm{PG}(2, q)$ if and only if $r = 2, q - 2$ or $q/2$.*

About the existence of irregular hyperovals the following is known.

Proposition 11.16. *In $\mathrm{PG}(2, q)$, $q = 2^n$, there exist irregular hyperovals if and only if $n \geq 4$.*

Proof. In PG(2, 2) each hyperoval is a complement of a line, hence projectively equivalent to the set of points $\{(1 : 0 : 0), (1 : 1 : 1), (0 : 0 : 1), (0 : 1 : 0)\}$, that is the conic with equation $X_0 X_2 = X_1^2$ and its nucleus. Thus each hyperoval is regular.

Let $\mathcal{H}$ be a hyperoval in PG(2, 4). Then, by a corollary of Pascal's theorem, through any subset of five points of the six points of $\mathcal{H}$ there is a conic $\mathcal{C}$ (see Exercise 11.4). The nucleus of $\mathcal{C}$ is uniquely determined and it must be the sixth point of $\mathcal{H}$, hence $\mathcal{H}$ is regular.

In PG(2, 8) the situation is a bit more complicated, and we only sketch the proof. If f is an o-polynomial, then $\deg f$ must be even (see Corollary 11.24), hence $\deg f = 2, 4$ or 6. If $\deg f = 2$, then the hyperoval is regular by definition. If $\deg f = 4$ or 6, then one can prove that the hyperoval is projectively equivalent to $\mathcal{D}(x^4)$ or $\mathcal{D}(x^6)$, respectively. As $4 = q/2$ and $6 = q - 2$, these hyperovals are regular by Corollary 11.15.

Now consider the case $n \geq 4$. It follows from Corollary 11.15 that irregular translation hyperovals arising from the monomial x^{2^h} exist if $n = 5$ or $n \geq 7$.

For $n = 4$ *Lunelli* and *Sce* [120] constructed irregular hyperovals by the aid of a computer in 1958. (Later on *Hall* [79] proved that there are only two projectively non-equivalent hyperovals in PG(2, 16), the regular and the Lunelli-Sce ones.) The corresponding o-polynomial is

$$f(x) = x^{12} + x^{10} + e^{11}x^8 + x^6 + e^2 x^4 + e^9 x^2,$$

where $e^4 = e + 1$.

For $n = 6$ there are several examples. The o-polynomial

$$\begin{aligned}
f(x) = {} & x^{62} + x^{30} + x^{24} + e^{21}(x^{60} + x^{58} + x^{54} + x^{52}) \\
& + e^{21}(x^{46} + x^{44} + x^{40} + x^{38} + x^{34} + x^{16} + x^{14} + x^{10} + x^8 + x^4) \\
& + e^{42}(x^{50} + x^{48} + x^{42} + x^{36} + x^{32} + x^{26} + x^{20} + x^{18} + x^{12} + x^6)
\end{aligned}$$

defines an irregular hyperoval if e is a primitive element of GF(64) satisfying $e^6 = e + 1$. Of course, for $n = 4$ and 6 it is highly non-trivial to verify that the given polynomial is an o-polynomial. □

In the following we will need to solve quadratic equations. The well-known Quadratic Formula which gives the solution over the reals also works in fields of odd characteristic, but the situation in fields of characteristic 2 is more complicated. First, we recall a definition from algebra.

Definition 11.17. *The* trace function *from* GF(q^n) *to* GF(q), *denoted by* Tr, *is defined as*

$$\mathrm{Tr}(x) = x + x^q + \ldots + x^{q^i} + \ldots + x^{q^{n-1}}.$$

The proof of the following lemma is straightforward, so we leave it as an exercise.

Lemma 11.18. *The trace function from* $\mathrm{GF}(q^n)$ *to* $\mathrm{GF}(q)$ *has the following properties:*

- Tr *is linear over* $\mathrm{GF}(q)$, *which means*

$$\mathrm{Tr}(\alpha x + \beta y) = \alpha \mathrm{Tr}(x) + \beta \mathrm{Tr}(y)$$

  for all $\alpha, \beta \in \mathrm{GF}(q)$ *and for all* $x, y \in \mathrm{GF}(q^n)$.

- *In particular,* Tr *is additive,* $\mathrm{Tr}(x + y) = \mathrm{Tr}(x) + \mathrm{Tr}(y)$.

- *The equation* $\mathrm{Tr}(x) = a$ *has* q^{n-1} *roots over* $\mathrm{GF}(q^n)$ *if* $a \in \mathrm{GF}(q)$, *and has no roots if* $a \notin \mathrm{GF}(q)$.

The next lemma describes when a quadratic equation over $\mathrm{GF}(2^n)$ has a solution.

Lemma 11.19. *If* $a \neq 0$, *then the quadratic equation*

$$aX^2 + bX + c = 0 \tag{11.3}$$

has no solution over $\mathrm{GF}(2^n)$ *if and only if*

$$\mathrm{Tr}\left(\frac{ac}{b^2}\right) = 1.$$

Proof. If $b = 0$, then $x = (c/a)^{\sigma^{-1}}$ is the unique root of Equation (11.3), because the mapping $\sigma : x \mapsto x^2$ is an automorphism of $\mathrm{GF}(2^n)$. Suppose that $b \neq 0$. Multiply Equation (11.3) by a/b^2 and introduce the new indeterminate $U = aX/b$. Then the equation can be written as $U^2 + U + ac/b^2 = 0$. Apply now the Tr function to both sides. As Tr is additive, $\mathrm{Tr}(0) = 0$ and $\mathrm{Tr}(U^2) = \mathrm{Tr}(U)$, we get

$$\mathrm{Tr}(U^2) + \mathrm{Tr}(U) + \mathrm{Tr}\left(\frac{ac}{b^2}\right) = \mathrm{Tr}(0), \quad \text{so} \quad \mathrm{Tr}\left(\frac{ac}{b^2}\right) = 0.$$

Hence if Equation (11.3) has a solution, then $\mathrm{Tr}\left(\frac{ac}{b^2}\right) = 0$.

The function Tr is linear over $\mathrm{GF}(2)$, thus there are exactly 2^{n-1} elements in $\mathrm{GF}(q)$ with 0 trace. On the other hand, the mapping $\tau : \mathrm{GF}(2^n) \to \mathrm{GF}(2^n)$, $\tau(U) = U^2 + U$, is also linear over $\mathrm{GF}(2)$. Its kernel has dimension 1, because $U^2 + U = 0$ implies $U = 0$ or $U = 1$, hence the image of τ has dimension $n - 1$. Thus $\mathrm{Tr}(U^2 + U) = 0$ implies that an element $s \in \mathrm{GF}(2^n)$ can be written in the form $U^2 + U$ if and only if $\mathrm{Tr}(s) = 0$. So Equation (11.3) has no solution if $\mathrm{Tr}\left(\frac{ac}{b^2}\right) = 1$. $\qquad\square$

Corollary 11.20. *The equation* $X^2 + X + 1 = 0$ *has a root in* $\mathrm{GF}(2^n)$ *if and only if* n *is even.*

Proof. In this case $a = b = c = 1$, hence $\mathrm{Tr}\left(\frac{ac}{b^2}\right) = \mathrm{Tr}(1) = n$. This equals to 0 if and only if n is even. $\qquad\square$

The next example, constructed by *Segre* [152] in 1962, was the first hyperoval other than translation ones.

Proposition 11.21 (Segre). *If $q = 2^n$, then $f(x) = x^6$ is an o-polynomial if and only if n is odd.*

Proof. If n is even, then $(6, 2^n - 1) = 3$, so condition (i) of Corollary 11.10 is not satisfied, hence x^6 cannot be an o-polynomial.

If n is odd, then $(6, 2^n - 1) = (5, 2^n - 1) = 1$, hence conditions (i) and (ii) of Corollary 11.10 are satisfied. Condition (iii) means that $g(x) = ((x+1)^6 + 1)/x$ is a permutation polynomial. Elementary calculation gives $g(x) = x^5 + x^3 + x$. Suppose that $t^5 + t^3 + t = s^5 + s^3 + s$ holds with $s \neq t$. After rearranging and dividing by $(s + t)$ we get

$$1 + s^2 + st + t^2 + s^4 + s^3t + s^2t^2 + st^3 + t^4 = 0.$$

Let $S = s^2 + t^2$ and $T = st$, then the equation can be written as

$$S^2 + S(T + 1) + T^2 + T + 1 = 0. \tag{11.4}$$

If $T = 1$, then it becomes $S^2 + 1 = 0$, hence $S = 1$. So $s^2 + t^2 = 1$ which means $s + t = 1$, so $t = s + 1$. Substituting it to $st = T = 1$ we get $s^2 + s + 1 = 0$. As n is odd, by Corollary 11.20, this equation has no solution. So we may assume that $T \neq 1$.

Dividing Equation (11.4) by $(T + 1)^2$ and introducing the new indeterminate $S/(T + 1) = U$ we get

$$U^2 + U + \frac{T^2 + T + 1}{(T + 1)^2} = U^2 + U + \frac{1}{T + 1} + \frac{1}{(T + 1)^2} + 1$$

$$= \left(U + \frac{1}{T + 1}\right)^2 + \left(U + \frac{1}{T + 1}\right) + 1 = 0.$$

Again, by Corollary 11.20, this equation has no solution, so $g(x)$ is a permutation polinomial. $\square$

A more technical criterion for o-polynomials is due to *Glynn* [74]. We skip its proof, but recall the key lemma of the proof and the criterion itself.

Lemma 11.22 (Glynn). *The following are equivalent:*

(i) *f is an o-polynomial,*

(ii) *$g_k(c) = \sum_x (f(x) + cx)^k = 0$, if $k < q - 1$, $g_{q-1}(c) = c^{q-1} + 1$,*

(iii) *$\deg f \geq 1$, and $\binom{k+l}{k} \sum_x f(x)^k x^l = 0$, for all $1 \leq k + l \leq q - 1$, $(k, l) \neq (q - 1, 0)$.*

Before considering Glynn's criterion, we define a partial ordering on a set of integers.

Let $q = 2^h$ and write the integers $0 \leq a, b < q$ in binary number system, let $a = \sum_{i=0}^{h-1} a_i 2^i$ and $b = \sum_{i=0}^{h-1} b_i 2^i$, where $a_i, b_i = 0$ or 1. We define $b \prec a$ if and only if $b_i \leq a_i$ for all i. By Lucas' theorem (Theorem 7.10), $b \prec a$ means that $\binom{a}{b}$ is odd.

Theorem 11.23 (Glynn's criterion). *Let $f(x)$ be a polynomial of degree at most $(q-2)$ for which $f(0) = 0$ and $f(1) = 1$. Then f is an o-polynomial if and only if the coefficient of x^c in $f(x)^b$ modulo $x^q - x$ is 0 for all pairs (b, c) satisfying $1 \leq b \leq c \leq q - 1$, $b \neq q - 1$ and $b \prec c$.*

Glynn's criterion is equivalent to the possibilities listed in Lemma 11.22. The next corollary was first proved by *Segre* and *Bartocci* [154].

Corollary 11.24 (Segre, Bartocci). *If f is an o-polynomial, then it consists of even degree terms only.*

Proof. If the coefficient of x^s in $f(x)$ is not zero and s is odd, then let $k = 1$ and $l = q - 1 - s$ in condition (iii) of Lemma 11.22. This gives $\binom{k+l}{k} \neq 0$, and $f(x)x^l$ contains a term x^{q-1}, hence $\sum_x f(x)x^l$ is not zero, a contradiction. $\square$

Glynn [73] has found two more examples of monomial o-polynomials. The proof of the next result is beyond the scope of this book.

Theorem 11.25 (Glynn). *Let $q = 2^n$, $n = 2e + 1$ odd,*

$$\sigma : x \mapsto x^{2^{e+1}} \quad and \quad \gamma : \begin{cases} x \mapsto x^{2^{(e+1)/2}}, & \text{if e is odd,} \\ x \mapsto x^{2^{(3e+2)/2}}, & \text{if e is even.} \end{cases}$$

Then $x^{\sigma+\gamma}$ and $x^{3\sigma+4}$ are o-polynomials.

Let us remark that in all of the previous examples the binary form of the exponent contains at most three 1's. Glynn conjectured that these are the only hyperovals with a monomial o-polynomial. *Cherowitzo* and *Storme* [45] proved the conjecture when the binary form of the exponent contains at most two 1's, and *Glynn* proved it in small planes, namely for $q \leq 2^8$. The conjecture is still open in general.

In the rest of this chapter we consider constructions producing families of hyperovals. These methods are related to q-clans of matrices, generalized quadrangles, flocks of quadratic cones, translation planes and the correspondences among them. For more details about these topics we refer to the web page of *Cherowitzo* [42].

Definition 11.26. *Let $f(x)$ and $g(x)$ be o-polynomials satisfying the extra condition that for a fixed $a \in \mathrm{GF}(q)$ with $\mathrm{Tr}(a) = 1$, for all $s \in \mathrm{GF}(q)$ the polynomial*

$$f_s(x) = \frac{f(x) + asg(x) + s^{1/2}x^{1/2}}{1 + as + s^{1/2}} \tag{11.5}$$

is also an o-polynomial. Then the set of $q+1$ hyperovals defined by the set of o-polynomials

$$\{f_s(x)\colon s \in \mathrm{GF}(q)\} \cup \{g(x)\}$$

is called a herd of hyperovals.

Definition 11.27. *A 2×2 matrix B is called* anisotropic *if $\mathbf{u}B\mathbf{u}^T = 0$ implies $\mathbf{u} = \mathbf{0}$.*

A q-element family of matrices

$$\mathcal{C} = \left\{ A_t = \begin{pmatrix} x_t & y_t \\ 0 & z_t \end{pmatrix} : t \in \mathrm{GF}(q) \right\}$$

is called a q-clan if for all $s \neq t$ the matrices $(A_s - A_t)$ are anisotropic.

The notion of q-clan was introduced by *Kantor* [104] and *Payne* [139] in relation with generalized quadrangles. The q-clans are defined for all q, but we consider the q even case only. Obviously, $\mathcal{C}$ is a q-clan if and only if $\mathcal{C} + A_0 = \{A_t + A_0 \colon A_t \in \mathcal{C}\}$ is a q-clan. Thus, we may assume without loss of generality that A_0 is the zero matrix. The quadratic form belonging to $A_t + A_s$ is

$$(x_t + x_s)X^2 + (y_t + y_s)XY + (z_s + z_t)Y^2.$$

As $(A_s + A_t)$ is anisotropic, this quadratic form is 0 if and only if $X = Y = 0$. Suppose that $Y \neq 0$ and let $U = X/Y$ be a new variable. Then the condition means that the equation

$$(x_t + x_s)U^2 + (y_t + y_s)U + (z_s + z_t) = 0$$

has no solution. Hence none of its coefficients is 0, so $x_s \neq x_t$, $y_s \neq y_t$ and $z_s \neq z_t$ for $s \neq t$. This gives that each of x_t, y_t and z_t must be a permutation of the elements of $\mathrm{GF}(q)$, and by Lemma 11.19,

$$\mathrm{Tr}((x_s + x_t)(z_s + z_t)/(y_s + y_t)^2) = 1 \tag{11.6}$$

for all $s \neq t$.

As y_t is a permutation, we can reparametrise $\mathcal{C}$ and normalize it so that $y_t = t^{1/2}$. We have already assumed that $A_0 = \begin{pmatrix} 0 & 0 \\ 0 & 0 \end{pmatrix}$, and now by multiplying elements of $\mathcal{C}$ by a fixed λ, we may assume $A_1 = \begin{pmatrix} 1 & 1 \\ 0 & a \end{pmatrix}$ with $\mathrm{Tr}(a) = 1$, because $A_1 + A_0$ is anisotropic. Let $f(t) = x_t$ and $g(t) = z_t/a$. Then $f(0) = g(0) = 0$ and $f(1) = g(1) = 1$. In this normalized form $\mathcal{C}$ is a q-clan if and only if

$$\mathrm{Tr}\left(\frac{a(f(s) + f(t))(g(s) + g(t))}{s + t}\right) = 1 \tag{11.7}$$

for all $s \neq t$.

Conversely, if the functions f and g satisfy Equation (11.7) and we define the matrices A_t for all $t \in \mathrm{GF}(q)$ as $A_t = \begin{pmatrix} f(t) & t^{1/2} \\ 0 & ag(t) \end{pmatrix}$ then we get a q-clan. The next theorem describes the connection between q-clans and herds of hyperovals. It is due to *Cherowitzo, Penttila, Pinneri* and *Royle* [44], although it was implicitly known probably in earlier work of *Payne*.

Theorem 11.28 (Cherowitzo, Penttila, Pinneri, Royle). *Let $f(x)$ and $g(x)$ be o-polynomials and consider the polynomials $f_s(x)$, $s \in \mathrm{GF}(q)$, given in Definition 11.26. If these polynomials determine a herd of hyperovals, then the set of matrices*

$$\mathcal{C} = \left\{ \begin{pmatrix} f(t) & t^{1/2} \\ 0 & ag(t) \end{pmatrix} : t \in \mathrm{GF}(q) \right\}$$

is a normalized q-clan.

Conversely, let $\mathcal{C}$ be a normalized q-clan (hence $f(0) = g(0) = 0$ and $f(1) = g(1) = 1$). Then starting with the two polynomials in the main diagonal and applying Formula (11.5) of Definition 11.26, the resulting set of $q + 1$ o-polynomials

$$\{f_s(x) \colon s \in \mathrm{GF}(q)\} \cup \{g(x)\}$$

corresponds to a herd of hyperovals.

Proof. First, suppose that the set $\{f_s(x) \colon s \in \mathrm{GF}(q)\} \cup \{g(x)\}$ of o-polynomials determines a herd of hyperovals. As f_s is an o-polynomial for all s, $x \neq y$ implies $f_s(x) \neq f_s(y)$, hence

$$f(x) + f(y) + s(ag(x) + ag(y)) + s^{1/2}(x^{1/2} + y^{1/2}) \neq 0$$

for all $s \in \mathrm{GF}(q)$. This can be considered as a quadratic equation for $s^{1/2}$, thus Lemma 11.19 gives (11.7) immediately.

In the opposite direction, suppose that $f(x)$ and $g(x)$ satisfy Equation (11.7) with some a, such that $\mathrm{Tr}(a) = 1$. This means that both f and g are permutations, otherwise (11.7) would be violated for some $s \neq t$. We claim that the points of the affine plane $(x, f(x))$, $(y, f(y))$ and $(z, f(z))$ are not collinear. Being on a line with slope m would mean that

$$(f(x) + f(y))/(x + y) = (f(x) + f(z))/(x + z) = (f(y) + f(z))/(y + z) = m.$$

By substituting it to (11.7) we get $\mathrm{Tr}(am(g(x) + g(y))) = 1$, $\mathrm{Tr}(am(g(x) + g(z))) = 1$ and $\mathrm{Tr}(am(g(y) + g(z))) = 1$. Summing up these three equations and applying the additivity of the Trace function we get

$$\begin{aligned}
1 &= \mathrm{Tr}(am(g(x) + g(y))) + \mathrm{Tr}(am(g(x) + g(z))) + \mathrm{Tr}(am(g(y) + g(z))) \\
&= \mathrm{Tr}(am(g(x) + g(y)) + am(g(x) + g(z)) + am(g(y) + g(z))) \\
&= \mathrm{Tr}(0) = 0,
\end{aligned}$$

giving a contradiction. Hence f is an o-polynomial. As the roles of f and g can be interchanged, we get that g is an o-polynomial, too.

Now, we show that f_s is also an o-polynomial for all $0 \neq s \in \mathrm{GF}(q)$. The conditions $f_s(0) = 0$ and $f_s(1) = 1$ are satisfied by definition. Let $b = a + s^{-1} + s^{-1/2}$. Then

$$\mathrm{Tr}(b) = \mathrm{Tr}(a + s^{-1} + s^{-1/2}) = \mathrm{Tr}(a) + \mathrm{Tr}(s^{-1} + s^{-1/2})$$
$$= \mathrm{Tr}\left((s^{-1/2})^2 + s^{-1/2}\right) = \mathrm{Tr}(a) = 1.$$

We claim that if the triple (f, g, a) satisfies Equation (11.7), then so does the triple (f, f_s, b). This is the key observation in the proof. As

$$bsf_s(x) = (a + s^{-1} + s^{-1/2})s \frac{f(x) + asg(x) + s^{1/2}x^{1/2}}{1 + as + s^{1/2}} = f(x) + asg(x) + s^{1/2}x^{1/2},$$

we have

$$bf_s(x) = ag(x) + \frac{f(x)}{s} + \frac{x^{1/2}}{s^{1/2}}. \tag{11.8}$$

The triple (f, f_s, b) satisfies condition (11.7) if and only if

$$\mathrm{Tr}\left(\frac{b(f(x) + f(y))(f_s(x) + f_s(y))}{x + y}\right) = 1$$

for all $x \neq y$. By substituting from Equation (11.8), applying the additive property of the Trace function, we get

$$\mathrm{Tr}\left(\frac{b(f(x) + f(y))(f_s(x) + f_s(y))}{x + y}\right)$$
$$= \mathrm{Tr}\left(\frac{f(x) + f(y)}{x + y}\left(ag(x) + \frac{f(x)}{s} + \frac{x^{1/2}}{s^{1/2}} + ag(y) + \frac{f(y)}{s} + \frac{y^{1/2}}{s^{1/2}}\right)\right)$$
$$= \mathrm{Tr}\left(\frac{a(f(x) + f(y))(g(x) + g(y))}{x + y}\right)$$
$$\quad + \mathrm{Tr}\left(\frac{1}{s}\frac{(f(x) + f(y))^2}{x + y} + \frac{1}{s^{1/2}}\frac{f(x) + f(y)}{(x + y)^{1/2}}\right).$$

The second term is of the form $\mathrm{Tr}(u^2 + u)$, hence it is 0. This proves

$$\mathrm{Tr}\left(\frac{b(f(x) + f(y))(f_s(x) + f_s(y))}{x + y}\right) = 1.$$

Thus if the triple (f, g, a) satisfies Equation (11.7), then the triple (f, f_s, b) also satisfies it. Hence we can repeat the first part of the proof of the opposite direction with the triple (f, f_s, b) and we get that f_s is an o-polynomial. $\square$

This theorem played a crucial role in the construction of new classes of hyperovals, because it is much easier to check the validity of condition (11.7) than the conditions of Theorem 11.9 about o-polynomials. The reason is that the additivity of the Trace function and the identity $\text{Tr}(u^2) = \text{Tr}(u)$ simplify the calculations. For example, the so-called *Subiaco hyperovals* (see in Table 11.1 at the end of the chapter) were constructed in this way.

Some other interesting geometric objects are also related to herds of hyperovals. We briefly mention some of them, but only pay more attention to one of them.

Definition 11.29. *Let $\mathcal{Q}$ be a quadratic cone with vertex V in $\text{PG}(3, q)$. A partition of $\mathcal{Q} \setminus \{V\}$ into q disjoint conics is called a* flock *of $\mathcal{Q}$.*

A flock can be given by the set of q planes cutting off the conics from the cone. Such a set of planes defines a flock if none of them contains V and the line of intersection of any two planes is disjoint from $\mathcal{Q}$.

If a line ℓ has no point in common with $\mathcal{Q}$, then the q planes through ℓ but not through V obviously give a flock. This type of flock is called a linear *flock.*

There is a close connection between q-clans and flocks. This is explained in the following theorem which is valid for all q.

Theorem 11.30 (Thas). *The set*

$$\left\{ \begin{pmatrix} x_t & y_t \\ 0 & z_t \end{pmatrix} : t \in \text{GF}(q) \right\}$$

is a q-clan if and only if the set of planes

$$\{x_t X_0 + z_t X_1 + y_t X_2 + X_3 = 0 : t \in \text{GF}(q)\}$$

gives a flock of the quadratic cone $\mathcal{Q}$ having equation $X_0 X_1 = X_2^2$.

Proof. The proof is mostly calculation. We show that Equation (11.7) is satisfied if and only if all of the pairwise intersection lines of the planes are skew to the cone. This happens if and only if the set of equations

$$x_s X_0 + z_s X_1 + y_s X_2 + X_3 = 0,$$
$$x_t X_0 + z_t X_1 + y_t X_2 + X_3 = 0,$$
$$X_0 X_1 = X_2^2$$

has no solution for $s \neq t$. Summing up the first two equations and then rearranging it, we get

$$(x_s + x_t)X_0 + (z_s + z_t)X_1 + (y_s + y_t)X_2 = 0,$$

$$X_2 = \frac{(x_s + x_t)X_0 + (z_s + z_t)X_1}{y_s + y_t}.$$

Substituting it to the third equation and dividing by X_1^2 gives

$$\frac{X_0}{X_1} = \frac{(x_s + x_t)^2 X_0^2 + (z_s + z_t)^2 X_1^2}{(y_s + y_t)^2 X_1^2}.$$

This is a quadratic equation for the new variable $Y = X_0/X_1$. By Lemma 11.3, it has no solution if and only if

$$\text{Tr}\left(\frac{(x_s + x_t)^2(z_s + z_t)^2}{(y_s + y_t)^2}\right) = 1,$$

which is obviously equivalent to Equation (11.6). This proves the statement.
□

Definition 11.31. *Let Q be a non-singular quadric in $\text{PG}(3, q)$. A flock $\mathcal{F}$ of Q is a maximum cardinality set of disjoint conic sections of Q. In particular, if $Q = \mathcal{H}_3$ is a hyperbolic quadric, then $|\mathcal{F}| = q + 1$ and the elements of $\mathcal{F}$ give a partition of the points of $\mathcal{H}_3$; if $Q = \mathcal{E}_3$ is an elliptic quadric, then $|\mathcal{F}| = q - 1$ and the two points of $\mathcal{E}_3$ on no conic of $\mathcal{F}$ are the carriers of $\mathcal{F}$.*
A flock of an ovoid and its carriers are defined in the same way as a flock and its carriers for an elliptic quadric.

Linear flocks exist for both non-singular quadrics. If a line ℓ has no point in common with Q, then the $q + 1$ planes through ℓ define a flock. In the elliptic case the point of contact of the two tangent planes through ℓ are the carriers.

If Q is an elliptic quadric, then *Orr* [136] (for q odd) and *Thas* (for q even) proved that each flock of Q is linear. If Q is a hyperbolic quadric and q is odd, then there exist non-linear flocks of Q, but their complete description is not known. The following nice proof of the linearity of flocks of non-singular quadrics for q even is due to *Thas*.

Theorem 11.32 (Thas). *Let $\mathcal{M}$ be an ovoid or a hyperbolic quadric in $\text{PG}(3, q)$. If q is even, then each flock of $\mathcal{M}$ is linear.*

Proof. Let $\mathcal{F} = \{C_1, C_2, \ldots, C_s\}$ be a flock and let S and T be the carriers of $\mathcal{F}$ in the ovoidal case. Let Π_i be the plane of C_i and P_i be the pole of the plane Π_i with respect to $\mathcal{M}$. Let us remark that, by Segre's theorem (Theorem 9.14), an ovoid defines a null polarity. Then P_i is the nucleus of C_i, because the polar planes of the points of C_i contain P_i. We claim that the set of $q + 1$ points,

$$\mathcal{P} = \begin{cases} \{P_1, P_2, \ldots, P_{q+1}\}, & \text{if } \mathcal{M} \text{ is hyperbolic quadric,} \\ \{P_1, P_2, \ldots, P_{q-1}\} \cup \{S, T\}, & \text{if } \mathcal{M} \text{ is ovoid,} \end{cases}$$

meets each plane of $\text{PG}(3, q)$.

If Π is the tangent plane to $\mathcal{M}$ at a point $P \in C_i$ then the line PP_i is a tangent line to C_i at P, because P_i is the nucleus of C_i. Hence $PP_i \subset \Pi$ which implies $P_i \in \Pi$. If Π is the plane of C_i for some i, or Π contains at least one of the points S and T, then the statement is obvious. If Π meets $\mathcal{M}$ in an oval $\mathcal{O}$, then there exists at least one element of $\mathcal{F}$ meeting $\mathcal{O}$ in exactly one point, because $q + 1$ is odd and the intersection $\mathcal{O} \cap C_i$ is contained in the line

$\Pi \cap \Pi_i$. We may assume without loss of generality that $|C_1 \cap \mathcal{O}| = 1$. This means that the line $\Pi \cap \Pi_1$ is tangent to both C_i and $\mathcal{C}$, hence $P_i \in \Pi_i \cap \Pi$.

Thus, by Proposition 9.52, the set of points $\mathcal{P}$ is a line p. So the polar line of p with respect to $\mathcal{M}$ is contained in each element of $\mathcal{F}$, which means that $\mathcal{F}$ is linear. $\qquad\square$

According to this theorem, the existence of non-linear flocks remained an open question for quadratic cones for all q and for hyperbolic quadrics if q is odd. For the sake of completeness we present examples in each of these cases. We omit the (far from trivial) proofs that these are really non-linear flocks.

Example 11.33 (Thas). Let $\mathcal{H}$ be a hyperbolic quadric in $\mathrm{PG}(3, q)$, q odd. Then each non-tangent plane meets $\mathcal{H}$ in a conic (see Corollary 4.63). Define a relation $\sim$ on the set of these conics. Let $C_1 \sim C_2$ if and only if there exists a conic C for which $|C \cap C_1| = |C \cap C_2| = 1$. One can prove that $\sim$ is an equivalence relation with two classes. Hence $\sim$ divides the non-tangent planes into two classes, too. Let ℓ be a line skew to $\mathcal{H}$ and ℓ^* be its polar line with respect to $\mathcal{H}$. With a suitable choice of the elements of the two equivalence classes of planes containing ℓ and ℓ^*, respectively, we get a non-linear flock. Let us remark that this flock consists of two subsets of two linear flocks.

The following examples show the existence of non-linear flocks of quadratic cones if q is odd (for q even, we have already seen such examples via hyperovals).

Example 11.34. Let the equation of the quadratic cone $\mathcal{Q}$ be $X_0 X_1 = X_2^2$.

- (*Fisher and Thas* [68], *Walker* [180]) Let the planes be given by the equations

$$tX_0 + 3t^3 X_1 + 3t^2 X_2 + X_3 = 0, \quad t \in \mathrm{GF}(q).$$

  If $q \equiv -1 \pmod 3$, then this set of q planes defines a flock of $\mathcal{Q}$. This flock is linear if and only if $q = 2$.

- (*Kantor* [103]) Let the planes be given by the equations

$$tX_0 - mt^\sigma + X_3 = 0, \quad t \in \mathrm{GF}(q).$$

  If q is odd, m is a fixed non-square element and σ is an automorphism of $\mathrm{GF}(q)$, then this set of q planes defines a flock of $\mathcal{Q}$.

Flocks are related to line spreads of $\mathrm{PG}(3, q)$ and via this to translation planes (see Example 5.20). The relationship between flocks and translation planes was known much earlier than the connection between q-clans and herds of hyperovals, and it was the starting point for investigations of these objects. It is due to *Walker* [180] and was discovered by *J. Thas* [167] independently.

Starting from a flock we can construct a spread in the following way. Let $\mathcal{F} = \{C_1, \ldots, C_s\}$ be a flock of $\mathcal{Q}$ with $s = q - 1, q$ or $q + 1$, according to whether $\mathcal{Q}$ is an elliptic quadric, a quadratic cone or a hyperbolic quadric. Let $\mathcal{K}_5$ be the Klein quadric in $\mathrm{PG}(5, q)$ and let us embed $\mathcal{Q}$ into $\mathcal{K}_5$. Let Σ_3 be a 3-dimensional subspace of $\mathrm{PG}(5, q)$ such that $\mathcal{Q} = \Sigma_3 \cap \mathcal{K}_5$. For $i = 1, 2, \ldots, s$ denote the plane of C_i by Π_i, the polar plane of Π_i with respect to $\mathcal{K}_5$ by Π_i^* and let $C_i^* = \Pi_i^* \cap \mathcal{Q}$. As the line $\ell_{ij} = \Pi_i \cap \Pi_j$ is external to $\mathcal{Q}$, it is also external to $\mathcal{K}_5$. Thus the polar subspace of ℓ_{ij} with respect to $\mathcal{K}_5$ is a 3-dimensional subspace ℓ_{ij}^* which contains the planes Π_i^* and Π_j^* and meets $\mathcal{K}_5$ in an elliptic quadric. In particular, the lines joining two points of the set $C_i^* \cup C_j^*$ cannot lie on $\mathcal{K}_5$. The set $C_1^* \cup C_2^* \cup \ldots \cup C_s^*$ and the carriers or the vertex of the cone will be a subset $\mathcal{O}$ of $\mathcal{K}_5$ with cardinality $q^2 + 1$. By the previous remark, no three of its points are collinear, hence $\mathcal{O}$ is an ovoid.

This means that under the Klein correspondence (see Corollary 5.3) the points of $\mathcal{O}$ go to pairwise skew lines of $\mathrm{PG}(3, q)$. As $|\mathcal{O}| = q^2 + 1$, these lines form a spread. This spread contains at least s reguli, because $\mathcal{O}$ is the union of conics. According to the construction given in Example 5.20, a translation plane Π of order q^2 arises from the spread. The plane Π is Desarguesian if and only if the spread is regular. It is not too difficult to show that it happens if and only if the flock $\mathcal{F}$ is linear, because in this case a 3-dimensional subspace cuts $\mathcal{O}$ from $\mathcal{K}_5$.

Flocks and generalized quadrangles are also related. This is how *Payne* [139] and *Kantor* [104] discovered some classes of generalized quadrangles and also their connections with q-clans. More details can be found in the book [140].

To end this chapter we survey all known infinite families of hyperovals. For the sake of completeness, the table also contains those hyperovals that have been discussed earlier in this chapter. Table 11.1 can be found in *Hirschfeld's* book [88]. An up-to-date version of the table can be found on the hyperoval web site of *Cherowitzo* [42].
The polynomials in the table are defined as follows:

$$P(X) = X^{1/6} + X^{3/6} + X^{5/6},$$

$$C(X) = X^{\sigma} + X^{\sigma+2} + X^{3\sigma+4},$$

$$S_1(X) = \frac{\omega^2(X^4 + X)}{X^4 + \omega^2 X^2 + 1} + X^{1/2},$$

$$S_2(X) = \frac{\delta^2 X^4 + \delta^5 X^3 + \delta^2 X^2 + \delta^3 X}{X^4 + \delta^2 X^2 + 1} + \left(\frac{X}{\delta}\right)^{1/2},$$

$$S_3(X) = \frac{(\delta^4 + \delta^2)X^3 + \delta^3 X^3 + \delta^2 X}{X^4 + \delta^2 X^2 + 1} + \left(\frac{X}{\delta}\right)^{1/2},$$

$$S(X) = \frac{T(\beta^m)(X + 1)}{T(\beta)} + \frac{T((\beta X + \beta^q)^m)}{T(\beta)(X + T(\beta)X^{1/2} + 1)^{m-1}} + X^{1/2}.$$

The Subiaco hyperovals really belong to one class. If $q = 2^h$ and $h \not\equiv 2$ (mod 4), then they are projectively equivalent to $\mathcal{D}(S_3)$, while if $h \equiv 2$ (mod 4), then they split into two classes, $\mathcal{D}(S_1)$ and $\mathcal{D}(S_2)$.

TABLE 11.1
Known infinite families of hyperovals

Name	o-polynomial	$q = 2^h$	Conditions
Regular	X^2		
Translation	X^{2^i}		$(h, i) = 1$
Segre	X^6	h odd	
Glynn I	$X^{3\sigma+4}$	h odd	$\sigma = 2^{(h+1)/2}$
Glynn II	$X^{\sigma+\lambda}$	h odd	$\sigma = 2^{(h+1)/2};$ $\lambda = 2^m$ if $h = 4m - 1;$ $\lambda = 2^{3m+1}$ if $h = 4m + 1$
Payne	$P(X)$	h odd	
Cherowitzo	$C(X)$	h odd	$\sigma = 2^{(h+1)/2}$
Subiaco	$S_1(X)$	$h = 4r + 2$	$\omega^2 + \omega + 1 = 0$
Subiaco	$S_2(X)$	$h = 4r + 2$	$\delta = \zeta^{q-1} + \zeta^{1-q}$ ζ primitive in $GF(q^2)$
Subiaco	$S_3(X)$	$h \neq 4r + 2$	$\mathrm{Tr}(1/\delta) = 1$
Adelaide	$S(X)$	h even, $h \geq 4$	$1 \neq \beta \in GF(q^2), \beta^{q+1} = 1,$ $T(X) = x + x^q$ $m \equiv \pm(q-1)/3 \pmod{q+1}$

There is only one known sporadic example not listed in the table. *O'Keefe* and *Penttila* [135] found an irregular hyperoval in PG(2, 32). The corresponding *o*-polynomial is

$$f(x) = x^{28} + x^{16} + x^4 + \beta^{11}(x^{26} + x^{22} + x^{18} + x^{14} + x^{10} + x^6)$$
$$+ \beta^{20}(x^{20} + x^8) + \beta^6(x^{24} + x^{12}),$$

where β is a primitive element of GF(32) satisfying the equation $\beta^5 = \beta^2 + 1$.

Exercises

11.1. Prove that the *o*-polynomials x^2, x^{q-2} and $x^{q/2}$ define the same regular hyperoval.

11.2. Prove the properties of the trace function listed in Lemma 11.18.

11.3. Let $\mathcal{C}$ be a conic in PG(2, q) containing the points $(1:0:0)$, $(0:1:0)$, $(0:0:1)$ and $(1:1:1)$. Show that the equation of $\mathcal{C}$ can be written as

$$\alpha X_0 X_1 + \beta X_0 X_2 + \gamma X_1 X_2 = 0$$

where α, β and γ are suitable non-zero elements of GF(q) satisfying the equation $\alpha + \beta + \gamma = 0$.

11.4. Let $\mathcal{C}$ be a conic in PG(2, 2^r) containing the points $(1:0:0)$, $(0:1:0)$, $(0:0:1)$ and $(1:1:1)$. Show that the equation of $\mathcal{C}$ can be written as

$$X_0 X_1 + \alpha X_0 X_2 + (1+\alpha)X_1 X_2 = 0$$

where $0 \neq \alpha \neq 1$ is a suitable element of GF(2^r).

11.5. Let $P_1, P_2, \ldots, P_5$ be five points in general position in PG(2, q). Prove that there exists a unique conic containing them.

11.6. Consider the plane PG(2, 8). Prove the following:

- There are 30 points whose all coordinates are non-zeros.
- These points can be divided into six conics, $\mathcal{C}_1, \mathcal{C}_2, \ldots, \mathcal{C}_6$, such that $\mathcal{C}_i \cap \mathcal{C}_j = \{(1:0:0), (0:1:0), (0:0:1), (1:1:1)\}$ for all $i \neq j$.
- Let $\mathcal{H}_1$ and $\mathcal{H}_\in$ be hyperovals. Show that they coincide if they have at least six points in common (see Exercise 6.7).
- Let $\mathcal{H}$ be an irregular hyperoval containing the points $(1:0:0)$, $(0:1:0)$, $(0:0:1)$ and $(1:1:1)$. Prove that $\mathcal{H} \cap \mathcal{C}_i$ contains exactly five points for all i.
- Show that the hyperoval $\mathcal{D}(x^4)$ is regular in PG(2, 8).

11.7. Show that if the authomorphism group of a hyperoval in PG(2, q) is transitive on irs points, then $q = 2, 4$ or 16.

11.8. Determine the projectively equivalent monomial hyperovals in PG(2, 2^h) for $h = 4, 5$ and 6.

12

Some applications of finite geometry in combinatorics

In this chapter we collect some problems and results in extremal combinatorics where the extremal structures are related to finite geometry. Our first example is a theorem of *De Bruijn* and *Erdős*. This will be formulated in a dual setting as it makes the connection with finite geometry more transparent.

Definition 12.1. *The set system* $\mathcal{L} = (\mathbf{P}, \mathbf{L})$ $(\mathbf{L} \subset 2^{\mathbf{P}})$ *is called a* linear space *if*

(i) *for any two points (elements of* $\mathbf{P}$*) there is a unique element of* $\mathbf{L}$*, which contains them,*

(ii) *each set in* $\mathbf{L}$ *contains at least two points,*

(iii) *there are at least two elements in* $\mathbf{L}$*.*

In case of linear spaces elements of $\mathbf{L}$ will be called *lines*. Perhaps the name "line space" would be better than linear space as it cannot be mixed with a vector space. We could have defined linear spaces as a point-line incidence geometry, however, because of (i) and (ii), two lines cannot be incident with the same set of points, hence lines can be identified by the set of points incident with them.

In what follows v denotes the number of points (that is $|\mathbf{P}|$) and b the number of elements of $\mathbf{L}$. The number of elements in $\mathbf{L}$ through a point p will be denoted by $\deg(p)$.

For any set $\mathbf{P}$, with cardinality at least three, one can define linear spaces. Pick a point w of $\mathbf{P}$ and define lines in the following way:

$$(\mathbf{P} \setminus \{w\}) \cup \{\{w, x\} : x \in \mathbf{P} \setminus \{w\}\}.$$

This trivial structure is called the *degenerate* linear space; axioms P3 and P4 of projective planes essentially exclude these structures (besides other, even more trivial ones). Let us mention that for a degenerate linear space $b = v$, however, the structure is not uniform (and also not regular).

Now let us see the basic relations between the parameters of linear spaces that can be obtained by elementary counting arguments.

Counting incident point-line pairs in two ways we get the fundamental equality, which is valid for hypergraphs not just for linear spaces:

$$\sum_{p \in \mathbf{P}} \deg(p) = \sum_{L \in \mathbf{L}} |L|. \tag{12.1}$$

Double counting triples consisting of two distinct points and the line joining them gives the following equality:

$$v(v-1) = \sum_{L \in \mathbf{L}} |L|(|L|-1). \tag{12.2}$$

Fixing a point p and applying (12.1) for the lines passing through p we get

$$v - 1 = \sum_{p \in L \in \mathbf{L}} (|L|-1). \tag{12.3}$$

Intuitively, this means that the point is joined to other points by a unique line.

Now we turn to Conway's proof of the inequality of De Bruijn-Erdős for the number of lines in linear spaces. The proof uses the following lemma.

Lemma 12.2. *Let* $\mathcal{L} = (\mathbf{P}, \mathbf{L})$ *be a linear space,* p *a point,* L *a line not through* p. *Then* $\deg(p) \geq |L|$ *and in case of equality every line passing through* p *meets* L.

Proof. The point p can be joined to each point of L and these lines are pairwise different, so $\deg(p) \geq |L|$. □

Theorem 12.3 (De Bruijn–Erdős). *In a linear space the number of lines is at least the number of points* ($b \geq v$). *If* $b = v$, *then the linear space is either a projective plane or a degenerate linear space.*

Proof. Let $\mathcal{L} = (\mathbf{P}, \mathbf{L})$ be a linear space on v points with b lines. Assume that $b \leq v$. We are going to show that $b = v$.

As $b \leq v$ and $\deg(p) \geq |L|$ for every non-incident point-line pair (p, L), we have

$$\frac{\deg(p)}{b - \deg(p)} \geq \frac{|L|}{v - |L|}. \tag{12.4}$$

Add up these inequalities for every non-incident point-line pair. Adding the left hand sides from point to point we get

$$\sum_{p \notin L} \frac{\deg(p)}{b - \deg(p)} = \sum_{p \in \mathbf{P}} \sum_{L: p \notin L} \frac{\deg(p)}{b - \deg(p)} = \sum_{p \in \mathbf{P}} \deg(p),$$

since for every point p there are exactly $b - \deg(p)$ lines not through p. Similarly, if we add up the right hand sides from line to line, then we get

$$\sum_{p:p\notin L} \frac{|L|}{v - |L|} = \sum_{L\in\mathbf{L}}|L|.$$

Since we added up the starting inequalities (12.4),

$$\sum_{L\in\mathbf{L}}\sum_{p\in\mathbf{P}} \deg(p) \geq \sum_{L\in\mathbf{L}}|L|$$

follows. Here we must have equality because of (12.1). This can only happen if $b = v$ and for each non-incident point-line pair (p, L) we have $\deg(p) = |L|$. This means that for $b = v$ every two lines intersect each other. If there is a line consisting of just two points then it is not difficult to see that our linear space is degenerate. If all lines contain at least three points, then our structure satisfies all the axioms of a projective plane. This completes the proof of the theorem of De Bruijn-Erdős. □

It should be noted that the original formulation of the theorem of De Bruijn-Erdős was about "dual linear spaces" or simply set systems in which the subsets intersect in precisely one point.

Let us state a sharpening of the De Bruijn-Erdős theorem, due to *Erdős, Mullin, T. Sós* and *Stinson* [63]. Intuitively, it means that when the number of points v of a linear space is close to the number of points of a projective plane of order n, then the least number of lines is obtained when we delete some points from a projective plane of order n (together with the lines that contain less than two points). The more precise statement of the result is the following.

Theorem 12.4 (Erdős, Mullin, T. Sós, Stinson). *Let $n^2 - n + 1 < v \leq n^2 + n + 1$ and define the function $B(v)$ as follows:*

$$B(v) = \begin{cases} n^2 + n - 1, & \text{if } v = n^2 - n + 2 \text{ and } v \neq 4, \\ n^2 + n, & \text{if } n^2 - n + 3 \leq v \leq n^2 + 1 \text{ or } v = 4, \\ n^2 + n + 1, & \text{if } n^2 + 2 \leq v. \end{cases}$$

Then for a non-degenerate linear space $b \geq B(v)$.

Note that this theorem is sharp if there is a projective plane of order n; deleting an appropriate point-set gives a linear space with $B(v)$ lines. *Metsch* [128] proved that a linear space on v points with $B(v)$ lines can almost always be embedded in a projective plane of order n. There is a unique exception on $v = 8$ points.

Before turning to the applications of finite geometry in graph theory let us summarize the standard notation. The set of vertices (or nodes) of a graph

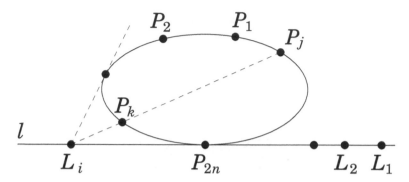

FIGURE 12.1
One-factorization arising from an oval

G is denoted by $V(G)$, the set of its edges by $E(G)$. The number of points and edges will be denoted by $v(G)$ and $e(G)$, respectively. If it does not cause confusion they will simply be denoted by v and e. The degree of a vertex $x \in V(G)$ will be denoted by $\deg(x)$. The equation $\sum_{x \in V(G)} \deg(x) = 2e(G)$, corresponding to (12.1), will also be used frequently.

Definition 12.5. *A* one-factor (*or* perfect matching) *of the graph G is a set $F \subset E(G)$ of pairwise disjoint edges of G such that every vertex $x \in V(G)$ is contained in exactly one of them. A* one-factorization $\mathcal{F} = \{F_1, F_2, \ldots, F_k\}$ *of G is a decomposition of $E(G)$ into edge-disjoint one-factors; that is every edge $e \in E(G)$ is contained in exactly one $F_i \in \mathcal{F}$.*

The one-factorizations of K_{2n}, the complete graph on $2n$ vertices, have an interesting application. Suppose that several soccer teams play against each other in a league. The competition can be represented by a graph with the teams as vertices and edges as games (the edge uv corresponds to the game between the two teams u and v). If every pair of teams plays exactly once, then the graph is complete. When several matches are played simultaneously, and every pair of teams must compete once, the set of games held at the same time is called a round. Thus a round of games corresponds to a one-factor of the underlying graph. The schedule of the games is the same as a one-factorization of K_{2n}. We shall construct one-factorizations of the complete graph from ovals and hyperovals of appropriate projective planes.

Example 12.6. Suppose that the projective plane Π_{2n-1} contains an oval $\mathcal{O} = \{P_1, P_2, \ldots, P_{2n}\}$. Take the points of $\mathcal{O}$ as the vertices of K_{2n}. Let E be an external point of $\mathcal{O}$. The one-factor F belonging to E consists of the edges $P_j P_k$ if the points P_j, P_k and E are collinear, and the edge $P_\ell P_m$ if the lines EP_ℓ and EP_m are the two tangent lines to $\mathcal{O}$ through E. It is clear that the set of edges F is a one-factor of the complete graph K_{2n}.

Let ℓ be the tangent line to $\mathcal{O}$ at the point P_{2n}, for $i = 1, 2, \ldots, 2n-1$, let e_i be the tangent line to $\mathcal{O}$ at the point P_i, L_i denote the point $e_i \cap \ell$ and let F_i be the one-factor belonging to L_i (see Figure 12.1).

Lemma 12.7. *The set of the one-factors* F_i, $i = 1, 2, \ldots, 2n-1$, *gives a one-factorization of* K_{2n}.

Proof. The edge $P_{2n}P_k$ belongs to F_k, and $L_i \neq L_j$ if $i \neq j$. If $i \neq 2n \neq j$, then there is a uniqe intersection point L_k of the lines P_iP_j and ℓ. Hence there is a unique one-factor $\mathcal{F}_k$ containing the edge P_iP_j. $\qquad \square$

Example 12.8. The following similar construction gives a one-factorization of K_{2n} if there exists a projective plane of order $2n-2$ containing a hyperoval $\mathcal{H} = \{P_1, P_2, \ldots, P_{2n}\}$.

Take the points $P_1, P_2, \ldots, P_{2n}$ as the vertices of K_{2n}. Let ℓ be an external line to $\mathcal{H}$ and let $L_1, L_2, \ldots, L_{2n-1}$ be the points on ℓ.

The one-factor F_i belonging to the point L_i is defined to consist of the edges P_jP_k if the points P_j, P_k and L_i are collinear. The union of the one-factors F_i is a one-factorization of K_{2n} because there is a unique point of intersection of the lines P_iP_j and ℓ (see Figure 12.2).

Note that one could construct schedules for some European soccer championships with the above mentioned model based on ovals and hyperovals. For example, the German Bundesliga has 18 teams, so one can start with an oval in PG(2, 17) or a hyperoval in PG(2, 16). Several championships (including the Premier League, Serie A, Primera Division) have 20 teams and in those cases the construction based on an oval of PG(2, 19) can be used.

Let us also mention the probably easiest one-factorization of K_{2n}. Consider the vertices of K_{2n} as the center C and the vertices $\{P_1, P_2, \ldots, P_{2n-1}\}$ of a regular $(2n-1)$-gon in the Euclidean plane. The edges of our complete graph are the radii, that is the pairs CP_i $(i = 1, 2, \ldots, 2n-1)$, and the sides and diagonals, that is the pairs P_iP_j, where $i, j = 1, 2, \ldots, 2n-1$, $i \neq j$, of the regular $(2n-1)$-gon. Let the i-th one-factor be

$$F_i = \{CP_i\} \cup \{P_{i-s}P_{i+s} : i = 1, 2, \ldots, n-1\},$$

where the indices are calculated modulo $2n-1$. Then $\mathcal{F} = \{F_1, F_2, \ldots, F_{2n-1}\}$ is a one-factorization of K_{2n}.

Our next example is about extremal graphs. *Turán* posed the problem of finding the maximum number of edges in a simple graph G on n vertices if G does not contain certain subgraphs. More precisely, let $H_1, \ldots, H_k$ be fixed graphs, the excluded subgraphs. We wish to determine $\text{ex}(n, H_1, \ldots, H_k)$, that is the maximum number of edges one can have in a simple graph on n vertices which does not contain a subgraph isomorphic to H_i, for any $i = 1, 2, \ldots, k$. We can even allow infinitely many excluded subgraphs. In what follows cycles and complete bipartite graphs will occur as excluded subgraphs. The cycle on r vertices will be denoted by C_r, the complete bipartite graph with s vertices in one of the classes and t in the other class will be denoted by $K_{s,t}$.

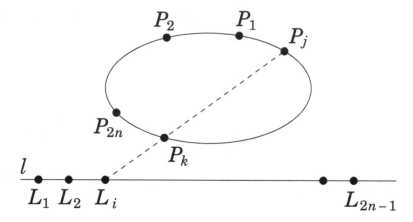

FIGURE 12.2
One-factorization arising from a hyperoval

Historically, the case of $H = K_3 = C_3$, that is triangle-free graphs, was investigated by *Mantel* [123] in 1907 (see Exercise 12.1). Turán's theorem determines the maximum number of edges if we exclude $H = K_{r+1}$ and determines the extremal graphs called r-partite Turán graphs. For more details on Turán's theorem, see [31].

Let us begin with the graph $C_4 = K_{2,2}$.

Theorem 12.9. *If a simple graph on n vertices does not contain a C_4, then it has at most*
$$e \le \frac{n}{4}(1 + \sqrt{4n - 3})$$
edges.

Proof. We are going to count in two ways pairs of adjacent edges (that is, edges that have a common endpoint). Counting them at their "free" edges we get at most $\binom{n}{2}$, since for any pair of points they can have at most one common neighbour, otherwise we would get a C_4. Counting adjacent edges at their common endpoint we get exactly
$$\sum_{v \in V(G)} \binom{\deg(v)}{2}.$$
Using $\sum_{v \in V(G)} \deg(v) = 2e$,
$$\sum_{v \in V(G)} \deg(v)^2 \le 2e + n(n-1)$$
follows. Applying the inequality between quadratic and arithmetic means gives
$$\sqrt{\frac{\sum_{v \in V(G)} \deg(v)^2}{n}} \ge \frac{\sum_{v \in V(G)} \deg(v)}{n},$$

and using $\sum_{v \in V(G)} \deg(v) = 2e$ again we get

$$\left(\frac{2e}{n}\right)^2 \cdot n \le 2e + n(n-1).$$

This is a quadratic inequality in e:

$$4e^2 - 2en - n^2(n-1) \le 0.$$

Solving the inequality gives the upper bound in the theorem. □

Essentially the same question appears in a paper by *Zarankiewicz* [182]. He posed the following problem: *what is the maximum number of '1'-s in an $m \times n$ 0-1 matrix if no $s \times t$ submatrix is the all '1' matrix?*

Of course, this problem is equivalent to a Turán type problem for bipartite graphs: what is the maximum number of edges in a bipartite graph with classes of size m and n if the graph cannot contain a subgraph $K_{s,t}$ in such a way that the s vertices of it are in the class of size m?

To fix notation, let us formally define Zarankiewicz numbers in the bipartite graph terminology.

Definition 12.10. *A bipartite graph with classes A, B and edges E is denoted as $G = (A, B, E)$. It is $K_{s,t}$-free if it does not contain s vertices in A and t vertices in B that span a subgraph isomorphic to $K_{s,t}$. We call $(|A|, |B|)$ the size of G. The maximum number of edges a $K_{s,t}$-free bipartite graph of size (m, n) can have is denoted by $Z_{s,t}(m, n)$, and is called the corresponding Zarankiewicz number.*

In the matrix formulation rows correspond to the vertices in the class of size m, columns to the vertices in the class of size n. Two vertices are joined by an edge if the element in the corresponding row and column is a 1. The non-existence of an all '1' minor of size $s \times t$ implies that this bipartite graph is $K_{s,t}$-free.

We shall concentrate on the particular case $s = t$ and $n = m$, but also mention some results about the general case. We continue with the case $s = t = 2$ and study the bipartite version for $m = n$ in detail.

Theorem 12.11 (Reiman). *There are at most $\frac{n}{2}(1 + \sqrt{4n - 3})$ '1'-s in an $n \times n$ 0-1 matrix if no 2×2 submatrix is the all '1' matrix. Hence*

$$Z_{2,2}(n, n) \le \frac{n}{2}(1 + \sqrt{4n - 3}).$$

Proof. First, let us see how the previous proof has to be modified for the bipartite case. Our graph has $2n$ vertices and the "free" vertices of adjacent edges must belong to the same class in the bipartition. So in our starting inequality we have $2\binom{n}{2}$ instead of $\binom{2n}{2}$. Copying the previous proof gives the inequality

$$e^2 - ne - n^2(n-1) \le 0,$$

and solving the inequality indeed implies $e \le \frac{n}{2}(1 + \sqrt{4n - 3})$. □

Let us also see the original proof by *Reiman* using the matrix formulation of the problem.

Proof. Let us denote the row vectors of the $0 - 1$ matrix A by $\mathbf{a}_i$, its column vectors by $\mathbf{a}_j^*$. Since there are no all "1" submatrices it implies $\mathbf{a}_i\mathbf{a}_j = 0$ or 1, and similarly $\mathbf{a}_i^*\mathbf{a}_j^* = 0$ or 1. Here the product denotes the inner product of the vectors. The inner product $\mathbf{a}_i\mathbf{a}_i = r_i$ is the number of '1'-s in the i-th row, $\mathbf{a}_i^*\mathbf{a}_i^* = c_i$ is the number of '1'-s in the i-th column, where $i = 1, \ldots, n$. Clearly, $\sum_{i=1}^{n} r_i = \sum_{i=1}^{n} c_i = e$, the total number of '1'-s in our matrix A.

Compute the inner product

$$(\mathbf{a}_1 + \mathbf{a}_2 + \ldots + \mathbf{a}_n)^2.$$

Squaring a vector means its inner product with itself. On the one hand, it is $c_1^2 + \ldots + c_n^2$; on the other hand, using properties of inner product it is

$$(\mathbf{a}_1^2 + \ldots + \mathbf{a}_n^2) + 2(\mathbf{a}_1\mathbf{a}_2 + \ldots + \mathbf{a}_{n-1}\mathbf{a}_n).$$

The first term here is just $r_1 + \ldots + r_n = e$, and the second term is at most $n(n - 1)$, since all summands in the second term are either 0 or 1. By the inequality between quadratic and arithmetic means, the sum $c_1^2 + \ldots + c_n^2$ is at least e^2/n. This implies

$$\frac{e^2}{n} \le e + n(n - 1),$$

which is just the inequality appearing in the previous proof.   □

The result can easily be extended to $m \times n$ matrices, giving the following upper bound.

Theorem 12.12 (Reiman). *Let A be a $0 - 1$ matrix with m rows and n columns. Assume that no 2×2 submatrix of A is the all '1' matrix. Then the total number e of '1'-s in the matrix and the Zarankiewicz number $Z_{2,2}(m, n)$ satisfy the inequalities*

$$e \le Z_{2,2}(m, n) \le \frac{1}{2}\left(n + \sqrt{n^2 + 4nm(m - 1)}\right). \tag{12.5}$$

The proof given for the case $n = m$ can be copied, so we leave it as Exercise 12.3. In general, it is not difficult to determine when the above bound is sharp, see Exercise 12.4.

To formulate the result for bipartite graphs has the advantage that the extremal graphs can be described easily. The incidence graph (Levi graph) of an incidence geometry was defined in Chapter 10, Definition 10.2. When the incidence geometry satisfies that every two distinct points are incident with at most one line, then the resulting incidence graph cannot contain a C_4.

Of course, for any bipartite graph with bipartition $L \cup U$ one can define an incidence geometry whose points are the vertices of U, lines are the vertices of L, and incidence is defined by the edges. So $u \in U$ is incident with $\ell \in L$ if and only if (u, ℓ) is an edge.

Theorem 12.13 (Reiman). *If an $n \times n$ 0−1 matrix contains no 2×2 submatrix consisting of four '1'-s, the matrix contains precisely $e = \frac{n}{2}(1 + \sqrt{4n - 3})$ '1'-s and $n > 3$, then it is the incidence matrix of a non-degenerate projective plane.*

Proof. The second proof of Theorem 12.11 implies that in case of equality every pair of distinct rows has inner product 1 (and the same holds for columns), moreover, every row and column have the same number of '1'-s. This implies that for the incidence geometry coming from the bipartite graph defined by our $0 - 1$ matrix, the first two axioms of projective planes are valid and lines have the same size. As $n > 3$, this implies that our bipartite graph is the incidence graph of a non-degenerate projective plane. □

We could have cited the theorem of De Bruijn and Erdős: our incidence geometry is a linear space, it has the same number of points and lines, and all lines are incident with the same number of points which is greater than 2 by $n > 3$. These imply again that our bipartite graph is indeed the incidence graph of a non-degenerate projective plane.

Let us remark that in the cases $n = 1$ and $n = 3$ we can have equality. The case $n = 1$ is trivial and if $n = 3$ then we get a degenerate linear space on 3 points.

A similar result can be deduced when $m = k^2$ and $n = k^2 + k$ with $k > 1$. In this case the maximum number of edges is $k^2(k + 1)$ and in the case of equality the graph is the incidence graph of an affine plane. We leave the proof as Exercise 12.5.

Let us repeat that, using the language of bipartite graphs, the extremal graph is the incidence graph of a projective plane; that is vertices in one class correspond to points, vertices in the other class correspond to lines and edges correspond to incident point-line pairs. This description suggests that we may "identify" the vertices of the two classes and this way get close(r) to the bound for the number of edges in a not necessarily bipartite C_4-free graph. Let Π be a projective plane of order n, π be a polarity of Π. Let the vertices of a graph G be the points of Π and let xy be an edge if $x \neq y$ and yIx^π. We will call this graph the *Erdős Rényi graph* as they introduced it first. As π is a polarity, the graph is undirected. One can also see that it is C_4-free. Assume to the contrary that x, y, z, u is a four-cycle in G. Then both y and u are common points of x^π and z^π. This is impossible because two lines intersect in a unique point.

The Erdős–Rényi graph is not regular, because self-conjugate points have one fewer neighbours than other points. The following theorem of Baer gives an estimate on the number of self-conjugate points of a polarity of an arbitrary projective plane.

Theorem 12.14 (Baer). *Let Π be a projective plane of order n and suppose that Π admits a polarity π. Then π has at least $n + 1$ self-conjugate points.*

Proof. List the points of the plane in some order $P_1, P_2, \ldots, P_{n^2+n+1}$ and let us order the lines in such a way that $\ell_i = P_i^\pi$. By the definition of a polarity (see Definition 4.20), $P_i \mathbin{I} \ell_j$ implies $P_i^\pi \mathbin{I} \ell_j^\pi$, that is $P_j \mathbin{I} \ell_i$. Thus the incidence matrix A of the plane corresponding to this order will be symmetric. This means that $AA^T = A^2$, hence, by Lemma 1.15, we have that the eigenvalues of A^2 are $(n+1)^2$ with multiplicity one, and n with multiplicity $n^2 + n$. The first eigenvalue corresponds to the all-1 vector as eigenvector. This implies that the eigenvalues of the symmetric matrix A can be $\pm(n+1)$ and $\pm\sqrt{n}$. The all-1 vector is an eigenvector of A and the corresponding eigenvalue is $(n+1)$. Since the eigenvalue $(n+1)^2$ of A^2 has multiplicity one, $-(n+1)$ cannot be an eigenvector of A. Assume that the eigenvector $\sqrt{n}$ has multiplicity m. Then $-\sqrt{n}$ has multiplicity $n^2 + n - m$. The number of self-conjugate points of π is just the trace of A, which is clearly a non-negative integer. This can also be computed using the eigenvalues and we get $\mathrm{Tr}(A) = n + 1 - (n^2 + n - 2m)\sqrt{n}$. This expression cannot be zero, so π has a self-conjugate point. Observe that for a non-square n the expression for $\mathrm{Tr}(A)$ only gives an integer value if the term containing $\sqrt{n}$ has coefficient 0, so we have exactly $n + 1$ self-conjugate points.

Let us now assume that n is a square and let P be a self-conjugate point of π. Let us consider the case when n is odd. Take a line e through P which is not the polar of P and introduce the following mapping π^* on the points of $e \setminus \{P\}$: for a point R let $\pi^*(R) = e \cap R^\pi$. The defining properties of π imply that π^* is an involutorial mapping. As n is odd, the line e has to contain a point fixed by π^*, that is, a self-conjugate point of π. Therefore, the number of self-conjugate points is at least $n + 1$. Essentially the same argument shows that for n even every line contains a self-conjugate point, hence by Lemma 6.26, the number of self-conjugate points is at least $n + 1$. Indeed, take a line e. If it is self-conjugate, then its pole is a self-conjugate point on e. If not, then define the above mapping π^* on the points of e. As it is involutorial and $n + 1$ is odd, there has to be a point fixed by π^*, that is, a self-conjugate point of π. $\qquad\square$

When the number of self-conjugate points is $n + 1$, a more careful analysis of the proof above shows that the self-conjugate points form an oval if n is odd and a line if n is even (see Exercise 12.6). It is also clear (see Exercise 12.7) that any self-conjugate line contains exactly one self-conjugate point, namely its pole. Hence the set of self-conjugate points has at least as many tangent lines as points, and this implies that the number of self-conjugate points is at most $n\sqrt{n} + 1$ (see Exercise 12.8, where the proof of Theorem 6.47 has to be modified slightly). According to this exercise, if a polarity has $n\sqrt{n} + 1$ self-conjugate points then they form a unital (in other words, each line contains either 1 or $\sqrt{n} + 1$ self-conjugate points).

According to this result, there are at least $n + 1$ points of degree n in G and the remaining points have degree $n + 1$. Thus the number of edges of G is at most $\frac{1}{2}n(n + 1)^2$. If we start from the plane $\mathrm{PG}(2, q)$ and π is an orthogonal polarity, then such a graph can be given explicitly. As prime numbers are a dense subset of natural numbers, this construction also shows that Theorem 12.9 gives the correct order of magnitude asymptotically. *Füredi* [70] also showed that for $n \geq 15$ a C_4-free graph on $N = n^2 + n + 1$ vertices contains at most $\frac{1}{2}n(n+1)^2$ edges. He also proved that in the case of extremal graphs only the above Erdős–Rényi graph occurs for $q > 13$.

The next theorem, due to *Erdős, Rényi* and *T. Sós,* is also related to polarities in finite planes.

Theorem 12.15 (Friendship theorem). *Let G be a simple graph on the vertex set V, $|V| = n$. Assume that every two vertices of G have precisely one common neighbour. Then n is odd and the graph is unique: one vertex v is joined to each other vertex and the remaining edges form a perfect matching (one-factor) on $V \setminus \{v\}$.*

Proof. We define an incidence geometry $(\mathcal{P}, \mathcal{L}, \mathrm{I})$ and show that it is a projective plane admitting a polarity. Let $\mathcal{P} = V$. For every $x \in V$ consider the set of neighbours L_x of x and call it a line, that is let $\mathcal{L} = \{L_x : x \in V\}$. The incidence is defined as set theoretical containment. If a and b are two vertices, then there is precisely one x such that $a, b \in L_x$, namely the common neighbour of a and b. This shows that the first axiom of projective planes is satisfied. Similarly, if L_x and L_y are two distinct lines, then they have a unique point in common, namely the common neighbour of x and y. So, our incidence geometry is a possibly degenerate projective plane. This incidence geometry is a degenerate linear space if and only if there is a vertex adjacent to all the remaining vertices. In this case the graph must clearly be the "windmill graph" described in the theorem. If the linear space is non-degenerate, then it is a projective plane, by the theorem of De Bruijn-Erdős, since it has the same number of points and lines. One can easily check that the mapping $x \mapsto L_x$, $L_u \mapsto u$ is a polarity of our projective plane. This polarity has no self-conjugate point, since $x \notin L_x$. This contradicts the Theorem of Baer. $\square$

This result has been rediscovered several times. For example, it appears in a paper by *Wilf*, where it is attributed to *Higman*. There are also several different proofs of the Friendship Theorem [2].

Let us now turn our attention to the general case of s and t. The proof, based on counting adjacent pairs of edges, can be extended to $K_{s,t}$-free graphs. Such a general result was first obtained by *Kővári, T. Sós* and *Turán* [107].

Theorem 12.16 (Kővári, T. Sós, Turán). *If $s \geq t$ are fixed, then a $K_{s,t}$-free simple graph contains at most*

$$\frac{1}{2}((s-1)^{1/t}n^{2-1/t} + (t-1)n)$$

edges.

Proof. We will count in two different ways the sets of t disjoint edges having a common point. For the sake of brevity we call them $K_{1,t}$-s. On the one hand, the number of $K_{1,t}$-s is at most $(s-1)\binom{n}{t}$, since for a set of t endpoints there are at most $s-1$ $K_{1,t}$-s; otherwise we would get a $K_{s,t}$ in G. On the other hand, counting these $K_{1,t}$-s at their common vertex we get exactly $\sum_{v \in V(G)} \binom{\deg(v)}{t}$. We are going to use Jensen's inequality for the real function

$$\binom{x}{t} = x(x-1)\ldots(x-t+1)/t!.$$

Unfortunately, this function is not convex for $0 < x < t-1$, so we modify it. Instead of f we use the convex function f^* for which $f^*(x) = f(x)$, when $x \geq t-1$ (or $x \leq 0$), and $f^*(x) = 0$, when $0 < x < t-1$. It is clear that $f^*(x)$ is a convex function. According to Jensen's inequality

$$\sum_{v \in V(G)} \binom{\deg(v)}{t} \geq nf^*\left(\sum_{v \in V(G)} \frac{\deg(v)}{n}\right).$$

When the value of f^* is 0, we get $e \leq \frac{1}{2}n(t-1)$, which is better than the upper bound in the theorem. So, it is sufficient to consider the case when

$$f^*\left(\sum_{v \in V(G)} \frac{\deg(v)}{n}\right) = \binom{\sum_{v \in V(G)} \deg(v)/n}{t}.$$

By substituting the number of edges in our inequality we get

$$(s-1)\binom{n}{t} \geq n \cdot \binom{2e/n}{t}.$$

The binomial coefficient on the left hand side can be bounded from above by $n^t/t!$. Similarly, the binomial coefficient on the right hand side can be bounded from below by $(\frac{2e}{n} - t + 1)^t/t!$. Multiplying the two sides by $t!$ and using the above bounds

$$(s-1)n^t \geq n\left(\frac{2e}{n} - t + 1\right)^t$$

follows. After taking t-th roots we get

$$(s-1)^{1/t}n^{1-1/t} \geq \frac{2e}{n} - t + 1,$$

which implies the estimate in the theorem after elementary calculations. □

The previous proofs (see Theorems 12.9, 12.11 and 12.13) always used the inequality between quadratic and arithmetic means. Instead of this, one can also use Jensen's inequality for convex functions as we did for the Theorem of Kővári, T. Sós and Turán. Let $I \subset \mathbb{R}$ be an interval and $f: I \to \mathbb{R}$ a convex function. Then for any n, and any $x_1, \ldots, x_n \in I$, $p_1, \ldots, p_n > 0$, $p_1 + p_2 + \cdots + p_n = 1$, we have

$$f(p_1 x_1 + \ldots + p_n x_n) \leq p_1 f(x_1) + \ldots + p_n f(x_n).$$

In particular, when all the p_i's are equal to $1/n$, we get

$$f\left(\frac{x_1 + \ldots + x_n}{n}\right) \leq \frac{f(x_1) + \ldots + f(x_n)}{n}. \qquad (12.6)$$

For example, the function $x \mapsto x^2$ is convex on $\mathbb{R}$, so (this special case of) Jensen's inequality implies the inequality between quadratic and arithmetic means. If the function is strictly convex (intuitively, its graph does not contain a line segment), then equality implies $x_1 = \ldots = x_n$.

For the next part the reader is referred to the papers by *Roman* [146], and by *Damásdi, Héger* and *Szőnyi* [51]. We follow the approach of the latter paper closely. Both the inequality between quadratic and arithmetic means and the special case of Jensen's inequality mentioned above has a sharpening for the case when $x_1, \ldots, x_n$ are integers. Note that we have a fixed arithmetic mean A and wish to minimize the sum on the right hand side. If the arithmetic mean A is not an integer, then $x_1 = \ldots = x_n$ cannot occur in Inequality (12.6). In this case, the minimum value of the quadratic mean is obtained when the x_i's are either $\lfloor A \rfloor$ or $\lceil A \rceil$. In other words, the difference between x_i and x_j is at most one. Indeed, if $x_i - x_j \geq 2$, then replacing x_i by $x_i - 1$ and x_j by $x_j + 1$ would not change the arithmetic mean and would lower the right hand side. This idea is used in the proof of Turán's theorem on K_{r+1}-free graphs.

Let us now state the sharpening of the special case (when $p_i = 1/n$) of Jensen's inequality mentioned above for integers. A convenient version to use the sharpening of Jensen's is the following result by *Roman*.

Theorem 12.17 (Roman's inequality [146]). *Let $f \colon \mathbb{R} \to \mathbb{R}$ be an increasing convex function, $n \in \mathbb{N}$, $x_1, \ldots, x_n \in \mathbb{Z}$ and $p \in \mathbb{Z}$. Then*

$$\sum_{i=1}^{n} x_i \leq \frac{\sum_{i=1}^{n} f(x_i)}{f(p+1) - f(p)} + n \cdot \frac{pf(p+1) - (p+1)f(p)}{f(p+1) - f(p)}.$$

Equality holds if and only if $x_i \in \{p, p+1\}$ for every $1 \leq i \leq n$, or $\{x_1, \ldots, x_n, p, p+1\} \subset I$ for an interval I on which f is linear.

Proof. Let

$$a = \frac{1}{f(p+1) - f(p)}, \quad c = \frac{pf(p+1) - (p+1)f(p)}{f(p+1) - f(p)}$$

and define the function $F \colon \mathbb{R} \to \mathbb{R}$ as $F(x) = af(x) - x + c$. Then $F(p) = F(p+1) = 0$. Since f is increasing and convex, $a > 0$, hence $F(x) = af(x) - x + c$ is also convex. Therefore either $F(x) = 0$ on an interval I containing $[p, p+1]$, or the only values for which $F(x) = 0$ are p and $p+1$, two consecutive integers. In both cases $F(x) \geq 0$ for every integer x.

Hence

$$0 \leq \sum_{i=1}^{n} F(x_i) = \sum_{i=1}^{n} (af(x_i) - x_i + c) = a \sum_{i=1}^{n} f(x_i) - \sum_{i=1}^{n} x_i + nc,$$

thus

$$\sum_{i=1}^{n} x_i \leq a \sum_{i=1}^{n} f(x_i) + nc.$$

Substituting the values of a and c gives the formula stated above. Equality holds if and only if $F(x_i) = 0$ for every $1 \leq i \leq n$, which implies one of the two stated cases. $\qquad\square$

Corollary 12.18 (Roman [146]). *Let* $f(x) = \binom{x}{k} = \frac{x(x-1)\cdots(x-k+1)}{k(k-1)\cdots 1}$, $x_i \geq k-1$, $x_i \in \mathbb{N}$ $(i = 1, 2, \ldots, n)$, $p \geq k-1$, $p \in \mathbb{N}$. *Then*

$$\sum_{i=1}^{n} x_i \leq \frac{\sum_{i=1}^{n} \binom{x_i}{k}}{\binom{p}{k-1}} + n \cdot \frac{(p+1)(k-1)}{k}.$$

Equality holds if and only if $x_i \in \{p, p+1\}$ *for all* $1 \leq i \leq n$.

Proof. Again, instead of f we use the function f^* for which $f^*(x) = f(x)$, when $x \geq k-1$, and $f^*(x) = 0$, when $x < k-1$. It is clear that $f^*(x)$ is increasing and convex, so we can use Roman's inequality which gives the statement at once. $\qquad\square$

The consequence of Roman's bound for Zarankiewicz numbers is the following.

Theorem 12.19 (Roman's bound [146]). *Let* $G = (A, B, E)$ *be a* $K_{s,t}$-*free bipartite graph of size* (m, n), *and let* $p \geq s-1$. *Then the number of edges in* G *satisfies the inequality*

$$|E| \leq \frac{(t-1)}{\binom{p}{s-1}} \binom{m}{s} + n \cdot \frac{(p+1)(s-1)}{s}.$$

Equality holds if and only if every vertex in B *has degree* p *or* $p+1$ *and every s-tuple in* A *has exactly* $t-1$ *common neighbours in* B.

Proof. Let d_i denote the degrees of the vertices in B for $i = 1, \ldots, n$. By adding edges, we may assume that $d_i \geq s-1$ for all $i \in \{1, \ldots, n\}$ while preserving G being $K_{s,t}$-free. Count the number of $K_{s,1}$ subgraphs in G (with the single node in B). On the one hand, we get that their exact number is $\sum_{i=1}^{n} \binom{d_i}{s}$; on the other hand, since G is $K_{s,t}$-free, the number is at most $(t-1)\binom{m}{s}$. Hence

$$\sum_{i=1}^{n} \binom{d_i}{s} \leq (t-1) \binom{m}{s}.$$

Corollary 12.18 with $f(x) = \binom{x}{s}$ yields

$$|E| = \sum_{i=1}^{n} d_i \le \frac{\sum_{i=1}^{n} \binom{d_i}{s}}{\binom{p}{s-1}} + n \cdot \frac{(p+1)(s-1)}{s}.$$

Combining these two inequalities, the proof is finished. The case of equality is straightforward. □

We shall use an abbreviation for the upper bound in the previous theorem.

Definition 12.20. *For* $s, t, m, n, p \in \mathbb{N}$, $p \ge s-1$, *let*

$$R(s, t, m, n, p) := \frac{(t-1)}{\binom{p}{s-1}} \binom{m}{s} + n \cdot \frac{(p+1)(s-1)}{s}.$$

Let us now return to the case $s = t = 2$. In this special case, we saw that affine and projective planes meet Reiman's general upper bound. It is natural to ask what happens for parameter sets between affine and projective planes. This is not known in general, but we shall discuss some results in this direction.

First of all, we need an embeddability result by *Metsch* [129]. The structures in the theorem are sometimes called partial projective planes.

Theorem 12.21 (Metsch). *Let* $n \ge 15$, $(\mathcal{P}_1, \mathcal{E}_1, I)$ *be an incidence geometry with* $|\mathcal{P}_1| = n^2 + n + 1$, $|\mathcal{E}_1| \ge n^2 + 2$ *such that every line in* $\mathcal{E}_1$ *is incident with* $n + 1$ *points of* $\mathcal{P}_1$ *and every two lines have at most one point in common. Then a projective plane* Π *of order* n *exists and* $(\mathcal{P}_1, \mathcal{E}_1, I)$ *can be embedded into* Π.

Actually, the conditions of Metsch's embeddability theorem can be relaxed so that it can be used directly for $K_{2,2}$-free incidence graphs.

Lemma 12.22. *Let* $n \ge 15$, $(\mathcal{P}, \mathcal{E}, I)$ *be an incidence geometry with* $|\mathcal{P}| = n^2 + n + 1$, $|\mathcal{E}| \ge n^2 + 2$ *such that every line in* $\mathcal{E}$ *is incident with at least* $n + 1$ *points of* $\mathcal{P}$, *and every two lines have at most one point in common. Then there exists a projective plane* Π *of order* n, *so that* $(\mathcal{P}, \mathcal{E}, I)$ *can be embedded into* Π. *In particular, every line in* $\mathcal{E}$ *is incident with exactly* $n + 1$ *points of* $\mathcal{P}$.

Proof. Let G be the incidence graph of the geometry. By deleting edges from G, we can obtain a graph $G' = (\mathcal{P}, \mathcal{E})$ in which the vertices of $\mathcal{E}$ have degree exactly $n + 1$. Then, by Theorem 12.21, G' is a subgraph of the incidence graph of a projective plane Π of order n. Suppose that there is a line $\ell \in \mathcal{E}$ whose degree in G is at least $n + 2$. This guarantees the existence of a point P such that ℓ is incident with P in G, but not in Π. Then each of the $n + 1$ lines passing through P in Π intersects ℓ in a point different from P. As $|\mathcal{E}| \ge n^2 + 1$, at least one of these lines is a line of G as well, but it intersects ℓ in at least two points in G, giving a contradiction. Hence every line has $n + 1$ points in G. □

Theorem 12.23 (Damásdi, Héger, Szőnyi). *Let $n \geq 15$ and $c \leq n/2$. Then*

$$Z_{2,2}(n^2 + n + 1 - c, n^2 + n + 1) \leq (n^2 + n + 1 - c)(n + 1).$$

Equality holds if and only if a projective plane of order n exists. Moreover, graphs giving equality are subgraphs of the incidence graph of a projective plane of order n.

Proof. If a projective plane of order n exists, delete c of its lines. This yields an incidence geometry with $n^2 + n + 1$ points and $n^2 + n + 1 - c$ lines, hence a graph on $(n^2 + n + 1 - c, n^2 + n + 1)$ vertices and $(n^2 + n + 1 - c)(n + 1)$ edges.

Suppose that $G = (A, B, E)$ is a $K_{2,2}$-free graph on $(n^2+n+1-c, n^2+n+1)$ vertices and $|E| \geq |A|(n + 1)$ edges. Let m be the number of vertices in A of degree at most n (low-degree vertices). Assume that $m \geq n-c$. Delete $(n - c)$ low-degree vertices to obtain a graph G' on $(n^2 + 1, n^2 + n + 1)$ vertices with at least $(n^2 + 1)(n + 1) + (n - c)$ edges. Roman's bound with $p = n$ (Theorem 12.19) implies $Z_{2,2}(n^2 + 1, n^2 + n + 1) \leq (n^2 + 1)(n + 1) + (n - 1)/2$, hence $n - c \leq (n - 1)/2$. This contradicts $c \leq n/2$, thus $m < n - c$ must hold.

Now delete all the low-degree vertices from A to obtain a graph G' on the vertex sets (A', B) with $|A'| \geq n^2 + 2$, $|B| = n^2 + n + 1$. Then every vertex in A' has degree at least $n + 1$, hence we can apply Lemma 12.22 to derive that G' can be embedded into a projective plane Π of order n. Therefore, every vertex in A' has degree $n + 1$ which, combined with $|E| \geq |A|(n + 1)$, yields that every vertex in A has degree $n+1$ (in G). Thus G itself can be embedded into Π. $\qquad\qquad\square$

Now we turn to results related to affine planes. The next observation was already made by *Guy* [76] in the sixties.

Theorem 12.24 (Guy). *Let $Z_{2,2}(M, N) \leq e = a(d + 1) + bd$, $a, b, d \in \mathbb{N}$ and $M = a + b$. Let $c \in \mathbb{N}$.*

  1. *If $b = 0$, then $Z_{2,2}(M + c, N) \leq (M + c)(d + 1)$.*

  2. *If $b > 0$ or $Z_{2,2}(M + 1, N) \leq e + d$, then $Z_{2,2}(M + c, N) \leq e + cd$.*

In both cases equality for some $c \geq 1$ implies that equality holds for all $c' \in \mathbb{N}$, $0 \leq c' < c$, too. Moreover, any $K_{2,2}$-free graph G of size $(M + c, N)$ having the maximum number of edges induces a graph of size $(M + c - 1, N)$ having the maximum number of edges.

Proof. We prove the second case by induction on c. If $c = 0$, the statement is trivial. Suppose $Z_{2,2}(M+c, N) \geq e+cd$, and let $G = G(A, B)$ be an extremal graph of size $(M + c, N)$. There is no vertex of degree $< d$ in A; otherwise removing such a vertex we would obtain a graph of size $(M + c - 1, N)$ with more than $e + (c - 1)d$ edges. This is impossible by the induction hypothesis. Now either by taking a subgraph on (M, N) vertices in the case of $b \geq 1$, or by taking a subgraph on $(M + 1, N)$ vertices and considering our assumption in the case of $b = 0$, we find a vertex in A of degree d. Removing this vertex we obtain a graph of size $(M + c - 1, N)$ with at least $e + (c - 1)d$ edges. The induction hypothesis guarantees equality here. Therefore, $Z_{2,2}(M + c - 1, N) = e + (c - 1)d$, and $Z_{2,2}(M + c, N) = e(G) = e + cd$.

The proof of the first case is very similar; we only have to see that all vertices of A have degree $\geq d + 1$, and that there is at least one vertex of that degree. □

In the case of projective planes the embeddability theorem of Metsch was crucial in our proof. In the case of affine planes a similar embeddability result is due to *Totten* [173]. Recall that an affine plane of order n is always embeddable into a projective plane of order n. The next result is about incidence structures that are projective planes minus two lines.

Theorem 12.25 (Totten [173]). *Let $S = (\mathcal{P}, \mathcal{E})$ be a finite linear space with $|\mathcal{P}| = n^2 - n$, $|\mathcal{E}| = n^2 + n - 1$, $2 \leq n \neq 4$, and every point having degree $n + 1$. Then S can be embedded into a projective plane of order n.*

In the next result the upper bounds are direct consequences of Theorem 12.24, Totten's result is only used for characterizing graphs attaining equality in (12.8).

Corollary 12.26 (Damásdi, Héger, Szőnyi). *Let $c \in \mathbb{N}$. Then*

$$Z_{2,2}(n^2 + c, n^2 + n) \leq n^2(n + 1) + cn, \tag{12.7}$$

$$Z_{2,2}(n^2 - n + c, n^2 + n - 1) \leq (n^2 - n)(n + 1) + cn, \tag{12.8}$$

$$Z_{2,2}(n^2 - 2n + 1 + c, n^2 + n - 2) \leq (n^2 - 2n + 1)(n + 1) + cn,$$
$$\text{if } n \geq 4. \tag{12.9}$$

Equality can be reached in all three inequalities if a projective plane of order n exists and $c \leq n + 1$, or $c \leq 2n$, or $c \leq 3(n - 1)$, respectively.

Moreover, if $c \leq n + 1$, or $c \leq 2n$ and $2 \leq n \neq 4$, then graphs reaching the bound in (12.7) or in (12.8), respectively, can be embedded into a projective plane of order n.

Proof. For the parameters of an affine plane of order n, the incidence graph is of size $(n^2, n^2 + n)$. We have equality in Reiman's, hence also in Roman's bound (see Exercise 12.5). Moreover, $Z_{2,2}(n^2 + 1, n^2 + n) \leq R(2, 2, n^2 + 1, n^2 + n, n) = n^2(n + 1) + n$, hence Theorem 12.24 can be applied with $a = n^2$, $b = 0$, $d = n$, to obtain the upper bound (12.7). In the case of (12.8) and (12.9), we can simply calculate that $R(2, 2, n^2 - n, n^2 + n - 1, n - 1) = n^3 - n$, and $R(2, 2, n^2 - n + 1, n^2 + n - 1, n - 1) = n^3$, which imply (12.8) using Theorem 12.24. Similarly, one can compute that $R(2, 2, n^2 - 2n + 1, n^2 + n - 2, n - 2) = (n^2 - 2n + 1)(n + 1) = e$, and $R(2, 2, n^2 - 2n + 2, n^2 + n - 2, n - 2) = e + n + 1/(n - 2)$, whence $Z_{2,2}(n^2 - 2n + 2, n^2 + n - 2) \leq e + n$.

By taking a projective plane of order n, and deleting one, two, or three of its lines and all but c of their points that are contained in one of the three lines only, we can reach equality in (12.7), (12.8) and (12.9), respectively.

In (12.7), Theorem 12.24 also provides an affine plane of order n as an induced subgraph in graphs attaining equality. Now the c extra points of degree n must be incident with non-intersecting lines to avoid C_4's in the graph; that is, they are the common points of c distinct parallel classes. Adding the missing $n+1-c$ ideal points and the line at infinity, we obtain a projective plane of order n.

In (12.8), Theorem 12.24 provides us with an extremal C_4-free subgraph $G = G(A, B)$ on (n^2-n, n^2+n-1) vertices and $(n^2-n)(n+1)$ edges in graphs reaching equality. As $R(2, 2, n^2 - n, n^2 + n - 1, n - 1) = (n^2 - n)(n + 1)$, every line in B has degree $n - 1$ or n, and any two distinct points are contained in a (unique) line. Let P be a point in A, and let α and β denote the number of lines of degree $n-1$ and n through P, respectively. Then counting the total number of points on the lines through P we obtain $n^2 - n = 1 + \alpha(n - 2) + \beta(n - 1)$, which has two relevant solutions, namely $\alpha = n$, $\beta = 1$ and $\alpha = 1$, $\beta = n - 1$. In both cases each point has degree at most $n+1$ and we have equality because of the number of edges. Thus, by Theorem 12.25, G is the incidence graph of a projective plane Π of order n minus two lines. As before, it is easy to see that the embedding extends to the c extra points as well. □

Let us now go back to the general case of $K_{s,t}$-free graphs. The most interesting case is $s = t$. It is not known in general whether the order of magnitude of the upper bound in the theorem of Kővári, T. Sós, Turán can be reached. In the particular case $s = t = 3$ there are some constructions and we are going to discuss this particular case in more detail. We first consider a construction by *Brown*, which has an intuitive idea from classical geometry behind it: if we join the points at distance 1 in the classical Euclidean 3-space, then this graph does not contain a $K_{3,3}$. Indeed, if we take three points, then the points at distance 1 from each of these points are on a line perpendicular to the plane of the three points and intersecting it in the centre of the circumscribed circle of the triangle. Therefore, there can be at most two such points, one on both sides of the plane containing the three points. The goal in the next construction is to translate this idea for finite 3-dimensional spaces.

Theorem 12.27 (Brown's construction). *Let k_1, k_2 be chosen so that the equation $X^2 + k_1 Y^2 + k_2 Z^2 = 1$ defines an elliptic quadric $\mathcal{E}$ of $AG(3, q)$, q odd. Define a graph on the points of $AG(3, q)$ in the following way: the points (x, y, z) and (a, b, c) are joined by an edge if and only if $(x - a)^2 + k_1(y - b)^2 + k_2(z - c)^2 = 1$.*

This graph has $n = q^3$ vertices, asymptotically $\sim \frac{1}{2}n^{5/3}$ edges and contains no $K_{3,3}$.

Proof. The quadric $\mathcal{E}$ meets the hyperplane at infinity in the conic $X^2 + k_1 Y^2 + k_2 Z^2 = 0$, so the number of its affine points is $q^2 + 1 - (q + 1) = q^2 - q$. The neighbours of a point $A = (a, b, c)$ are on a translate of $\mathcal{E}$, hence every vertex in our graph has degree $q^2 - q$. Thus the number of edges is $e = \frac{1}{2}q^3(q^2 - q)$. Since $n = q^3$, the number of edges indeed satisfies $e \sim \frac{1}{2}n^{5/3}$.

Let $A = (a, b, c)$, $D = (d, e, f)$ and $G = (g, h, i)$ be three distinct points. Denote by $\mathcal{E}_A$, $\mathcal{E}_D$ and $\mathcal{E}_G$ the three translates of $\mathcal{E}$, that is the set of neighbours of A, D and G. The equation of these quadrics is

$$(X - a)^2 + k_1(Y - b)^2 + k_2(Z - c)^2 = 1,$$

$$(X - d)^2 + k_1(Y - e)^2 + k_2(Z - f)^2 = 1,$$

$$(X - g)^2 + k_1(Y - h)^2 + k_2(Z - i)^2 = 1,$$

respectively. Subtract the first equation from the second and third one. We get two linear equations,

$$(a-d)X + k_1(b-e)Y + k_2(c-f)Z + (d^2 + k_1e^2 + k_2f^2 - a^2 - k_1b^2 - k_2c^2)/2 = 0,$$

$$(a-g)X + k_1(b-h)Y + k_2(c-i)Z + (g^2 + k_1h^2 + k_2i^2 - a^2 - k_1b^2 - k_2c^2)/2 = 0.$$

It is not difficult to check that these equations cannot coincide. This means that the planes defined by them either meet each other in a line ℓ, or they are disjoint. Since $\mathcal{E}_A$ contains no line, ℓ intersects it in at most two points. Hence the number of common neighbours of A, D and G is at most two, so the graph cannot contain a $K_{3,3}$. $\qquad\square$

Brown's construction shows the sharpness of the bound of Kővári, T. Sós, Turán for $s = t = 3$ regarding its order of magnitude. *Füredi* [69] proved that Brown's construction gives an asymptotically exact result, that is, not just the exponent but also the constant in the main term is correct. To show this, he had to improve on the upper bound. We follow his proof closely. First, we need a lemma which is in some sense a converse of Jensen's inequality for a particular function f, like Roman's inequality.

Lemma 12.28 (Füredi). *Let v be an integer, $y, x_1, x_2, \ldots, x_v \geq 0$ be real numbers. If $\sum_{i=1}^{v} \binom{x_i}{3} \leq \binom{y}{3}$, then*

$$\sum_{i=1}^{v} x_i \leq yv^{2/3} + 2v.$$

Proof. Let $S = \sum_{i=1}^{v} x_i$. Our assertion is trivial for $S \leq 2v$. For $S > 2v$; similarly to the proof of the theorem of Kővári, T. Sós, Turán, we can apply Jensen's inequality for the function $x \mapsto \binom{x}{3}$, which is convex for $x \geq 2$. This gives $v\binom{S/v}{3} \leq \sum_{i=1}^{v} \binom{x_i}{3}$, and by our condition it is at most $\binom{y}{3}$. Therefore,

$$v \leq \frac{x_0}{S/v} \frac{x_0 - 1}{S/v - 1} \frac{x_0 - 2}{S/v - 2} \leq \left(\frac{x_0}{S/v - 2}\right)^3.$$

After expressing S we get the required result. $\qquad\square$

Theorem 12.29 (Füredi). *Let G be a $K_{3,3}$-free simple graph on n vertices. Then the number of edges e satisfies*

$$e \le \frac{1}{2}n^{5/3} + n^{4/3} + \frac{1}{2}n.$$

Proof. The set of neighbours of x is denoted by $N(x)$, and the set of common neighbours of x and y by $N(x,y)$. Let $\deg(x,y) = |N(x,y)|$. The idea is to compute the sum $\sum_{x \in V(G)} \sum_{y \ne x} \deg(x,y)$ in two different ways.

On the one hand, count triples $\{z_1, z_2, z_3\} \subset N(x)$. For every triple there is at most one y such that $\{z_1, z_2, z_3\} \subset N(x,y)$, since the graph does not contain a $K_{3,3}$. Hence $\sum_{y \ne x} \binom{\deg(x,y)}{3} \le \binom{\deg(x)}{3}$. By the previous lemma, this implies $\sum_{y \ne x} \deg(x,y) \le n^{2/3}\deg(x) + 2n$. Now add up these inequalities for every vertex $x \in V(G)$. This gives

$$\sum_{x \in V(G)} \sum_{y \ne x} \deg(x,y) n^{2/3} \sum_{x \in V(G)} \deg(x) + 2n^2 \le n^{2/3}2e + 2n^2.$$

On the other hand,

$$\sum_{x \in V(G)} \sum_{y \ne x} \deg(x,y) = \sum_{x \in V(G)} \sum_{y \ne x} \sum_{z \in N(x,y)} 1 = \sum_{z \in V(G)} \sum_{y \in N(z)} \sum_{x \in N(z), x \ne y} 1$$

$$= \sum_{z \in V(G)} \deg(z)(\deg(z)) - 1),$$

which can be bounded from below by using the inequality between quadratic and arithmetic means. Thus

$$\sum_{x \in V(G)} \sum_{y \ne x} \deg(x,y) = \sum_{x \in V(G)} (\deg(x)^2 - \deg(x)) \ge \left(\frac{\sum \deg(x)}{n}\right)^2 \cdot n - 2e.$$

Substituting the upper bound and the number of edges we get the inequality

$$\frac{(2e)^2}{n} - 2e \le 2en^{2/3} + 2n^2.$$

This implies

$$e \le \frac{1}{2}n^{5/3} + \frac{1}{2}\frac{n^3}{e} + \frac{1}{2}n.$$

If the number of edges is larger than $\frac{1}{2}n^{5/3}$, then the second term on the right hand side is at most $n^{4/3}$, so the bound in the assertion follows. $\qquad\square$

Observe that the upper bound by *Füredi* shows that Brown's construction gives the correct main term. Actually, *Alon, Rónyai* and *Szabó* [3] have a different construction for $K_{3,3}$-free graphs showing that the second largest term is also correct. Füredi [69] has extended the above proof for arbitrary s and t. Fifteen years later *Nikiforov* [133] improved the smaller order terms in Füredi's bound and proved the following theorem.

Theorem 12.30 (Nikiforov). *For the number of edges of a $K_{s,t}$-free graph on n vertices we have*

$$e \leq \frac{1}{2}(s-t+1)^{1/t} n^{2-1/t} + \frac{1}{2}(t-2)n + \frac{1}{2}(t-1)n^{2-2/t}.$$

It is important to note that for $s > t! + 1$ *Kollár, Rónyai* and *Szabó* [109] showed that the upper bound of Kővári, T. Sós, Turán is asymptotically sharp. Although their construction is not coming directly from finite geometry, the construction has some geometric flavour. They had to estimate the number of solutions of systems of algebraic equations over finite fields. Later *Alon, Rónyai* and *Szabó* generalized the construction and also improved the above bound relating s and t. We refer the reader to the book by *Ball* [8], Chapter 6, in particular 6.8, for more results about the case of bipartite excluded graphs in Turán's problem.

A natural generalization of quadrilateral-free graphs is the investigation of graphs containing no cycles of length $5, 6, 7, \ldots$. Much less is known about these graphs but the next remark fits naturally in our presentation.

Theorem 12.31. *The incidence graph of a generalized quadrangle of order (q,q) is a graph on $n = 2(q^3 + q^2 + q + 1)$ vertices which contains no cycles of length $3, 4, 5, 6, 7$.*

Proof. By Lemma 10.21, the GQ has $q^3 + q^2 + q + 1$ points and the same number of lines. Thus the number of vertices in the incidence graph is indeed $n = 2(q^3 + q^2 + q + 1)$. The non-existence of cycles listed in the assertion follows either from the fact that the graph is bipartite (contains no odd cycles) or from the definition of generalized quadrangles. $\square$

Note that Example 10.17 shows the existence of a generalized quadrangle with these parameters if q is a power of a prime.

Another topic in graph theory, connected to finite geometry at least in a special case, is the theory of cages (and the degree/diameter problem). Recall that the girth of a graph is the length of the shortest cycle in it.

Definition 12.32. *A (k,q)-graph is a k-regular graph of girth g. A (k,g)-cage is a (k,g)-graph with as few vertices as possible. We denote the number of vertices of a (k,g)-cage by $c(k,g)$.*

It is not obvious that there exist (k,g)-graphs for every $k \geq 2$ and $g \geq 3$, but *Erdős* and *Sachs* found a recursive construction in the 1960's. This implies that the definition of (k,g)-cages is meaningful.

The study of cages was probably initiated by *Kárteszi* around 1960. About the same time *Hoffman* proposed the degree/diameter problem. In this case we look for the largest possible number of vertices of a k-regular graph of diameter d. We shall focus on the cage problem; the reader can find a brief introduction and some exercises about the degree/diameter problem at the end of this chapter.

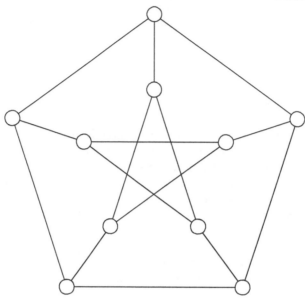

FIGURE 12.3
Petersen graph

We shall concentrate on the case $g = 6$, the other cases will only be mentioned.

Finding the cages for $g = 3$ and $g = 4$ are easy exercises (see Exercises 12.9 and 12.10). For $g = 5$ let us start with a small example: if $k = 3$, then $n \geq 10$ and in case of equality the graph is the *Petersen graph*. This implies $c(3, 5) = 10$ (see Figure 12.3).

Start from a vertex and consider its neighbours and then the neighbours of neighbours. These vertices are all distinct, since the graph contains no cycles of length less than 5. This implies immediately that the number of vertices in a k-regular graph of girth 5 is at least $1 + k + k(k-1)$, that is at least $1 + k^2$. It is a natural question whether there is graph of girth 5 on this many vertices. *Hoffman* and *Singleton* [94] proved that this is not possible in general, only in cases $k = 2, 3, 7$ and possibly 57. For $k = 2$ the 5-cycle, for $k = 3$ the Petersen graph is the extremal graph. It is an easy exercise to prove the uniqueness of the extremal graphs in these cases. For $r = 7$, *Hoffman* and *Singleton* [94] constructed an r-regular graph of girth 5; for $r = 57$, the existence is still open. The proof uses linear algebra (applied for the theory of strongly regular graphs), which can be found in the book by *Aigner* and *Ziegler* [2].

The simple proof above can be copied for odd g without essentially modifying it; for k-regular graphs of girth g on n vertices it gives

$$n \geq 1 + k + k(k - 1) + \ldots + k(k - 1)^{(g-3)/2}. \tag{12.10}$$

Let us now turn to the case $g = 6$.

Theorem 12.33. *A k-regular graph having girth 6 has at least*

$$2(1 + (k - 1) + (k - 1)^2)$$

vertices. In the case of equality the graph is the incidence graph of a projective plane.

Proof. This time start from an edge and consider the neighbours of the two endpoints of this edge. Continue with the neighbours of these neighbours. All these vertices are pairwise distinct, otherwise a cycle of length less than 6 would appear in the graph. Both endpoints have $k-1$ neighbours, each of them has another $k-1$ neighbours, so in total we see at least $2(1+(k-1)+(k-1)^2)$ vertices indeed.

Let us colour red and blue the two endpoints of the edge we started with. Continue with a good colouring, that is colour the neighbours of a red vertex blue and vice versa. We can colour the $2(1 + (k - 1) + (k - 1)^2)$ vertices red and blue in such a way that every edge joins a red and a blue vertex. If our graph has no more vertices, then the remaining edges must also join a red and a blue vertex. One can easily check that an edge joining two points of the same colour would result in a cycle of length less than 6. To sum up, in the case of equality in the lower bound, we must have a bipartite graph with the same number of vertices in the two colours.

As our graph is C_4-free, the results about Zarankiewicz's problem can be used. According to that, our graph has at most $(1+(k-1)+(k-1)^2)k$ edges. Since our graph is k-regular, it has exactly this many edges. Reiman's theorem then gives that our graph has to be the incidence graph of a projective plane of order $k - 1$. □

Note that the proof can be completed without referring to Reiman's theorem and Zarankiewicz's problem. Here we only sketch the argument; details are left for the reader as an exercise. We have to show that two vertices of the same colour (say, red) have a unique common neighbour. Take one of the points and any edge emanating from it as the starting edge in the first part of the proof. Then the other vertex is either among the neighbours of the neighbours of the red endpoint of the starting edge, or it is among the neighbours of the blue endpoint of the (new) starting edge. In both cases the two red points have a common neighbour. Using this it is not difficult to show that the graph is the incidence graph of a projective plane indeed.

Further elementary results on cages can be found in Chapters 4.2 and 4.3 of the book by *Kárteszi* [105]. More general results can be found in Ball's book [8] or in the dynamic surveys [65] and [131].

Copying the previous proof for even g gives the lower bound

$$n \geq 2(1 + (k - 1) + (k - 1)^2 + \ldots + (k - 1)^{g/2-2}). \tag{12.11}$$

Sometimes this lower bound and the bound in Inequality 12.10 are called the *Moore bound*; even *Moore* considered the degree-diameter problem instead of cages (see the end of this chapter).

It is not difficult to show that equality in this bound is closely related to the existence of generalized n-gons. The details of the proof of the next theorem is left as an exercise.

Theorem 12.34. *The incidence graph of a generalized quadrangle of order* $(k-1, k-1)$ *is a* $(k, 8)$*-cage. Conversely, equality in the Moore bound implies that the graph is the incidence graph of an appropriate generalized quadrangle.*

Proof. The number of points and lines in a GQ of order $(k-1, k-1)$ is $1 + (k-1) + (k-1)^2 + (k-1)^3$, hence the number of vertices in its incidence graph meets the Moore bound. The graph cannot contain a cycle of length less than 8, because of the definition of generalized quadrangles (see axiom GQ2 and Lemma 10.6).

In the other direction, we can copy the first part of the proof for $g = 6$ and deduce that the graph is bipartite. Similar to that proof, by the trick of changing the starting edge in the construction of the vertices of our graph, one can also show that for any two points of distance three there is a unique path of length three joining them. □

The number of vertices in a (k, g)-cage can also be bounded from above using the fact that the distance of any two vertices in it is at most $g - 1$. This gives the upper bound

$$n \leq 1 + k + k(k-1) + \ldots + k(k-1)^{g-2} < \frac{k}{k-1}(k-1)^g. \qquad (12.12)$$

The proof is left as an exercise and can be found in Lovász's book [115] as Exercises 10.11-13. Further details can also be found in Bollobás's book [31].

One can extend the previous approach for larger even girth and show that the Moore bound is attained for incidence graphs of generalized n-gons. However, they do not exist in general, so we only have a similar result for $g = 12$. Several other ideas can also be copied, for example the use of polarities that we saw in the construction of the Erdős-Rényi graph. This was done by *Lazebnik, Ustimenko* and *Woldar*. For larger girth the extremal graphs are obtained by combining algebraic and probabilistic methods and not related directly to incidence geometries. More details can be found in [31].

Let us briefly mention the other source of similar extremal graphs, the so-called degree/diameter problem. It was *Moore* who proposed the following problem around 1960: *Assume that a graph is* Δ*-regular and its diameter is* d. *What is the maximum number of vertices such a graph can have?* Let us recall that the distance of two vertices is just the length of the shortest path joining them, and the diameter of a graph is the maximum distance that occurs between its vertices. The ideas used for the cage problem (more precisely, to build the graph from a vertex) immediately imply an upper bound on the number of vertices n, namely

$$n \leq 1 + \Delta + \Delta(\Delta - 1) + \ldots + \Delta(\Delta - 1)^{d-1}.$$

The first term counts the starting vertex and the $(i+1)$-st term on the right hand side gives an upper bound on the number of points at distance i from the starting vertex. One can also note that assuming the regularity of the graphs is not necessary; the same upper bound applies if all the degrees are at most Δ. It is very rare that the upper bound (also called the Moore bound) can be attained. As an illustration let us recall a theorem of *Hoffman* and *Singleton* who investigated the case $d = 2$. In this case the upper bound is just $n \leq 1 + \Delta^2$. This is the same as the lower bound for the number of vertices of a $(\Delta, 5)$-cage, and we have already seen that they can only exist for $\Delta = 2, 3, 7, 57$. If a graph has girth g, g odd, then it contains vertices at distance $d = (g - 1)/2$. For this d, the lower and upper bounds are the same, so this shows that the two problems are connected. The reader can find more details in the dynamic surveys by *Exoo, Jajcay* [65] (on cages) and *Miller, Širáň* [131] (on the degree/diameter problem).

Exercises

12.1. (Mantel [123]) Show that if a simple graph on n vertices contains no triangle, then it has at most $\left\lfloor \frac{n^2}{4} \right\rfloor$ edges.

12.2. (Korchmáros, Pace, Sonnino [112]) Let n be odd and $\mathcal{O}$ be an oval in a projective plane Π_n. Then the complete graph K_{n+1} has a natural representation on $\mathcal{O}$ where vertices are the points of $\mathcal{O}$ and edges are the secants to $\mathcal{O}$. If P and R are distinct points of $\mathcal{O}$, then with the secant PR there is associated an external point E_{PR} and vice versa so that E_{PR} is the point of intersection of the tangents to $\mathcal{O}$ through P and R. Show that

- every one-factor of K_{n+1} is represented by a set $\mathcal{E}$ consisting of $\frac{n+1}{2}$ external points to $\mathcal{O}$ such that every tangent to $\mathcal{O}$ meets $\mathcal{E}$ in exactly one point;

- a one-factorization of K_{n+1} is represented by a partition $\mathcal{E}_1, \mathcal{E}_2, \ldots, \mathcal{E}_n$ of the set of external points to $\mathcal{O}$ such that $\mathcal{E}_i$ has size $\frac{n+1}{2}$ for all i and every tangent to $\mathcal{O}$ meets each $\mathcal{E}_i$ in exactly one point;

- if ℓ is an external line with respect to $\mathcal{O}$, then the $\frac{n+1}{2}$ external points on ℓ give a one-factor;

- if PR is a secant to $\mathcal{O}$, then the union of the $\frac{n-1}{2}$ external points on PR and E_{PR} gives a one-factor $\mathcal{F}_{PR}$ of K_{n+1};

- if P is a point on $\mathcal{O}$ and $PR_1, PR_2, \ldots, PR_n$ are the n secant lines through P, then $\cup_{i=1}^n \mathcal{F}_{PR_i}$ is a one-factorization of K_{n+1}.

12.3. Prove Theorem 12.12.

12.4. Let $G = (A, B, E)$ be a $K_{2,2}$-free bipartite graph of size (m, n). Show that equality holds in Inequality (12.5) if and only if every vertex in B has degree $\frac{e}{n}$ and every pair of vertices in A has exactly one common neighbour in B.

12.5. Let $G = (A, B, E)$ be a $K_{2,2}$-free bipartite graph of size $(k^2, k^2 + k)$ which contains $k^2(k+1)$ edges with $k > 1$. Show that G is the incidence graph of an affine plane of order k.

12.6. Prove that if a polarity of a projective plane of order n has $n + 1$ self-conjugate points, then they form an oval for n odd and a line for n even.

12.7. Show that any self-conjugate line of a polarity of a projective plane contains exactly one self-conjugate point, namely its pole.

12.8. Show that if a set of points $\mathcal{P}$ in a projective plane of order n has at least $|\mathcal{P}|$ tangent lines then $|\mathcal{P}| \leq n\sqrt{n} + 1$.

12.9. Show that for $k > 2$ the $c(k, 3)$-cage is unique; it is the complete graph K_{k+1}.

12.10. Show that for $k > 1$ the $c(k, 4)$-cage is unique; it is the complete bipartite graph $K_{k,k}$.

12.11. Prove the uniqueness of the $c(3, 5)$-cage; show that it is isomorphic to the Petersen graph.

12.12. Show that if $g > 2$ even, then a (k, g)-cage attaining the Moore bound exists if and only if a generalized $\frac{g}{2}$-gon of order $(k - 1, k - 1)$ exists.

12.13. Prove Inequality (12.12) on the number of vertices in a (k, g)-cage.

13

Some applications of finite geometry in coding theory and cryptography

In this chapter, some constructions for codes, secret sharing schemes and authentication systems will be given. All the constructions are based on finite geometry, more precisely on particular substructures discussed earlier in this book.

Coding theory (or rather a chapter of coding theory on channel coding) deals with the following communication model. A *sender* (Alice) wants to send data through a certain channel to the *receiver* (Bob). During transmission parts of the data might be changed. How can the receiver find out which part of the message is changed and how can he correct the data? Of course, this is an oversimplified model and we are going to define this model more precisely. For more realistic models and more details about different types of codes, the reader is referred to the textbooks [147, 176]. Our discussion is mostly based on Chapters 5 and 6 of [21].

We shall assume that the message $\mathbf{c}$ sent by the sender is an element of the vector space $V = \big(\mathrm{GF}(q)\big)^n$. In most cases $q = 2$, in other words, the message is a binary string of length n. We shall only consider correcting errors that come from changing a symbol to another one. This means that the length of the message remains the same (no erasures, no extra symbols).

So our problem is the following: during transmission an error vector $\mathbf{e}$ is added to $\mathbf{c}$ and the receiver gets the message $\mathbf{x} = \mathbf{c} + \mathbf{e}$. Our aim is to decode $\mathbf{x}$, that is to find the error vector and the original message $\mathbf{c}$. In our simplified model intuitively, errors are independent random errors, so we just want to find the message that differs from the transmitted message in the least number of symbols.

Definition 13.1. *Let* $\mathbf{v} = (v_1, v_2, \ldots, v_n)$ *and* $\mathbf{w} = (w_1, w_2, \ldots, w_n)$ *be two elements of* V. *The* Hamming distance *is the number of those coordinates in which* $\mathbf{v}$ *and* $\mathbf{w}$ *differ, that is*

$$d_H(\mathbf{v}, \mathbf{w}) = |\{i : v_i \neq w_i\}|.$$

The Hamming distance d_H defined above will simply be denoted by d in this chapter. It is well-known (and easy to check) that the Hamming distance is a metric, which means that it is symmetric, and satisfies the triangle inequality $d(\mathbf{x}, \mathbf{y}) + d(\mathbf{y}, \mathbf{z}) \geq d(\mathbf{x}, \mathbf{z})$, and $d(\mathbf{x}, \mathbf{y}) = 0$ if and only if $\mathbf{x} = \mathbf{y}$. We shall also use that the Hamming distance is translation invariant, that is

$$d(\mathbf{x}, \mathbf{y}) = d(\mathbf{x} + \mathbf{u}, \mathbf{y} + \mathbf{u}).$$

The ball with centre $\mathbf{c} \in V$ and radius r is defined as

$$S_r(\mathbf{c}) = \{\mathbf{v} \in V : d(\mathbf{c}, \mathbf{v}) \leq r\}.$$

Definition 13.2. *Let t be a positive integer. Let V be the vector space $V = (GF(q))^n$. A subset $C \subset V$ is called a t-error correcting code if for any two distinct vectors $\mathbf{v}, \mathbf{w} \in C$ we have*

$$d(\mathbf{v}, \mathbf{w}) \geq 2t + 1.$$

The elements of C are called codewords, *the elements of V are called* words. *The number of non-zero coordinates of a word $\mathbf{v}$ is called the* weight *of v and denoted by $w(v)$. The quantity*

$$d(C) = min\{d(\mathbf{v}, \mathbf{w}) : \mathbf{v}, \mathbf{w} \in C, \mathbf{v} \neq \mathbf{w}\} \geq 2t + 1$$

is called the minimum distance *of C.*

Lemma 13.3. *Let C be a t-error correcting code. Then for every word $\mathbf{v} \in V$ there is at most one $\mathbf{c} \in C$ codeword for which $d(\mathbf{c}, \mathbf{v}) \leq t$.*

Proof. The triangle inequality for the Hamming distance implies that if C is a t-error correcting code, then balls of radius t whose centers are the codewords are pairwise disjoint. This is just the assertion of the lemma. $\square$

This lemma explains the name t-error correcting code. If the sender transmits codewords and during transmission at most t errors occur, then the receiver can decode the message. Indeed, he just finds the ball around a codeword that contains the received message and the center of the ball will be the original message. Of course, this decoding procedure implicitly uses some assumptions about the errors but we are not going into details. For us decoding will simply mean finding the nearest codeword. More details can be found in the coding theory textbooks mentioned at the beginning of this chapter.

If one wants to use a code in practice, then codewords have to be stored or generated, one has to determine how many errors the code can correct (or find the minimum distance of the code) and it is crucial to have an efficient decoding algorithm. For a class of codes some of this is simpler. From now on we shall focus on these codes, the so-called linear codes.

Definition 13.4. *The code $C \subset V$ is* linear *if C is a linear subspace of the vector space V. If C has dimension k, then we call it a* linear $[n, k]$ code*. If also the underlying field plays a role then the notation $[n, k]_q$ is used and in most cases also the minimum distance d of the code is mentioned, that is we speak about a linear $[n, k, d]_q$ code.*

If the vectors $\mathbf{c}_1, \mathbf{c}_2, \dots, \mathbf{c}_k$ form a basis of C then the $k \times n$ matrix G, whose i-th row is the vector $\mathbf{c}_i$, $(i = 1, 2, \dots, k)$, is called the generator matrix *of the code C.*

In case of linear codes one can simplify the definiton of the minimum distance and replace it with distance from the origin.

Lemma 13.5. *The minimum distance of a linear code is equal to the minimum weight of non-zero codewords.*

Proof. If C is a linear code, then $\mathbf{0} \in C$. Therefore,

$$d(C) = \min\{d(\mathbf{v}, \mathbf{w}) : \mathbf{v}, \mathbf{w} \in C, \mathbf{v} \neq \mathbf{w}\}$$
$$\leq \min\{d(\mathbf{v}, \mathbf{0}) : \mathbf{v} \in C, \mathbf{v} \neq \mathbf{0}\} = w(\mathbf{v}),$$

and hence the minimum distance is at most the smallest weight of a non-zero codeword. On the other hand, if $\mathbf{v}$ and $\mathbf{w}$ are vectors having distance $d(C) = d(\mathbf{v}, \mathbf{w})$, then by linearity of the code $\mathbf{v} - \mathbf{w} \in C$. As

$$w(\mathbf{v} - \mathbf{w}) = d(\mathbf{v} - \mathbf{w}, \mathbf{0}) = d(\mathbf{v} - \mathbf{w}, \mathbf{w} - \mathbf{w}) = d(\mathbf{v}, \mathbf{w}) = d(C),$$

there is a codeword whose weight is equal to the minimum distance. $\square$

In case of linear codes there is a seemingly simple method for decoding. To describe it some definitions are needed.

Definition 13.6. *Let $C \subset V$ be an arbitrary code. The* dual code *of C is the linear code*

$$C^{\perp} = \{\mathbf{v} \in V : \mathbf{v} \cdot \mathbf{c} = 0 \ \forall \, \mathbf{c} \in C\},$$

where $\cdot$ denotes the inner product.

We know from linear algebra that for a linear $[n, k]$ code C its dual code, $C^{\perp}$, is a linear $[n, n - k]$ code. A generator matrix H of $C^{\perp}$ is called a *parity check matrix* of C.

The minimum distance of the code can be found using the parity check matrix.

Theorem 13.7. *Let C be a linear code and H be a parity check matrix of it. The minimum distance of C is $d + 1$ if and only if every d columns of H are linearly independent, but there are $d + 1$ columns that are dependent.*

Proof. By Lemma 13.5, we can show the assertion for minimum weight instead of minimum distance. For a codeword $\mathbf{c} \in C$, we have $\mathbf{c}H^T = \mathbf{0}$. This gives a linear combination of the columns of H yielding the zero vector. The number of non-zero coefficients in this linear combination is the weight of our codeword $\mathbf{c}$. This observation proves the theorem. $\square$

Let us remark that finding the minimum distance of a code is seemingly simple but it is actually computationally difficult.

Definition 13.8. *If $C \subset V$ is a linear code with parity check matrix H, then for any $\mathbf{v} \in V$ the vector $s(\mathbf{v}) = \mathbf{v}H^T$ is called the* syndrom *of $\mathbf{v}$.*

The next lemma is straightforward from the definition of syndroms.

Lemma 13.9. *If $C \subset V$ is a linear code with parity check matrix H, then*

- $C = \{\mathbf{v} \in V : s(\mathbf{v}) = 0\}$;

- *for any $\mathbf{v}, \mathbf{w} \in V$ we have $s(\mathbf{v}) = s(\mathbf{w}) \iff \mathbf{v} + C = \mathbf{w} + C$. In other words, the syndrom of a vector only depends on the coset of C containing it.*

In the case of a t-error correcting linear code small weight codewords in a coset of C have a distinguished role.

Lemma 13.10. *If $C \subset V$ is a t-error correcting linear code, then each coset of C can contain at most one vector whose weight is at most t.*

Proof. Assume that there are vectors $\mathbf{v} \neq \mathbf{w} \in V$ of weight at most t in the same coset of C. Then $\mathbf{v} - \mathbf{w} \in C, w(\mathbf{v}) \leq t$ and $w(\mathbf{w}) \leq t$. As C is t-error correcting, Lemma 13.5 implies $2t + 1 \leq w(\mathbf{v} - \mathbf{w})$. Using the linearity of S and the triangle inequality for the Hamming distance we get

$$2t + 1 \leq w(\mathbf{v} - \mathbf{w}) = d(\mathbf{v}, \mathbf{w}) \leq d(\mathbf{v}, \mathbf{0}) + d(\mathbf{w}, \mathbf{0}) = w(\mathbf{v}) + w(\mathbf{w}) \leq t + t.$$

This contradiction proves the lemma.   □

Based on the previous lemma one can formulate the so-called syndrom decoding procedure for t-error correcting linear codes:

1. determine the syndroms of vectors $\mathbf{v} \in V$ with weight $w(\mathbf{v}) \leq t$;

2. compute the syndrom of $s(\mathbf{x})$ of the received vector $\mathbf{x}$;

3. pick the unique vector $\mathbf{v}$ of weight at most t in the coset of $\mathbf{x}$; that is the vector for which $s(\mathbf{x}) = s(\mathbf{v})$;

4. the original message is $\mathbf{c} = \mathbf{x} - \mathbf{v}$.

If there are at most t errors, then the previous lemma guarantees the correctness of the decoding procedure. Again, our method is a simplified version. Normally, one chooses a vector of smallest weight in each coset, called the coset leader. With this more general method one can decode the received message also when the number of errors exceeds t.

This method is not very efficient (we need all the coset leaders). It is usually applied when q^{n-k} is small and the coset leaders can be stored in a table. More generally, decoding linear codes is difficult. The NP-completeness of this problem was proved by *Berlekamp, McEliece* and *van Tilborg* [19].

"Good codes" can correct many errors. In the case of linear codes the error correcting capability and the dimension of the code work against each other if the length is given. This is the content of the next theorem.

Theorem 13.11 (Singleton's bound). *If d denotes the minimum distance of a linear C $[n, n - k]$ code, then*

$$d \leq k + 1.$$

Proof. By Lemma 13.5, it is enough to show that C contains a non-zero codeword with weight at most $k + 1$. If G is a generator matrix of C, then we can transform G, using elementary transformations, into the form

$$G = I_{(n-k) \times (n-k)} G^*_{(n-k) \times k},$$

which is also a generator matrix of C. Here I_{n-k} is the $(n - k) \times (n - k)$ identity matrix. Since there are at most k non-zero elements in each row of G^*, the weight of the codeword corresponding to any row of G^* is at most $k + 1$. □

Definition 13.12. *If the minimum distance of a linear $[n, n-k]$ code is d and the inequality in Singleton's bound is satisfied with equality, that is $d = k + 1$, then the code is called an* MDS *code.*

Note that MDS stands for maximum distance separable. According to Theorem 13.11, MDS codes have the largest minimum distance for a given dimension. In the case of a given length and minimum distance MDS codes have the largest dimension. This essentially means that MDS codes are optimal in many respects. However, for a given q and a pair (n, k) there does not exist an MDS code with these parameters in general. The next result gives a connection of MDS codes and arcs in higher dimensions.

Theorem 13.13. *Let us fix q, the size of the underlying field. For given natural numbers n and k there is a linear MDS code with parameters $[n, n-k]$ if and only if there is an n-arc in the projective space PG$(k - 1, q)$.*

Proof. Assume that there is a linear $[n, n - k, k + 1]_q$ code C. Let H denote a parity check matrix of C. Then any k columns of H are linearly independent by Theorem 13.7.

Consider the columns as points in PG$(k - 1, q)$. The linear independence of any k columns of H implies that no k of these points are in a hyperplane. Hence these points form an arc (consisting of n points).

Conversely, assume that there is an n-arc in PG$(k - 1, q)$. Let H be the matrix of size $k \times n$ whose columns are representative vectors of the points of our n-arc. Then H is a parity check matrix of a linear $[n, n - k]$ code C.

Any set of k columns of H is independent, since the points form an arc. So the minimum distance of the code is at least $k + 1$. By Singleton's bound we have equality here, hence C is an MDS code. □

Theorem 13.14. *The dual of an MDS code is an MDS code.*

Proof. Let C be an MDS code of length n and dimension $n - k$. Let us denote by H a parity check matrix of C. It is the generator matrix of the dual code $C^\perp$. To prove that $C^\perp$ is MDS we have to show that any linear combination of the rows of H has weight at least $n - k + 1$. Replacing an appropriate row of H by this linear combination of the rows one can see that it is enough to check this for an arbitrary row of H. As any k columns of H are independent, no row of H can contain k 0-s. Thus the weight of any row is at least $n - k + 1$ and it cannot be larger by Singleton's bound. □

Let us now see the usual description of *Reed–Solomon codes*. They essentially correspond to normal rational curves defined in Chapter 9, Example 9.26. The codes are called generalized Reed–Solomon (GRS) codes. The reason we need the slight generalization will be clear after the result on duals of such codes.

Definition 13.15. *Let $\mathbf{a} = (\alpha_1, \ldots, \alpha_n)$ be a vector consisting of n pairwise distinct elements of $\mathrm{GF}(q)$. Moreover, let $\mathbf{v} = (v_1, \ldots, v_n)$ be a fixed vector having no zero coordinates, that is $v_i \neq 0$ for every $i = 1, \ldots, n$. Define the code $\mathrm{GRS}_k(\mathbf{a}, \mathbf{v})$ by*

$$\mathrm{GRS}_k(\mathbf{a}, \mathbf{v}) = \{(v_1 f(\alpha_1), \ldots, v_n f(\alpha_n)) : f \in \mathrm{GF}(q)[x], \deg f < k\}.$$

This code is called a generalized Reed–Solomon code. *In other words, we evaluate polynomials of degree less than k at $\alpha_1, \ldots, \alpha_n$, and multiply the result coordinatewise by the vector $\mathbf{v}$.*

As polynomials of degree less than k form a vector space of dimension k, we immediately get that $\mathrm{GRS}_k(\mathbf{a}, \mathbf{v})$ is a linear $[n, k]$ code. Assume that $\mathrm{GRS}_k(\mathbf{a}, \mathbf{v})$ contains a codeword of weight less than $n - k + 1$. This must come from a polynomial f, which is zero for at least k of the α_i's. By the fact that a polynomial has at most as many roots as its degree, such a polynomial f must be the zero polynomial. This contradiction shows that the minimum weight, and hence the minimum distance, of $\mathrm{GRS}_k(\mathbf{a}, \mathbf{v})$ is at least $n - k + 1$. So the GRS code is MDS by Singleton's bound. To describe the code explicitly, we can choose the basis $1, x, \ldots, x^{k-1}$ in the vector space of polynomials of degree less than k and consider the corresponding vectors in $\mathrm{GRS}_k(\mathbf{a}, \mathbf{v})$. They are consecutive powers of the elements α_i multiplied by v_i. Clearly, these vectors form a basis of the GRS code. Putting them in generator matrix we get a Vandermonde type matrix, where the columns are multiplied by non-zero field elements.

Proposition 13.16. $\mathrm{GRS}_k(\mathbf{a}, \mathbf{v})^\perp = \mathrm{GRS}_{n-k}(\mathbf{a}, \mathbf{v}')$, *for a suitable vector $\mathbf{v}'$.*

Proof. The vector $\mathbf{v}'$ turns out to be independent of k. Hence we find it for $k = n - 1$ and then show that it is suitable for any k.

The code $\mathrm{GRS}_{n-1}(\mathbf{a}, \mathbf{v})$ is MDS, hence its dual $\mathrm{GRS}_{n-1}(\mathbf{a}, \mathbf{v})^{\perp}$ is also MDS. This dual code has dimension $1 = n - (n-1)$ and its minimum distance is $n - 1 + 1 = n$. If $\mathbf{v}' = (v'_1, \ldots, v'_n)$ generates this 1-dimensional subspace, then $v'_i \neq 0$, for every i. As $\mathbf{v}'$ is orthogonal to the standard basis of $\mathrm{GRS}_{n-1}(\mathbf{a}, \mathbf{v})$ we get

$$0 = \sum_{i=1}^{n} v_i v'_i \alpha_i^j, \quad 0 \leq j < n - 1.$$

In the j-th equation the polynomial x^j was evaluated at the coordinates of $\mathbf{a}$. The same system of equations immediately implies that $(v_1 \alpha_1^s, \ldots, v_n \alpha_n^s)$ is orthogonal to the vector $(v'_1 \alpha_1^t, \ldots, v'_n \alpha_n^t)$ if $s + t < n - 1$. In our case $t \leq n - k - 1$, $s \leq k - 1$, proving the assertion. $\square$

This theorem shows why one generalizes Reed–Solomon codes: it is not true that the dual code of an RS code is an RS code but it is true for GRS codes.

The previous results on the dual of MDS and GRS codes immediately imply the following.

Theorem 13.17. *Assume that there is an r-arc in $\mathrm{PG}(d, q)$. Then there is an r-arc also in $\mathrm{PG}(r - d - 2, q)$.*

Proof. This follows immediately from the fact that an r-arc in $\mathrm{PG}(d, q)$ corresponds to an MDS code with parameters $[r, d+1, r-d]_q$ whose parity check matrix consists of the points of the arc as columns. The dual code, which is also MDS, has parameters $[r, r-d-1, d+2]_q$, hence the columns of its parity check matrix give an r-arc in $\mathrm{PG}(r - d - 2, q)$. $\square$

This result implies the following corollary.

Corollary 13.18. *There are $(q + 2)$-arcs in $\mathrm{PG}(q - 2, q)$ if q is even. If the maximum size of an arc in $\mathrm{PG}(n, q)$ is $(q + 1)$, then the same is true for $\mathrm{PG}(q - n, q)$.*

Proof. The dual of a hyperoval of $\mathrm{PG}(2, q)$ will be a $(q+2)$-arc in $\mathrm{PG}(q-2, q)$. If there were a $(q + 2)$-arc in $\mathrm{PG}(q - n, q)$, then, by the previous theorem, its dual would be a $(q + 2)$-arc in $\mathrm{PG}(n, q)$. By assumption, it cannot exist. $\square$

Since normal rational curves correspond to generalized Reed–Solomon codes, we also have a similar theorem about $(q + 1)$-arcs and normal rational curves.

Corollary 13.19. *Assume that every $(q + 1)$-arc is a normal rational curve in $\mathrm{PG}(n, q)$. Then the same is true for $\mathrm{PG}(q - 1 - n, q)$.*

These results explain the duality principle for arcs, in particular Theorem 9.43 in Chapter 9.

Several other important codes have a representation using finite geometry. We only mention three of them. A detailed description of the codes mentioned here can be found in most introductory textbooks, for example, in the book [176] by *van Lint*.

Example 13.20. Let r be an arbitrary positive integer, $n = 2^r - 1$, H be the $(r \times n)$ matrix whose columns are the non-zero $0 - 1$ vectors. The *binary Hamming code* is the linear code with parity check matrix H.

It is easy to see that Hamming codes have minimum distance 3, since any two columns of H are independent and there are three columns which are dependent. In other words, this means that the binary Hamming codes correct 1 error. The dimension of binary Hamming codes is $2^r - r - 1$. Since

$$2^{2^r - 1 - r}(n + 1) = 2^{2^r - 1 - r} \cdot 2^r = 2^n,$$

the spheres of radius 1 around the codewords are disjoint and cover the entire vector space. Such codes are called perfect; more precisely, a t-error correcting code is called *perfect* if the spheres of radius t around the codewords cover the entire vector space. Thus binary Hamming codes are perfect. Perfect codes are very important but they are rare. All perfect codes correcting at least 2 errors are known (at least in our case when the words are elements of a vector space).

If $r = 1$ or 2, then the Hamming-code is trivial. For $r = 3$, the Hamming code of length 7 contains the following 16 codewords:

0000000	1111111
1110000	0001111
1001100	0110011
1000011	0111100
0101010	1010101
0100101	1011010
0011001	1100110
0010110	1101001

Let us identify the points of the Fano plane (see Figure 1.2) with the elements of $\{A, B, D, C, G, E, F\}$. The codewords of the Hamming code of length 7 are then just the characteristic vectors of the lines and their complements, together with $\mathbf{0}$ and the all-one vector $\mathbf{1}$. From geometric properties of the Fano plane it follows immediately that the Hamming code of length 7 is perfect and 1-error correcting. Indeed, if we consider words of length 7 as subsets of points and a subset contains at most one or at least six points, then only $\emptyset$ and the entire plane is at distance 1 from it. If the subset contains two or five points then it is at distance 1 either from a line (the line joining the two points) or from the complement of a line (the complement of the line joining the two points of the complementary set). When we have three non-collinear points, then it is at distance at least 2 from any line but it is at distance 1 from the complement of the line meeting the sides of the triangle in one-one point distinct from the points of the triangle (in the case of the Fano plane these three points are on a line). If four points are in general position, then their complement is a line. If not, then three of the points are collinear and this is the unique line which is at distance 1 from the set of four points. All the other lines and complements of lines are at distance at least 2.

Let us now extend the definition of Hamming codes to arbitrary fields GF(q).

Definition 13.21. *List the points of* PG($r - 1, q$) *and choose a fixed representative vector for each point. These vectors are* $\mathbf{v}_1, \ldots, \mathbf{v}_n$, *where* $n = (q^r - 1)/(q - 1)$. *Put these vectors in an* $r \times n$ *matrix* H *as columns. Let* C *be the code with parity check matrix* H. *This code* C *is the* q-*ary Hamming code.*

It is again clear that any two columns of H are independent, so C has minimum distance at least 3. One can easily find three columns that are dependent (three points on a line), hence the minimum distance is exactly 3. Essentially the same computation as before shows that the code is a 1-error correcting perfect code. Indeed, the number of words in a sphere of radius 1 is $1 + n(q - 1) = 1 + (q^r - 1) = q^r$, the number of codewords is q^{n-r}, so their product is q^n.

The dual code of any Hamming code also has interesting properties that can be proven by geometric arguments.

Theorem 13.22. *The dual code of the* q-*ary Hamming code is a* $[(q^r - 1)/(q - 1), r, q^{r-1}]$ *code. Moreover, any two distinct codewords have distance* q^{r-1}.

Such codes are called *equidistant* and the dual of the Hamming code is sometimes called a *simplex code*.

Proof. Instead of pairwise distances of codewords we just consider the weights of non-zero codewords. Let H be the parity check matrix of the Hamming code defined above. The columns of H correspond to points in PG($r - 1, q$) and the homogeneous coordinates in this projective space will be denoted by $X_1, \ldots, X_r$. Coordinates simply correspond to rows of H. In order to determine the weight of the first row of H we have to solve $X_1 = 0$. This defines a hyperplane of PG($r - 1, q$), so there are $(q^{r-1} - 1)/(q - 1)$ points satisfying it. In other words, the first row of H contains this many 0's, so the weight of the first row is $(q^r - 1)/(q - 1) - (q^{r-1} - 1)/(q - 1) = q^{r-1}$. An arbitrary codeword of the dual code is a linear combination $\mathbf{c} = c_1\mathbf{h}_1 + \ldots + c_r\mathbf{h}_r$, where $\mathbf{h}_1, \ldots, \mathbf{h}_r$ denote the rows of H. To determine the weight of $\mathbf{c}$ we have to solve the equation $c_1 X_1 + \ldots + c_r X_r = 0$. This is the equation of a hyperplane, so the weight of $\mathbf{c}$ is q^{r-1} as for the first row. This implies that our dual code is indeed equidistant. □

The other widely used class of codes, *Reed–Muller codes* can also be interpreted similarly to Reed–Solomon codes. In this description multilinear functions in several variables are evaluated at some points (in the binary case on the points of a hypercube).

Definition 13.23. *Let $H = \{0,1\}^m$ be the m-dimensional hypercube, $F = \mathrm{GF}(2)$, $V : F[x_1, \ldots, x_m] \to F^H$, which maps a polynomial F to the vector of its values on the points of the hypercube. Let*

$$\mathrm{RM}_{m,k} = \{V(f) : \ \deg f \leq k, \deg_{x_i} f \leq 1\}.$$

This code is called the binary Reed–Muller code *of order k.*

In the definition of Reed–Muller codes only multilinear polynomials were evaluated. This is enough since only the values $0, 1$ are substituted. It is easy to show that $\mathrm{RM}_{m,k}$ is linear and has dimension $\sum_{i=0}^{k} \binom{m}{i}$. To see this, note that the multilinear monomials $x_{i_1} \cdots x_{i_s}$ are independent. Clearly, any multilinear polynomial is the sum of monomials and the vector space of functions $H \to F$ has dimension 2^m. The functions which are 1 in just one vertex of the hypercube and 0 elsewhere generate this vector space and such a function can be written as a multilinear polynomial. As the total number of monomials is also 2^m, they form a basis of this vector space. This implies that $\mathrm{RM}_{m,k}$ has dimension $\sum_{i=0}^{k} \binom{m}{i}$, since the monomials generating it are independent. Also the minimum distance of Reed–Muller codes can be determined relatively easily.

Proposition 13.24. *The minimum distance of $\mathrm{RM}_{m,k}$ is 2^{m-k}.*

Proof. On the one hand, the weight of $V(x_1 \cdots x_k)$ is exactly 2^{m-k}. We have to show that there are no codewords of smaller weight. This is done by a simultaneous induction on k and m. For $k = 0$ and any m, the minimum weight is 2^m. Assume that we already know the assertion when either the degree is less than k or the number of variables is less than m. Let f be a multilinear polynomial of degree k and write it as $f = x_m g(x_1, ..., x_{m-1}) + h(x_1, ..., x_{m-1})$, where $\deg g \leq k-1$, $\deg h \leq k$. If $h = 0$, then $V(f) = 1$ if and only if $x_m = 1$ and $V(g) = 1$. As $V(g)$ is a codeword of $\mathrm{RM}_{m-1,k-1}$, our induction assumption gives that the weight of $V(g)$, and hence that of $V(f)$, is at least 2^{m-k}. The situation is similar if $g + h = 0$. In this case $f = (x_m + 1)g$ and we can copy the previous induction argument. Let us now substitute $x_m = 0$ in f. By induction, $V(h)$ has weight at least 2^{m-1-k}, so we see at least this many coordinates of $V(f)$, where the value is 1 and $x_m = 0$ for these coordinates. If we now substitute $x_m = 1$, the argument can be copied and again we see at least 2^{m-1-k} coordinates of $V(f)$, where the value is 1 and $x_m = 1$ for these coordinates. In total, the weight of $V(f)$ is at least $2^{m-1-k} + 2^{m-1-k} = 2^{m-k}$, as we had to prove. $\qquad\square$

Example 13.25. Let the points of the affine space $\mathrm{AG}(m, 2) = \mathcal{A}$ be $P_1, P_2, \ldots, P_{2^m}$. Identify a set M of points of $\mathcal{A}$ by its characteristic vector $\chi(M)$, so let

$$\chi(M) = (a_1, a_2, \ldots, a_{2^m}), \quad \text{where} \quad a_i = \begin{cases} 1, & \text{if } P_i \in M, \\ 0, & \text{if } P_i \notin M. \end{cases}$$

Let us define a linear code C generated by the characteristic vectors of $(m - k)$-dimensional subspaces of $\mathcal{A}$. Note that the characteristic vector of a subspace of dimension at least $m - k$ is a linear combination of characteristic vectors of subspaces of dimension exactly $m - k$, since an affine subspace of dimension $m - k + 1$ is the disjoint union of two subspaces of dimension $m - k$.

This is an alternative, geometric description of $\mathrm{RM}_{m,k}$ as the next proposition shows.

Proposition 13.26. *Let C be the linear code defined in the previous example. Then it is the binary Reed–Muller code $\mathrm{RM}_{m,k}$.*

Proof. Consider a subspace S of dimension at least $m - k$. It is the intersection of at most k hyperplanes. The equation of such a hyperplane H_i is a linear polynomial

$$a_{i,1}X_1 + a_{i,2}X_2 + \ldots + a_{i,m}X_m + a_{i,m+1}X_{m+1} = 0.$$

Let f denote the product of these equations. It is a polynomial of degree at most k and the characteristic vector of S is just the vector $V(f)$. This shows that the code C is contained in the Reed–Muller code $\mathrm{RM}_{m,k}$. Conversely, $\mathrm{RM}_{m,k}$ is generated by the vectors $V(x_{i_1} \ldots x_{i_s})$, where $s \leq k$, and all these vectors are characteristic vectors of subspaces of dimension at least $m - k$. Hence the two codes coincide. $\qquad\square$

While error correcting codes are used to correct random errors when transmitting data, authentication codes are used against attacks of a third person. Our model is the following:

A *sender* (Alice) transmits messages to a *receiver* (Bob) through a channel. During transmissions, a third party, called *Attacker* can see the transmitted messages and tries to do two types of attacks:

- tries to include his own messages among the transmitted ones (impersonation),

- tries to modify the original message (substitution).

Authentication systems defend the transmission against these attacks. Before transmission, Alice and Bob agree on a secret key k. Using an authentication algorithm F and the secret key k, Alice transforms the data d that has to be transmitted to a message $c = F_k(d)$ (this is the message that is actually transmitted). Let D and K denote the set of data and the set of keys, respectively. An authentication algorithm means that for any data $d' \in D$ and any key $k' \in K$, the message $c = F_{k'}(d')$ is uniquely determined. In other words, F is a mapping from $D \times K$ to C, the set of messages. When Bob gets a message c' (possibly modified by the *Attacker*), he checks whether there is a data $d_0 \in D$ for which $c' = F_{k'}(d_0)$. If this is true, then he accepts that Alice transmitted the message d_0, and if not, he rejects the message. In our oversimplified model it is assumed that *Attacker* knows the set K of keys and also the algorithm F. Of course, *Attacker* does not know the secret key k upon which Alice and Bob agreed. We also assume that for any message c and any key k there is precisely one data $d \in D$ so that $c = F_k(d)$. The security of the authentication system is the probability that Bob does not accept a message modified by *Attacker*. This seems to depend on the algorithm F, however, it turns out that it only depends on the size of keys, $|K|$.

If there are κ keys, then *Attacker* can make Bob accept a modified message with probability $1/\kappa$, because the message will be considered authentic for precisely one key. One can also prove that the security of an authentication system cannot be arbitrarily close to 1, if the number of keys is fixed. This is the content of the next theorem. The proof of this and other theorems in the rest of this chapter can be found in Chapter 6 of the book [21].

Theorem 13.27 (Gilbert, MacWilliams, Sloane). *If the number of keys in an authentication system is κ, the Attacker can crack the system with probability at least $1/\sqrt{\kappa}$.*

An authentication system is called *perfect* if there are κ keys and the probability of cracking the system is only $1/\sqrt{\kappa}$. In this sense perfect authentication systems are the most secure. In the next example a perfect authentication system is constructed from a projective plane.

Example 13.28. Let Π be a projective plane of order n and e be a line of Π. Let the data D be the set of points of e, the set K of keys be the points not on e. The messages are lines of Π, namely, for a point $P \in e$ and the key $Q \notin e$, the message c will be the line PQ.

The number of data in this system is $n + 1$, the number of keys is n^2, and the number of messages belonging to a given key is also $n + 1$. Let us see what are the chances of *Attacker* cracking the system.

If *Attacker* wants to put his own message, then for any data, that in any point $E \in e$, he can choose n messages, namely, one of the n lines through E different from e. Only one of these messages is authentic, so the chance of success for the *Attacker* is $1/n$.

If he tries to modify a message, then he knows a line $f \neq e$ and knows the algorithm. So he knows that the key is a point on f different from $e \cap f$. The number of possible keys (points on f) is again n, hence *Attacker* has chance $1/n$ to guess the key and this is the probability that the modified message is authentic.

So this authentication system is perfect. In our example the number of keys is $\kappa = n^2$ and

- for every data there are $\sqrt{\kappa}$ different messages;

- every message is authentic for $\sqrt{\kappa}$ keys;

- for any two messages there is one and only one key so that both messages are authentic.

One can prove that any perfect authentication system with κ keys satisfies the three properties above.

The authentication system of the previous example can be generalized. We only consider some of the points of e as data and also the messages are the lines through these points. If we delete the line e from Π, then in the resulting affine plane the set of messages is the union of some parallel classes of lines. Such a structure is called a *net*.

Definition 13.29. *A triple $(\mathcal{P}, \mathcal{E}, \mathrm{I})$, where $\mathcal{P}$ and $\mathcal{E}$ are disjoint finite sets, $\mathrm{I} \subset \mathcal{P} \times \mathcal{E}$ is an incidence relation, is called a* finite net *if the following axioms are satisfied:*

N1. *For any two distinct elements of $\mathcal{P}$ there is at most one element of $\mathcal{E}$ which is in relation I with both of them. In other words, two points are joined by at most one line.*

N2. *If $P \in \mathcal{P}$ is not in relation I with $e \in \mathcal{E}$, then there is precisely one element of $\mathcal{E}$, which is in relation I with P but is not in relation I with any element of $\mathcal{P}$ incident with e. In other words, for a non-incident points-line pair there is a unique line through the point which does not meet the line.*

By axiom N2, it is true that the lines of a finite net can be divided into classes consisting of pairwise non-intersecting lines. Two lines are called *parallel* if they are either disjoint or coincide. This relation of parallelism is clearly an equivalence relation. Since N2 is the axiom of parallelism, this can be seen as similar to the case of affine planes. It can also be proven that the existence of a finite net with $k + 2$ parallel classes is equivalent to the existence of k pairwise orthogonal Latin squares. Again, the proof is essentially the same as that of Theorem 1.29.

There are finite nets that cannot be obtained from an affine plane by removing some parallel classes of lines. For example, one can produce a finite net with 36 points in the following way. The Cayley table of the cyclic group of order 6 is a Latin square of order 6. Together with the horizontal and vertical lines (rows and columns of the Latin square) it gives a 3-net on 36 points but there is no affine plane of order 6 (see Theorem 1.32).

Starting from a finite net $\mathcal{N} = (\mathcal{P}, \mathcal{E}, \mathrm{I})$ we can construct an authentication system $A(\mathcal{N})$ similar to Example 13.28.

Example 13.30. Let the set of data D be the set of parallel classes of $\mathcal{N}$ and the keys be the elements of $\mathcal{P}$, that is the points of $\mathcal{N}$. The messages are lines of $\mathcal{N}$, namely the message of a parallel class d with key at point P is the (unique) line of the parallel class passing through P.

It is easy to show that the authentication system $A(\mathcal{N})$ is perfect. Moreover, the converse is also true as the next theorem shows.

Theorem 13.31. *Every perfect authentication system can be represented as $A(\mathcal{N})$ for an appropriate finite net $\mathcal{N}$.*

Proof. Let A be a perfect authentication system. We define the incidence structure $\mathcal{N} = (\mathcal{P}, \mathcal{E}, \mathrm{I})$ in the following way:

- the points of $\mathcal{N}$, that is the set $\mathcal{P}$, is the set of keys K;

- the lines of $\mathcal{N}$, that is the set $\mathcal{E}$, is the set of messages of A;

- a point is incident with a line if and only if the corresponding message is authentic with respect to the key.

Since A is perfect, for any messages belonging to two distinct data there is precisely one key, for which both messages are authentic. This means that two lines of $\mathcal{N}$ have at most one point in common. This implies N1.

If the number of keys is κ, then it follows from the perfectness of A that every message is authentic for $\sqrt{\kappa}$ keys and for any data there are $\sqrt{\kappa}$ different messages. In other words, every line of $\mathcal{N}$ is incident with $\sqrt{\kappa}$ points and the $\sqrt{\kappa}$ lines corresponding to the $\sqrt{\kappa}$ messages belonging to one data form a parallel class consisting of $\sqrt{\kappa}$ lines. This is only possible if there is a unique line from each parallel class through a given point. This implies N2.   □

Another branch of cryptography where finite geometric structures can be used is the theory of *secret sharing schemes*.

The simplest such scheme is that we have two locks on a safe and the two (different) keys are distributed between two people. Then they cannot open the safe individually but together they can open it.

The problem can be modelled in the following way: We have a secret data X and it is split into pieces X_i, called *shadows*. A set $\mathcal{X} = \{X_1, X_2, \ldots, X_k\}$ of shadows is called a *t-threshold scheme* if X can be reconstructed from any t elements of $\mathcal{X}$, but it cannot be reconstructed from $t - 1$ elements.

We assume that a person (or oracle) knows the secret and also splits into its shadows. We are not considering the problem of how this can be done.

Example 13.32. Let $\Pi = \mathrm{PG}(t, q)$ be a projective space of dimension t. Fix a line e in Π. Let the secret data be a point P, $P \in e$. Let H be a hyperplane of Π through P, not containing e. Let G be a $(q + 1)$-arc in H – for example a normal rational curve – containing P, and let the shadows be the points of G different from P.

If t shadows are known, then the hyperplane H is determined, because $t - 1$ points in general position determine a hyperplane of Π. If we know H, then the secret P will just be $e \cap H$. Knowing less than t shadows is clearly not enough to determine the secret. We can say more: if we know $k \le t - 1$ shadows, then the probability that we can guess the secret P is the same as the probability if no shadow is known. Namely, k points of G determine a subspace U of dimension k, which does not intersect e. For every point $E \in e$, the subspace U and the point E together generate a subspace that intersects e. So, even if we know k shadows, the probability of guessing the secret P is $1/(q + 1)$, which is clearly the chance of guessing a point of the line e.

A secret sharing scheme is *perfect* if the probability of guessing the secret is the same, no matter how many shadows are known provided that we do not know enough shadows to determine the secret. In other words, if we do not know enough shadows – perhaps nothing –, there is no better way to find the secret than random guessing. In our previous example, we have seen a perfect threshold scheme in which the probability of (illegal) guessing of the secret can be arbitrary small if q tends to infinity.

Another type of secret sharing schemes is that of the multilevel secret sharing schemes. In these schemes shadows differ in weight. In general, we require the following: when the total weight of a set of shadows is large enough, then they determine the secret; otherwise they do not. The simplest of such schemes is a $(2, s)$-*scheme*. Here the shadows are divided into two groups, S and T. It is required that the secret be determined by any 2 elements of T, any s elements of S, and also from one element of T together with $s - 1$ shadows from S, but not from any other subset of shadows. We now give an example for the simplest case $s = 3$. It was originally presented by *Simmons* [156, 157].

Example 13.33. Let e be a line of $\mathrm{PG}(3, q)$, the secret data be a point P on e. Let H be a plane of $\mathrm{PG}(3, q)$, which does not contain e, and passes through P. Let G be a k-arc in H which contains P, and let f be a tangent to G at P. Let the shadows in T be some points of the line f, the shadows in S be some points of G, different from P. These two sets of points have to be chosen in such a way that the line joining two distinct points of S does not pass through a point of T.

The secret can be determined if we either know the line f or the plane H, because these meet e in P. The line f is determined by any two of its points and the plane H is determined by any three non-collinear points of it. Hence the secret can be reconstructed from any two elements of T and any three elements of S. Since the lines joining two points of S meet the line f in a point not belonging to T, two points of S and one point of T also determine the plane H uniquely. If fewer shadows are known, then our chances are the same as knowing no shadows at all. Namely, even if we know a point of f, or know two points of H, or know a point of f and a point of H, nothing is known about the common point of these objects and e. Therefore, in all these

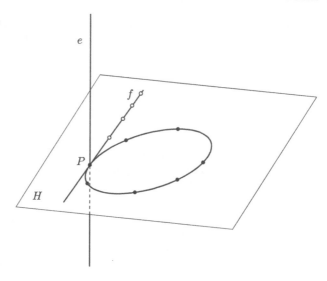

FIGURE 13.1
Construction of a 2-level secret sharing scheme

cases the chance of guessing the secret is $1/(q+1)$ implying that the scheme is perfect.

In our example, the elements of S were chosen from the points of a k-arc and the elements of T were on a tangent line to this k-arc in such a way that no two points of S and a point of T be collinear. Our goal is to construct the most efficient scheme in the sense that $|S| + |T|$ is as large as possible. The more points we choose from the arc, the fewer points are available on the tangent. There are $k-1$ secants to a k-arc through any of its points, and these lines meet f in distinct points, hence we cannot increase the sizes of S and T arbitrarily. This leads to the following definition.

Definition 13.34. *Let K be a k-arc and ℓ be an exterior line to K. Then K is called* very sharply focused, *or* hyperfocused *on ℓ if the secants of K cover exactly $k-1$ points of ℓ. The arc K is called* sharply focused *on ℓ if the secants of K cover exactly k points of ℓ.*

If we consider the tangent to our arc as the line at infinity of an affine plane, then maximizing the number of shadows indicates that we wish to select "many" points of the arc that determine "few" directions. A similar problem was considered in Chapter 6 when we studied Rédei type blocking sets.

Let K be a hyperfocused k-arc on ℓ in the plane $PG(2, q)$. Let A_0, A_1, A_2 be three points of K and $G = \{B_1, B_2, \ldots, B_{q+2-k}\}$ be the set of points on ℓ not covered by the secants of K. Then the points $A_1, A_2, A_3 \in K$ have the property that each line through at least one of them meets $G \cup \{A_0, A_1, A_2\}$ in exactly two points. Thus, by the results of Bichara and Korchmáros (Proposition 6.20), we get the following theorem.

Theorem 13.35 (Bichara, Korchmáros). *If a hyperfocused k-arc ($k > 2$) exists in the plane $PG(2, q)$, then q is even.*

Sharply focused arcs, instead, exist for both q even and q odd. For the characterization of sharply focused and very sharply focused sets we need the notion of *affinely regular polygons* and *generalized affinely regular polygons*.

Definition 13.36. *Let $\mathcal{A}$ be an abitrary affine plane, $P_0, P_1, \ldots, P_{n-1}$ be a set of n points in the plane. The sequence $P_0 P_1 \ldots P_{n-1}$ is called an* affinely regular polygon, *if there is a bijection γ mapping $\{P_0, P_1, \ldots, P_{n-1}\}$ onto the set of vertices of a regular n-gon in the classical Euclidean plane such that $P_i P_j$ and $P_k P_l$ are parallel in the plane $\mathcal{A}$ if and only if $\gamma(P_i)\gamma(P_j)$ and $\gamma(P_k)\gamma(P_l)$ are parallel in the classical Euclidean plane.*

Affinely regular polyons of the finite affine plane $AG(2, q)$, q odd, have been classified by *Korchmáros*, see [110], [113]. There are three classes of such polygons as described in the next theorem.

Theorem 13.37. *Let $\mathbb{K} = GF(q)$, $q = p^r$, p an odd prime. Every affinely regular polygon of the affine plane $\mathcal{A} = AG(2, \mathbb{K})$ can be mapped by an affinity to one of the affinely regular polygons described in the three classes below:*

1. *Let G be a multiplicative subgroup of $\mathbb{K}^*$. If $|G| = n$, then there is an affinely regular n-gon in $\mathcal{A}$, which is inscribed in the hyperbola with equation $XY = 1$. If $z \in G$ is a generator and the coordinates of P_j are (z^j, z^{-j}), then $P_0 P_1 \ldots P_{n-1}$ is an affinely regular n-gon.*

2. *Let P_j be the point with coordinates (j, j^2), where $j = 0, 1, \ldots, p - 1$. Then $P_0 P_1 \ldots P_{p-1}$ is an affinely regular p-gon inscribed in the parabola with equation $Y = X^2$.*

3. *Let $\widehat{\mathbb{K}} = \mathbb{K}(i)$ be a quadratic extension of $\mathbb{K}$ with the root of $x^2 - k = 0$, where k is a non-square in $\mathbb{K}$. The elements of $\widehat{\mathbb{K}}$ can be written as $a + bi$, $a, b \in \mathbb{K}$, and can be identified with the points of $\mathcal{A}$ us described in Lemma 1.24. The elements of $\widehat{\mathbb{K}}$ satisfying $a^2 - kb^2 = 1$ form a multiplicative subgroup H of order $q + 1$ of $\widehat{\mathbb{K}}^*$. Let G be a subgroup of H. If $|G| = n$, then there is an affinely regular n-gon in $\mathcal{A}$, inscribed in the ellipse with equation $X^2 - kY^2 = 1$. If $z \in G$ is a generator, $z^j = x_j + y_j i$, and P_j is the point with coordinates $(x_j; y_j)$, then $P_0 P_1 \ldots P_{n-1}$ is an affinely regular n-gon.*

Definition 13.38. *Let $\mathcal{P}$ be the parabola with equation $Y = X^2$ in $AG(2, q)$. Let $u \in GF(q)$ and $H = \{h_0, h_1, \ldots, h_{n-1}\}$ be a subgroup of the additive group of $GF(q)$. If $P_i = (u + h_i, (u + h_i)^2)$, for $i = 0, 1, \ldots, n - 1$, then the n-gon $P_0 P_1 \ldots P_{n-1}$ is called a* generalized affinely regular polygon.

Wettl [181] gave the complete classification of sharply focused sets in the plane PG$(2, q)$, q odd.

Theorem 13.39 (Wettl). *Let S be a sharply focused k-arc on ℓ in the plane PG$(2, q)$, $q = p^r$ odd. Then k divides $q + 1, q - 1$, or q, and S is an affinely regular k-gon if p^2 does not divide k. If p^2 divides k, then S is a generalized affinely regular k-gon.*

He proved an analogous theorem for planes of even order, too. Before citing this result we have to introduce the notion of *Pascal line of an oval* which comes from the classical Theorem of Pascal. A hexagon $\mathcal{H}$ of an oval $\mathcal{O}$ is an ordered set $(A_1, A_2, A_3, B_1, B_2, B_3)$ of points of $\mathcal{O}$ such that $A_i \neq A_j$, $B_i \neq B_j$ and $A_i \neq B_i$ for $i, j = 1, 2, 3$. The three points $A_i B_j \cap A_j B_i$, $i \neq j$, (in the case $A_i = B_j$ the line $A_i B_j$ defined as the tangent to $\mathcal{O}$ at A_i) are called the diagonal points of $\mathcal{H}$. A line ℓ is called a Pascal line with respect to $\mathcal{O}$ if each hexagon of $\mathcal{O}$ which has two diagonal points on ℓ also has the third diagonal point on ℓ.

Theorem 13.40 (Wettl). *Let S be a sharply focused k-arc on ℓ in the plane PG$(2, q)$, q even. If S is a subset of an oval $\mathcal{O}$ and the line ℓ is a Pascal line of $\mathcal{O}$, then either k divides $q \pm 1$ or $k + 1$ divides q.*

These results of Wettl come from more general results about internal nuclei. Let $\mathcal{K}$ be a k-set in PG$(2, q)$. A point $P \in \mathcal{K}$ is called an *internal nucleus* of $\mathcal{K}$ if each line through P meets $\mathcal{K}$ in at most two points including P. This notion was introduced for $k = q + 2$ by *Bichara* and *Korchmáros* [22] and generalized by *Wettl* [181]. The set of internal nuclei of $\mathcal{K}$ is denoted by $IN(\mathcal{K})$. If S is sharply focused on ℓ and $\mathcal{N}$ denotes the set of points of ℓ not covered by a chord of S, then $\mathcal{K} = S \cup \mathcal{N}$ is a set of $q + 1$ points and the points of S are internal nuclei of $\mathcal{K}$. If one starts from a very sharply focused set, then the corresponding $\mathcal{K}$ is a $(q + 2)$-set.

From our point of view the main results of *Bichara, Korchmáros*, and *Wettl* are the following.

Theorem 13.41 (Bichara, Korchmáros [22]). *If $\mathcal{K}$ is a set of $q + 2$ points in PG$(2, q)$ and $|IN(\mathcal{K})| \geq 3$, then q is even. If q is even and $|IN(\mathcal{K})| > q/2$, then $\mathcal{K}$ is a hyperoval.*

Theorem 13.42 (Wettl [181]). *If $\mathcal{K}$ is a set of $q + 1$ points in PG$(2, q)$, q odd, and $|IN(\mathcal{K})| > (q + 1)/2$, then $\mathcal{K} = IN(\mathcal{K})$, that is $\mathcal{K}$ is an oval.*

At an internal nucleus P, a line t is a tangent to $\mathcal{K}$ if it meets $\mathcal{K}$ in just P. For the set of tangents, Segre's Lemma of tangents (Lemma 6.17) can be used and for $|\mathcal{K}| = q + 1$, q odd, it gives that $IN(\mathcal{K})$ is contained in a conic (see Exercise 13.2). Affinely regular $(q + 1)/2$-gons together with the ideal points not covered by the chords show that the bound $(q + 1)/2$ in the theorem of Wettl is sharp. Similarly, for q even, generalized affinely regular $q/2$-gons show the sharpness of the bound in the theorem of Bichara and Korchmáros. For q

even, the above trick with Segre's Lemma of tangents gives that $(q+1)$-sets are contained in $(q+2)$-sets whose internal nuclei set cannot be smaller. More general embedding theorems were presented by *Szőnyi* [161] and by *Beato, Faina* and *Giulietti* [18]. *Szőnyi* [161] also proved that for $q > 121$, $(q+1)$-sets in PG$(2,q)$, q odd, having $(q+1)/2$ internal nuclei are projectively equivalent to $\mathcal{K} = \mathcal{S} \cup \mathcal{N}$, where $\mathcal{S}$ is an affinely regular $(q+1)/2$-gon, and the points of $\mathcal{N}$ are collinear.

Hyperfocused arcs contained in a conic were investigated by *Wen-Ai Jackson* [96]. Chapter 5 of her PhD thesis contains several results on sharply focused and hyperfocused sets, such as the next one.

Theorem 13.43 (Jackson). *Let $\mathcal{S}$ be a hyperfocused arc on ℓ in* PG$(2,q)$, *q even. Suppose that $\mathcal{S}$ is contained in a conic. Then $|\mathcal{S}| = k$ divides q, and $\mathcal{S}$ is a generalized affinely regular k-gon.*

It was already noticed by *Wettl* [181] that there were hyperfocused arcs not contained in a conic. *Cherowitzo* and *Holder* [43] constructed hyperfocused arcs that were contained in a hyperoval or a subplane. Recently, *Giulietti* and *Montanucci* [71] constructed hyperfocused translation arcs that were not contained in a hyperoval or a subplane. Moreover, these arcs are complete in the sense that every point not on the distinguished line ℓ belongs to some chord of the arc.

Finally, let us mention that the geometric questions for the analogue of Simmons' model for $s = 4$ were investigated by *Korchmáros, Lanzone* and *Sonnino* [111]. In this case, the secret data is a point P on a line e in PG$(4,q)$. As in Example 13.33, let H be a hyperplane of PG$(4,q)$, which does not contain e, and passes through P. Let f be a line of H through P. Let the shadows in T be some points of the line f, the shadows in S be some points of H, different from P. These two sets of points have to be chosen in such a way that S is a k-arc disjoint from f such that no point from $T \cup \{P\}$ is cut out by the plane determined by a triangle inscribed in S. Points on f which are coplanar with three points of S are called foci, the set consisting of all foci is called the *focus set* and it is denoted by $\mathcal{F}$. The scheme is efficient if $\mathcal{F}$ is small. The trivial lower bound on the cardinality of $\mathcal{F}$ is $k - 2$. The k-arc S is called a *spatial hyperfocused arc* if $|\mathcal{F}| = k - 2$ and it is called a *spatial sharply focused arc* if $|\mathcal{F}| = k - 1$. It is still unknown whether non-trivial spatial hyperfocused arcs or spatial sharply focused arcs do exist.

Exercises

13.1. Verify that the examples listed in Theorem 13.37 are indeed affinely regular polygons.

13.2. Define "combinatorially" the tangents of an affinely regular polygon and show that one can apply the Lemma of tangents for them. Conclude that affinely regular polygons are inscribed in a conic.

13.3. Let S be a sharply focused/hyperfocused k-arc on ℓ and $\mathcal{D}$ denote the set of points on ℓ collinear with two points of S. Show that $S \cup (\ell \setminus \mathcal{D})$ is a $(q+1)$-set/$(q+2)$-set whose internal nuclei are the points of $\mathcal{D}$.

13.4. Verify Theorem 13.43.

13.5. Show that one can indeed apply the Lemma of tangents for the tangents of a $(q+1)$-set at its internal nuclei.

13.6. Prove Theorem 13.42 using the previous exercise.

13.7. Let $\mathcal{O}$ be an oval, ℓ be a Pascal line of $\mathcal{O}$ and U be a point in the point set $\mathcal{O}' = \mathcal{O} \setminus \ell$. We define a binary operation, denoted by $*$, on $\mathcal{O}'$ in the following way. If $A, B \in \mathcal{O}'$, then let $H_{AB} = AB \cap \ell$, and

$$A * B = \begin{cases} U, & \text{if } H_{AB}U \text{ is the tangent to } \mathcal{O} \text{ at } U, \\ D, & \text{if } H_{AB}U \cap \mathcal{O} = \{D, U\}. \end{cases}$$

Show that $(\mathcal{O}', *)$ is an abelian group G.

Suppose that $\mathcal{O}$ is a conic in the plane $\mathrm{PG}(2, q)$. Prove that G is isomorphic to the

- additive group of $\mathrm{GF}(q)$ if ℓ is a tangent to $\mathcal{O}$,
- multiplicative group of $\mathrm{GF}(q)$ if ℓ is a secant to $\mathcal{O}$.

Bibliography

[1] R. W. Ahrens and G. Szekeres. On a combinatorial generalization of 27 lines associated with a cubic surface. *J. Austral. Math. Soc.*, 10:485–492, 1969.

[2] M. Aigner and G. M. Ziegler. *Proofs from The Book.* Springer-Verlag, New York, Berlin, Heidelberg, 1998.

[3] N. Alon, L. Rónyai, and T. Szabó. Norm-graphs: variations and applications. *J. Combin. Theory Ser. B*, 76:280–290, 1999.

[4] B. Bagchi and N. S. N. Sastry. Ovoidal packings in PG$(3, q)$ for even q. *Discrete Math.*, 313:2217–2217, 2013.

[5] R. D. Baker and G. L. Ebert. On Buekenhout–Metz unitals of odd order. *J. Combin Theory Ser. A*, 60:67–84, 1992.

[6] S. Ball. Multiple blocking sets and arcs in finite planes. *J. London Math. Soc.*, 54:581–593, 1996.

[7] S. Ball. On sets of vectors of a finite vector space in which every subset of basis size is a basis. *J. Europ. Math. Soc.*, 14:733–748, 2012.

[8] S. Ball. *Finite Geometry and Combinatorial Applications.* Cambridge University Press, Cambridge, 2015.

[9] S. Ball and A. Blokhuis. On the incompleteness of (k, n)-arcs in Desarguesian planes of order q where n divides q. *Geom. Dedicata*, 74:325–332, 1999.

[10] S. Ball, A. Blokhuis, and F. Mazzocca. Maximal arcs in desarguesian planes of odd order do not exist. *Combinatorica*, 17:31–47, 1997.

[11] S. Ball and J. De Beule. On sets of vectors of a finite vector space in which every subset of basis size is a basis II. *Des. Codes Cryptogr.*, 65:5–14, 2012.

[12] S. Ball and J. W. P. Hirschfeld. Bounds on (n, r)-arcs and their applications to linear codes. *Finite Fields Appl.*, 11:326–336, 2005.

[13] S. Ball and M. Lavrauw. Planar arcs. *J. Combin Theory Ser. A*, 160:261–287, 2018.

[14] J. Bamberg and T. Penttila. Completing Segre's proof of Wedderburn's little theorem. *Bull. London Math. Soc.*, 47:483–492, 2015.

[15] A. Barlotti. *Some Topics in Finite Geometrical Structures*, University of North Carolina, Mimeo Series, no. 439, 1965.

[16] A. Barlotti. Un'osservazione sulle k-calotte degli spazi lineari finiti di dimensione tre. *Boll. Un. Mat. Ital.*, 11:248–252, 1956.

[17] S. Barwick and G. Ebert. *Unitals in Projective Planes*. Springer-Verlag, New York, Berlin, Heidelberg, 2008.

[18] A. Beato, G. Faina, and M. Giulietti. Arcs in Desarguesian nets. *Contrib. Discrete Math.*, 3:96–108, 2008.

[19] E. R. Berlekamp, R. J. McEliece, and H. C. A. van Tilborg. On the inherent intractability of certain coding problems. *IEEE Trans. Inform. Theory*, 24:384–386, 1978.

[20] A. Beutelspacher. Blocking sets and partial spreads in finite projective spaces. *Geom. Dedicata*, 9:425–449, 1980.

[21] A. Beutelspacher and U. Rosenbaum. *Projective Geometry: From Foundations to Applications*. Cambridge University Press, Cambridge, 1998.

[22] A. Bichara and G. Korchmáros. Note on $(q+2)$-sets in a Galois plane of order q. *Ann. Discrete Math.*, 14:117–122, 1982.

[23] J. Bierbrauer and Y. Edel. Bounds on affine caps. *J. Combin. Des.*, 10:111–115, 2002.

[24] A. Blokhuis. On the size of a blocking set in PG$(2,p)$. *Combinatorica*, 14:111–114, 1994.

[25] A. Blokhuis. Blocking sets in Desarguesian planes. In *Combinatorics, Paul Erdős is Eighty*, pages 133–154. János Bolyai Math. Soc., Budapest, 1996.

[26] A. Blokhuis, A. E. Brouwer, and H. A. Wilbrink. Blocking sets in $PG(2,p)$ for small p, and partial spreads in $PG(3,7)$. *Advances in Geom.*, suppl.:245–253, 2003.

[27] A. Blokhuis, A. A. Bruen, and J. A. Thas. Arcs in PG(n,q), MDS-codes and three fundamental problems of B. Segre – Some extensions. *Geom. Dedicata*, 35:1–11, 1990.

[28] A. Blokhuis and P. Sziklai. On planes of order p^2 in which every quadrangle generates a subplane of order p. *Geom. Dedicata*, 79:341–347, 2000.

[29] A. Blokhuis, P. Sziklai, and T. Szőnyi. Blocking sets in projective spaces. In *Current Research Topics in Galois Geometriy (ed.: J. De Beule, L. Storme)*, pages 63–86. Nova Science Publishers, Inc., 2014.

[30] A. Blokhuis and H. A. Wilbrink. A characterization of exterior lines of certain sets of points in PG$(2, q)$. *Geom. Dedicata*, 23:253–254, 1987.

[31] B. Bollobás. *Extremal Graph Theory*. Dover Publications, Inc., Mineola, New York, 2004.

[32] E. Boros and T. Szőnyi. On the sharpness of a theorem of B. Segre. *Combinatorica*, 6:261–268, 1986.

[33] M. Braun, A. Kohnert, and A. Wassermann. Construction of (n, r)-arcs in PG$(2, q)$. *Innov. Incidence Geom.*, 1:133–141, 2005.

[34] A. E. Brouwer and A. Schrijver. The blocking number of an affine space. *J. Combin. Theory Ser. A*, 24:251–253, 1978.

[35] A. A. Bruen and M. J. de Resmini. Blocking sets in affine planes,. *Ann. Discrete Math.*, 18:169–175, 1981.

[36] A. A. Bruen, J. A. Thas, and A. Blokhuis. On M.D.S. codes, arcs in PG(n, q) with q even, and a solution of three fundamental problems of B. Segre. *Invent. Math.*, 92:441–459, 1988.

[37] F. Buekenhout. Existence of unitals in finite translation planes of order q^2 with a kernel of order q. *Geom. Dedicata*, 5:189–194, 1976.

[38] K. A. Bush. Orthogonal arrays of index unity. *Ann. Math. Statist.*, 23:426–434, 1952.

[39] L. R. A. Casse. A solution to Beniamino Segre's "Problem $I_{r,q}$" for q even. *Atti Accad. Naz. Lincei Rend. Cl. Fis. Mat. Natur.*, 46:13–20, 1969.

[40] L. R. A. Casse and D. G. Glynn. The solution to Beniamino Segre's problem $I_{r,q}$, $r = 3$, $q = 2^h$. *Geom. Dedicata*, 13:157–164, 1982.

[41] J. M. Chao and H. Kaneta. Classical arcs in PG(r, q) for $23 \leq q \leq 29$. *Discrete Math.*, 226:377–385, 2001.

[42] W. E. Cherowitzo. Project Website: http://math.ucdenver.edu/ wcherowi/research/hyperoval/hypero.html.

[43] W. E. Cherowitzo and L. D. Holder. Hyperfocused arcs. *Simon Stevin*, 12:685–696, 2005.

[44] W. E. Cherowitzo, T. Penttila, I. Pinneri, and G. F. Royle. Flocks and ovals. *Geom. Dedicata*, 60:17–37, 1996.

[45] W. E. Cherowitzo and L. Storme. α-flocks with oval herds and maximal hyperovals. *Finite Fields Appl.*, 4:185–194, 1998.

[46] S. Chowla, P. Erdős, and E. G. Straus. On the maximal number of pairwise orthogonal Latin squares of a given order. *Canad. J. Math.*, 12:204–208, 1960.

[47] A. Cossu. Su alcune proprietà dei $\{k, n\}$-archi di un piano proiettivo sopra un corpo finito. *Rend. Mat. e Appl.*, 20:271–277, 1961.

[48] H. S. M. Coxeter. *Projective Geometry.* Springer-Verlag, New York, Berlin, Heidelberg, 1987.

[49] E. S. Croot, V. F. Lev, and P. P. Pach. Progression-free sets in $\mathbb{Z}_4^n$ are exponentially small. *Ann. of Math.*, 185:331–337, 2017.

[50] B. Csajbók and T. Héger. Double blocking sets of size $3q-1$ in PG$(2, q)$. *arXiv*, 1805.01267:1–19, 201x.

[51] G. Damásdi, T. Héger, and T. Szőnyi. The Zarankiewicz problem, cages, and geometries. *Annales Univ. Sci. Eötvös Loránd*, LVI:3–37, 2013.

[52] A. A. Davydov and A. L. Tombak. Quasiperfect linear binary codes with distance 4 and complete caps in projective geometry. *Probl. Inform. Transm.*, 25:265–275, 1989.

[53] J. De Beule, T. Héger, T. Szőnyi, and G. Van de Voorde. Blocking and double blocking sets in finite planes. *Electron. J. Combin.*, 23:Paper # P2.5, 2016.

[54] M. de Finis. On semiovals in projective planes. *Ars Combin.*, 24:65–70, 1987.

[55] P. Dembowski. *Finite Geometries.* Springer-Verlag, New York, Berlin, Heidelberg, 1968.

[56] P. Dembowski and D. R. Hughes. On finite inversive planes. *J. London Math. Soc.*, 40:171–182, 1965.

[57] R. H. F. Denniston. Some maximal arcs in finite projective planes. *J. Combin. Theory*, 6:317–319, 1969.

[58] J. W. Di Paola. On mimimum blocking coalitions in small projective plane games. *SIAM J. Appl. Math.*, 17:378–392, 1969.

[59] G. L. Ebert. Partitioning projective geometries into caps. *Canad. J. Math.*, 37:1163–1175, 1985.

[60] Y. Edel. Extensions of generalized product caps. *Des. Codes Cryptogr.*, 31:5–14, 2004.

[61] Y. Edel and J. Bierbrauer. 41 is the largest size of a cap in PG(4, 4). *Des. Codes Cryptogr.*, 16:151–160, 1999.

[62] J. S. Ellenberg and D. C. Gijswijt. On large subsets of $\mathbb{F}_q^n$ with no three-term arithmetic progression. *Ann. of Math.*, 185:339–343, 2017.

[63] P. Erdős, R. C. Mullin, V. T. Sós, and D. R. Stinson. Finite linear spaces and projective planes. *Discrete Math.*, 47:49–62, 1983.

[64] G. Ewald. Beispiel einer Möbiusebene mit nichtisomorphen affinen Unterebenen. *Arch. Math.*, 11:146–150, 1960.

[65] G. Exoo and R. Jajcay. Dynamic Cage Survey. *Electron. J. Combin.*, Dynamic Survey DS16, 2008.

[66] W. Feit and G. Higman. The nonexistence of certain generalized polygons. *J. Algebra*, 1:114–131, 1964.

[67] J. C. Fisher, J. W. P. Hirschfeld, and J. A. Thas. Complete arcs in planes of square order. *Ann. Discrete Math.*, 30:243–250, 1986.

[68] J. C. Fisher and J. A. Thas. Flocks in $PG(3, q)$. *Math. Z.*, 169:1–11, 1979.

[69] Z. Füredi. An upper bound on Zarankiewicz' problem. *Combin. Probab. Comput.*, 5:29–33, 1996.

[70] Z. Füredi. On the number of edges of quadrilateral-free Graphs. *J. Combin. Theory Ser. B*, 68:1–6, 1996.

[71] M. Giulietti and E. Montanucci. On hyperfocused arcs in PG(2, q). *Discrete Math.*, 306:3307–3314, 2006.

[72] A. M. Gleason. Finite Fano planes. *Amer. J. Math.*, 78:797–807, 1956.

[73] D. G. Glynn. Two new sequences of ovals in finite Desarguesian planes of even order. In *Combinatorial Mathematics, X (Adelaide, 1982), Springer Lecture Notes in Mathematics Series 1036*, pages 217–229. Springer-Verlag, New York, Berlin, Heidelberg, 1983.

[74] D. G. Glynn. A condition for the existence of ovals in $PG(2, q)$, q even. *Geom. Dedicata*, 32:247–252, 1989.

[75] P. Goevarts and L. Storme. The classification of the smallest nontrivial blocking sets in PG(n, 2). *J. Combin. Theory Ser. A*, 113:1543–1548, 2006.

[76] R. K. Guy. The many faceted problem of Zarankiewicz. In *The Many Facets of Graphs Theory, Springer Lecture Notes in Mathematics Series 110*, pages 129–148. Springer-Verlag, New York, Berlin, Heidelberg, 1969.

[77] M. Hall Jr. *The Theory of Groups*. The MacMillan Company, New York, 1959.

[78] M. Hall Jr. Affine generalized quadrilaterals. In *Studies in Pure Mathematics*, pages 113–116. Academic Press, London, 1971.

[79] M. Hall Jr. Ovals in the Desarguesian plane of order 16. *Ann. Mat. Pura Appl.*, 102:159–176, 1975.

[80] M. Hall Jr., J. D. Swift, and R. J. Walker. Uniqueness of the projective plane of order eight. *Math. Tables Aids Comput.*, 10:186–194, 1956.

[81] O. Heden. A greedy search for maximal partial spreads in $PG(3,7)$. *Ars Combin.*, 32:253–255, 1991.

[82] U. Heim. Blockierende Mengen in endlichen projektiven Räumen. *Mitt. Math. Sem. Giessen*, (1996), Dissertation, Justus-Liebig-Universität, Giessen:pp. 1–82, 1995.

[83] R. Hill. On the largest size of a cap in $S_{5,3}$. *Atti Accad. Naz. Lincei Rend.*, 54:378–384, 1973.

[84] R. Hill. Caps and codes. *Discrete Math.*, 22:111–137, 1978.

[85] R. Hill and J. M. Mason. On (k,n)-arcs and the falsity of the Lunelli-Sce conjecture. In *Finite Geometries and Designs, LMS Lecture Note Series 49 (ed.: Cameron, P. J., Hirschfeld, J. W. P., Hughes, D. R.)*, pages 153–168. Cambridge University Press, Cambridge, 1981.

[86] J. W. P. Hirschfeld. Ovals in Desarguesian planes of even order. *Ann. Mat. Pura Appl.*, 102:79–89, 1975.

[87] J. W. P. Hirschfeld. *Finite Projective Spaces of Three Dimensions*. Clarendon Press, Oxford, 1985.

[88] J. W. P. Hirschfeld. *Projective Geometries over Finite Fields, second ed.* Clarendon Press, Oxford, 1998.

[89] J. W. P. Hirschfeld and G. Korchmáros. On the embedding of an arc into a conic in a finite plane. *Finite Fields Appl.*, 2:274–292, 1996.

[90] J. W. P. Hirschfeld, G. Korchmáros, and F. Torres. *Algebraic Curves over a Finite Field*. Princeton University Press, 2008.

[91] J. W. P. Hirschfeld and L. Storme. The packing problem in statistics, coding theory and finite projective spaces. In *Finite Geometries, Developments of Mathematics, Isle of Thorns*, pages 201–246. Kluwer, 2001.

[92] J. W. P. Hirschfeld and J. A. Thas. Linear independence in finite spaces. *Geom. Dedicata*, 23:15–31, 1987.

[93] J. W. P. Hirschfeld and J. A. Thas. *General Galois Geometries*. Clarendon Press, Oxford, 1991.

[94] A. J. Hoffman and R. R. Singleton. On Moore graphs with diameters 2 and 3. *IBM J. Res. Develop.*, 4:497–504, 1960.

[95] D. R. Hughes and F. C. Piper. *Projective Planes*. Springer-Verlag, New York, Berlin, Heidelberg, 1973.

[96] W. A. Jackson. *On designs which admit specific automorphisms*. Ph.D. Thesis, Royal Holloway and Bedford New College, University of London, 1989.

[97] R. E. Jamison. Covering finite fields with cosets of subspaces. *J. Combin. Theory Ser. A*, 22:253–266, 1977.

[98] D. Jungnickel. Difference sets. In *Contemporary Design Theory*, pages 241–324. John Wiley and Sons, New York, 1992.

[99] J. Kahn. Inversive planes satisfying the bundle theorem. *J. Combin. Theory Ser. A*, 29:1–19, 1980.

[100] J. Kahn. Finite inversive planes which satisfy the bundle theorem. *Geom. Dedicata*, 12:171–187, 1982.

[101] H. Kaneta and T. Maruta. An elementary proof and an extension of Thas' theorem on k-arcs. *Math. Proc. Cambridge Philos. Soc.*, 105:459–462, 1989.

[102] W. M. Kantor. Generalized quadrangles associated with $G_2(q)$. *J. Combin. Theory Ser. A*, 29:212–219, 1980.

[103] W. M. Kantor. Generalized quadrangles and translation planes. *Algebras, Groups Geom.*, 2:313–322, 1985.

[104] W. M. Kantor. Some generalized quadrangles with parameters q^2, q. *Math. Z.*, 192:45–50, 1986.

[105] F. Kárteszi. *Introduction to Finite Geometries*. North-Holland Publishing Co., Amsterdam, Oxford, 1976.

[106] B. C. Kestenband. Unital intersections in finite projective planes. *Geom. Dedicata*, 11:107–117, 1981.

[107] T. Kőváry, V. T. Sós, and P. Turán. On a problem of K. Zarankiewicz. *Colloq. Math.*, 3:50–57, 1954.

[108] J. H. Kim and V. H. Vu. Small complete arcs in projective planes. *Combinatorica*, 23:311–363, 2003.

[109] J. Kollár, L. Rónyai, and T. Szabó. Norm-graphs and bipartite Turán numbers. *Combinatorica*, 16:399–406, 1996.

[110] G. Korchmáros. Poligoni affin-regolari dei piani di Galois d'ordine dispari. *Atti Accad. Naz. Lincei Rend.*, 56:690–697, 1974.

[111] G. Korchmáros, V. Lanzone, and A. Sonnino. Projective k-arcs and 2-level secret-sharing schemes. *Des. Codes Cryptogr.*, 64:3–15, 2012.

[112] G. Korchmáros, N. Pace, and A. Sonnino. One-factorisations of complete graphs arising from ovals in finite planes. *J. Combin. Theory Ser. A*, 160:62–83, 2018.

[113] G. Korchmáros and T. Szőnyi. Affinely regular polygons in an affine plane. *Contrib. Discrete Math.*, 3:20–38, 2008.

[114] C. W. H. Lam, L. Thiel, and S. Swiercz. The non-existence of finite projective planes of order 10. *Canad. J. Math.*, 41:1117–1123, 1989.

[115] L. Lovász. *Combinatorial Problems and Exercises, second ed.* AMS Chelsea Publishing, 2007.

[116] G. Lunardon. Normal spreads. *Geom. Dedicata*, 75:245–261, 1999.

[117] G. Lunardon, P. Polito, and O. Polverino. A geometric characterisation of linear k-blocking sets. *J. Geom.*, 74:120–122, 2002.

[118] G. Lunardon and O. Polverino. Linear blocking sets: a survey. In *Finite Fields and Applications (ed.: Jungnickel, D. and Niederreiter, H.)*, pages 356–362. Springer, Berlin, Heidelberg, 2001.

[119] H. Lüneburg. *Translation Planes.* Springer-Verlag, New York, Berlin, Heidelberg, 1980.

[120] L. Lunelli and M. Sce. k-archi completi nei piani proiettivi desarguesiani di rango 8 e 16. *Centro di Calcoli Numerici, Politecnico di Milano, Milan*, page 11 pp, 1958.

[121] L. Lunelli and M. Sce. Considerazioni arithmetiche e risultati sperimentali sui $\{K; n\}_q$-archi. *Ist. Lombardo Accad. Sci. Rend. A*, 98:3–52, 1964.

[122] C. R. MacInness. Finite planes with less than eight points on a line. *Amer. Math. Monthly*, 14:171–174, 1907.

[123] W. Mantel. Problem 28 (Solution by H. Gouwentak, W. Mantel, J. Teixeira de Mattes, F. Schuh and W. A. Wythoff). *Wiskundige Opgaven*, 10:60–61, 1907.

[124] R. Mathon. New maximal arcs in Desarguesian planes. *J. Combin. Theory Ser. A*, 97:353–368, 2002.

[125] F. Mazzocca and O. Polverino. Blocking sets in $PG(2, q^n)$ from cones of $PG(2n, q)$. *J. Algebraic Combin.*, 24:61–81, 2006.

[126] F. Mazzocca, O. Polverino, and L. Storme. Blocking sets in $PG(r, q^n)$. *Des. Codes Cryptogr.*, 44:97–113, 2007.

[127] R. Meshulam. On subsets of finite abelian groups with no 3-term arithmetic progressions. *J. Combin. Theory Ser. A*, 71:168–172, 1995.

[128] K. Metsch. *Linear Spaces with Few Lines.* Springer-Verlag, New York, Berlin, Heidelberg, 1991.

[129] K. Metsch. On the maximum size of a maximal partial plane. *Rend. Mat. Appl.*, 12:345–355, 1992.

[130] R. Metz. On a class of unitals. *Geom. Dedicata*, 8:125–126, 1979.

[131] M. Miller and J. Širáň. Moore Graphs and Beyond: A survey of the Degree/Diameter Problem. *Electron. J. Combin.*, Dynamic Survey DS14, 2005.

[132] G. P. Nagy and T. Szőnyi. Caps in finite projective spaces of odd order. *J. Geom.*, 59:103–113, 1997.

[133] V. Nikiforov. A contribution to the Zarankiewicz problem. *Linear Algebra Appl.*, 432:1405–1411, 2010.

[134] I. Niven, H. S. Zuckerman, and H. L. Montgomery. *An Introduction to the Theory of Numbers, 5th Edition.* John Wiley & Sons, New York, 1991.

[135] C. M. O'Keefe and T. Penttila. A new hyperoval in $PG(2, 32)$. *J. Geom.*, 44:117–139, 1992.

[136] W. J. Orr. A characterization of subregular spreads in finite 3-space. *Geom. Dedicata*, 5:43–50, 1976.

[137] S. E. Payne. A geometric representation of certain generalized hexagons in $PG(3, s)$. *J. Combin. Theory*, 11:181–191, 1971.

[138] S. E. Payne. A complete determination of translation ovoids in finite desarguesian planes. *Atti Accad. Naz. Lincei Rend.*, 51:328–331, 1971.

[139] S. E. Payne. A new infinite family of generalized quadrangles. *Congr. Numer.*, 49:115–128, 1985.

[140] S. E. Payne and J. A. Thas. *Finite Generalized Quadrangles, second ed.* European Mathematical Society Publishing House, Zürich, 2009.

[141] G. Pellegrino. Sul massimo ordine delle calotte in $S_{4,3}$. *Matematiche (Catania)*, 25:1–9, 1970.

[142] P. Polito and O. Polverino. On small blocking sets. *Combinatorica*, 18:133–137, 1998.

[143] O. Polverino. Small minimal blocking sets in PG$(2, p^3)$. *Des. Codes Cryptogr.*, 20:319–324, 2000.

[144] G. Pólya and R. C. Read. *Combinatorial Enumeration of Groups, Graphs, and Chemical Compounds*. Springer-Verlag, New York, Berlin, Heidelberg, 1987.

[145] L. Rédei. *Lacunary Polynomials over Finite Fields*. North-Holland, American Elsevier, 1973.

[146] S. Roman. A problem of Zarankiewicz. *J. Combin. Theory Ser. A*, 18:187–198, 1975.

[147] S. Roman. *Coding and Information Theory*. Springer-Verlag, New York, Berlin, Heidelberg, 1992.

[148] P. Scherk. On the intersection number of two plane curves. *J. Geom.*, 10:57–68, 1977.

[149] G. Schild. *Coniques hermitiennes et unitaux*. Mémoire de Licence, Université de Bruxelles, 1974.

[150] B. Segre. On complete caps and ovaloids in three-dimensional Galois spaces of characteristic two. *Acta Arith.*, 5:315–332, 1959.

[151] B. Segre. *Lectures on Modern Geometry*. Edizioni Cremonese, Roma, 1961.

[152] B. Segre. Ovali e curve σ nei piani di Galois di caratteristica due. *Atti Accad. Naz. Lincei Rend.*, 32:785–790, 1962.

[153] B. Segre. Introduction to Galois geometries. *Atti Accad. Naz. Lincei Mem.*, 8:133–236, 1967. (edited by J. W. P. Hirschfeld.)

[154] B. Segre and U. Bartocci. Ovali ed altre curve nei piani di Galois di caratteristica due. *Acta Arith.*, 8:423–449, 1971.

[155] A. Seidenberg. *Elements of the Theory of Algebraic Curves*. Addison-Wesley, Reading, Mass., 1968.

[156] G. J. Simmons. How to (really) share a secret. *Advances in Cryptology – CRYPTO 88*, LNCS 403:390–448, 1989.

[157] G. J. Simmons. Sharply focused sets of lines on a conic in PG$(2, q)$. *Cong. Numer.*, 73:181–204, 1990.

[158] R. P. Stanley. *Enumerative Combinatorics, Volume 2*. Cambridge University Press, Cambridge, 2001.

[159] L. Storme and J. A. Thas. Complete k-arcs in PG(n,q), q even. *Discrete Math.*, 106/107:455–469, 1992.

[160] L. Storme, J. A. Thas, and S. K. J. Vereecke. New upper bounds for the sizes of caps in finite projective spaces. *J. Geom.*, 73:176–193, 2002.

[161] T. Szőnyi. k-sets in PG$(2,q)$ having a large set of internal nuclei. In *Combinatorics '88, Proceedings of the International Conference on Incidence Geometries and Combinatorial Structures, Ravello (1988), (ed.: A. Barlotti et al.)*, pages 449–458. Mediterranean Press, Research and Lecture Notes in Mathematics, Vol. 2, 1991.

[162] T. Szőnyi. Blocking sets in desarguesian affine and projective planes. *Finite Fields Appl.*, 3:187–202, 1997.

[163] P. Sziklai. On small blocking sets and their linearity. *J. Combin. Theory Ser. A*, 115:1167–1182, 2008.

[164] M. Tallini Scafati. Archi completi in un $S_{2,q}$, con q pari. *Atti Accad. Naz. Lincei Rend.*, 37:48–51, 1964.

[165] J. A. Thas. Normal rational curves and k-arcs in Galois spaces. *Rend. Mat.*, 1:331–334, 1968.

[166] J. A. Thas. Some results concerning $\{(q+1)(n-1);n\}$-arcs and $\{(q+1)(n-1);n\}$-arcs in finite projective planes of order q. *J. Combin. Theory Ser. A*, 19:228–232, 1975.

[167] J. A. Thas. Generalized quadrangles and flocks of cones. *European J. Combin.*, 8:441–452, 1987.

[168] J. A. Thas. Solution of a classical problem on finite inversive planes. In *Finite Geometries, Buildings, and Related Topics*, pages 145–159. Oxford University Press, Oxford, 1990.

[169] J. A. Thas. On k-caps in PG(n,q) with q even and $n \geq 3$. *Discrete Math.*, 341:1459–1471, 2018.

[170] J. A. Thas. On k-caps in PG(n,q) with q even and $n \geq 4$. *Discrete Math.*, 341:1072–1077, 2018.

[171] J. A. Thas, P. J. Cameron, and A. Blokhuis. On a generalization of a theorem of B. Segre. *Geom. Dedicata*, 43:299–305, 1992.

[172] J. Tits. Sur la trialité et certains groupes qui s'en déduisent. *Inst. Hautes Etudes Sci. Publ. Math.*, 2:13–60, 1959.

[173] J. Totten. Embedding the complement of two lines in a finite projective plane. *J. Austral. Math. Soc. Ser. A*, 22:27–34, 1976.

[174] T. Tsuzuku. *Finite Groups and Finite Geometries*. Cambridge University Press, Cambridge, 1982.

[175] B. L. van der Waerden and L. Smid. Eine Axiomatik der Kreisgeometrie und der Laguerre-Geometrie. *Math. Ann.*, 110:753–776, 1935.

[176] J. H. van Lint. *Introduction to Coding Theory*. Springer-Verlag, New York, Berlin, Heidelberg, 1992.

[177] H. Van Maldeghem. *Generalized Polygons*. Birkhäuser, Basel, 1998.

[178] J. F. Voloch. Arcs in projective planes over prime fields. *J. Geom.*, 38:198–200, 1990.

[179] J. F. Voloch. Complete arcs in Galois planes of nonsquare order. In *Advances in Finite Geometries and Designs, (Isle of Thorns)*, pages 401–406. Oxford University Press, Oxford, 1990.

[180] M. Walker. A class of some translation planes. *Geom. Dedicate*, 5:135–146, 1976.

[181] F. Wettl. On the nuclei of a pointset of a finite projective plane. *J. Geom.*, 30:157–163, 1987.

[182] K. Zarankiewicz. Problem of P101. *Colloq. Math.*, 2:301, 1951.

Index

Milton Keynes UK
Ingram Content Group UK Ltd.
UKHW031143141024
449569UK00024B/1121